Textile Science and Clothing Technology

Series Editor

Subramanian Senthilkannan Muthu, SgT Group & API, Hong Kong, Kowloon, Hong Kong

This series aims to broadly cover all the aspects related to textiles science and technology and clothing science and technology. Below are the areas fall under the aims and scope of this series, but not limited to: Production and properties of various natural and synthetic fibres; Production and properties of different yarns, fabrics and apparels; Manufacturing aspects of textiles and clothing; Modelling and Simulation aspects related to textiles and clothing; Production and properties of Nonwovens; Evaluation/testing of various properties of textiles and clothing products; Supply chain management of textiles and clothing; Aspects related to Clothing Science such as comfort; Functional aspects and evaluation of textiles; Textile biomaterials and bioengineering; Nano, micro, smart, sport and intelligent textiles; Various aspects of industrial and technical applications of textiles and clothing; Apparel manufacturing and engineering; New developments and applications pertaining to textiles and clothing materials and their manufacturing methods; Textile design aspects; Sustainable fashion and textiles; Green Textiles and Eco-Fashion; Sustainability aspects of textiles and clothing; Environmental assessments of textiles and clothing supply chain; Green Composites; Sustainable Luxury and Sustainable Consumption; Waste Management in Textiles; Sustainability Standards and Green labels; Social and Economic Sustainability of Textiles and Clothing.

More information about this series at http://www.springer.com/series/13111

Mohammad Shahid · Ravindra Adivarekar
Editors

Advances in Functional Finishing of Textiles

Editors
Mohammad Shahid
Department of Fibers and Textile Processing Technology
Institute of Chemical Technology
Mumbai, India

Ravindra Adivarekar
Department of Fibers and Textile Processing Technology
Institute of Chemical Technology
Mumbai, India

ISSN 2197-9863 ISSN 2197-9871 (electronic)
Textile Science and Clothing Technology
ISBN 978-981-15-3671-7 ISBN 978-981-15-3669-4 (eBook)
https://doi.org/10.1007/978-981-15-3669-4

This Springer imprint is published by the registered company Springer Nature Singapore Pte Ltd.
The registered company address is: 152 Beach Road, #21-01/04 Gateway East, Singapore 189721, Singapore

Contents

Surface Modification of Textiles with Nanomaterials for Flexible Electronics Applications 1
Dinesh Kumar Subbiah, Selva Balasubramanian, Arockia Jayalatha Kulandaisamy, K. Jayanth Babu, Apurba Das and John Bosco Balaguru Rayappan

Functional Finishing of Cotton Textiles Using Nanomaterials 43
N. Vigneshwaran and A. Arputharaj

Environmental Profile of Nano-finished Textile Materials: Implications on Public Health, Risk Assessment, and Public Perception 57
Luqman Jameel Rather, Qi Zhou, Showkat Ali Ganie and Qing Li

Biotechnology: An Eco-friendly Tool of Nature for Textile Industries 85
Shahid Adeel, Shagufta Kamal, Tanvir Ahmad, Ismat Bibi, Saima Rehman, Amna Kamal and Ayesha Saleem

Application of Enzymes in Textile Functional Finishing 115
Shrabana Sarkar, Karuna Soren, Priyanka Chakraborty and Rajib Bandopadhyay

Recent Advances in Development of Antimicrobial Textiles 129
Shagufta Riaz and Munir Ashraf

Advances of Textiles in Tissue Engineering Scaffolds 169
Pallavi Madiwale, Girendra Pal Singh, Santosh Biranje and Ravindra Adivarekar

Fabrication of Superhydrophobic Textiles 195
Munir Ashraf and Shagufta Riaz

Self-cleaning Finishes for Functional and Value Added Textile Materials 217
Subhankar Maity, Kunal Singha and Pintu Pandit

Insights into Phosphorus-Containing Flame Retardants and Their Textile Applications 231
Mohd Yusuf

From Smart Materials to Chromic Textiles 257
Tawfik A. Khattab and Meram S. Abdelrahman

Plasma Treatment Technology for Surface Modification and Functionalization of Cellulosic Fabrics 275
Nabil A. Ibrahim and Basma M. Eid

Cationization as Tool for Functionalization of Cotton 289
Ashwini Patil, Saptarshi Maiti, Aranya Mallick, Kedar Kulkarni and Ravindra Adivarekar

Role of Radiation Treatment as a Cost-Effective Tool for Cotton and Polyester Dyeing 313
Shahid Adeel, Fazal-ur-Rehman, Tanvir Ahmad, Nimra Amin, Shahzad Zafar Iqbal and Mohammad Zuber

Developments in Textile Continuous Processing Machineries 349
Kedar S. Kulkarni and Ravindra Adivarekar

Surface Modification of Textiles with Nanomaterials for Flexible Electronics Applications

Dinesh Kumar Subbiah, Selva Balasubramanian, Arockia Jayalatha Kulandaisamy, K. Jayanth Babu, Apurba Das and John Bosco Balaguru Rayappan

Abstract In the recent past, wearable electronics has emerged as one of the significant products of nanoscience and nanotechnology initiatives. Multi-functional textiles achieved by means of surface modification with nanomaterials are a promising strategy by which flexible and wearable electronics, also called as "next generation electronics" could be developed. Nanostructured metal and metal oxide materials, metal organic frameworks (MOF), nanostructured carbon derivatives and conducting polymers are the major kinds of materials being used to modify the surface of textiles for the fabrication of wearable devices. In this chapter, modification or functionalization of textile surface with nanomaterials through predominantly used techniques like Physical Vapour Deposition (PVD), Chemical Vapour Deposition (CVD) and other chemical techniques have been reviewed. Also, applications of wearable textiles with functionalized surface for energy harvesting, electromagnetic interference filter with special focus on UV, and wearable sensors for healthcare and environment quality monitoring, etc., have been highlighted.

Keywords Multifunctional textiles · Surface modification techniques · Metal organic frameworks · Nanomaterials · Flexible electronics · Wearable sensors

1 Introduction

Nanomaterials are highly functional as well as versatile and can be utilized for a variety of applications. Nanomaterials are like alchemists; different combinations of

---

Dinesh Kumar Subbiah and Selva Balasubramanian: equally contributed.

D. K. Subbiah · S. Balasubramanian · A. J. Kulandaisamy · K. Jayanth Babu · J. B. B. Rayappan (✉)
School of Electrical & Electronics Engineering, Centre for Nanotechnology & Advanced Biomaterials (CeNTAB), SASTRA Deemed University, Thanjavur 613 401, India
e-mail: rjbosco@ece.sastra.edu

A. Das
Department of Textile & Fibre Engineering, Indian Institute of Technology Delhi, New Delhi 110 016, India

M. Shahid and R. Adivarekar (eds.), *Advances in Functional Finishing of Textiles*, Textile Science and Clothing Technology, https://doi.org/10.1007/978-981-15-3669-4_1

chemical precursors and physical conditions can result in versatile nanostructures that finds utility in a range of applications [1]. In recent years, materials scientists have taken a paradigm shift and concentrated on engineering nanomaterials, where they intentionally design and develop synthetic materials with the desired physiochemical properties for a targeted purpose or function [2]. Nanomaterials have greatly enhanced the functional properties of matter, making them the most sought-after candidate for biocompatible applications like drug delivery [3], nanosensors [4] and nanomaterial modified textiles [5].

The demand for highly durable and functional garments has resulted in a growing necessity for textiles modified with nanomaterials. Nanomaterials offer enhanced functionalities to textiles such as enhancement of oil/water repellence [6], ultraviolet (UV) blocking ability [7], reduction of wrinkles [8], elimination of static charge build up [9], continuous monitoring of bodily functions and metabolism [10], rehabilitation, toxicity reduction [11], long term durability, and environmental impact without compromising their flexibility or comfort. This new approach has opened up several windows in the wearable and flexible technologies including garments, which is capable of sensing and responding to environmental stimuli including mechanical, chemical, electrical, thermal, optical, or magnetic sources. Studies have been conducted on electronic and photonic nanomaterials [12] integrated with textiles especially, to validate their plausible potential in sensing, optical displays [13], and drug delivery applications where they are examined in terms of their performance, durability, and connectivity.

Cotton is widely preferred by textile manufacturers towards realizing wearable textiles owing to its high absorbing capacity, chemically adaptable surface and flexibility. Notwithstanding these merits, the fibres of cotton fabrics have lower strength and easily flammable nature. As a result, such exacerbating properties of natural cotton imparts greater limitations in (i) electronic applications (where repeatability and durability at end user); (ii) antibacterial actions of cotton relatively decreases as a consequence of laundering [14]. To overcome these challenges with cotton, synthetic fibres have emerged as the other alternatives. It has been proved to have better anti-microbial and stain-resistance properties but at the cost of comfort. Thus, nanoengineered fabrics would plausibly stand as a conglomeration of the merits of both natural and synthetic fibres, while furnishing novel functionalities.

Chemical vapour deposition (CVD) and physical vapour deposition (PVD) techniques offer several advantages for surface modification of textiles [15]. Among the two techniques, PVD has proven to furnish high adherent coatings. PVD is a thin film coating technique, which involves condensation of vaporized thin film materials over the substrate/textile. PVD is generally conducted in vacuum, thereby offering high purity and uniformly coated thin films owing to the increased mean free path of the sputtered atoms. PVD techniques include cathode arc deposition, pulsed laser deposition, electron beam evaporation, evaporative deposition, sputtering, ion plating, thermal evaporation, and enhanced sputtering. The general mechanism in PVD techniques involve evaporation of the solid thin film materials to be coated onto the substrate by heat or ion bombardment (sputtering) [16]. Simultaneously, a reactive species in the form of gas is introduced into the vacuum chamber to form a compound with the target and the reactive species, which is subsequently coated on the surface of the substrate/textile as a thin film. Such deposited thin films are generally used in applications that demands high purity and repeatability in its functionality. On the other hand, thin films preparation through CVD techniques like spray coating, dip coating were carried out at atmospheric environment, where the influence of other gases in the atmosphere might influence and modify the prepared samples. These modifications limit the potential applications of the prepared samples. However, preparation of high purity thin films using PVD techniques fulfil the required demands and hence could be potentially employed in many applications like aircraft parts, biomedical instrumentations, optical devices, surface modification of textiles, etc.

Textile has proven to be a versatile substrate for the integration of nanostructured materials owing to its universality and the same has been exploited for the fabrication of wearable electronics devices. Laundering greatly decreases the effects imparted by functionalization of the fabrics thereby reducing its long-term usage. This poses a significant challenge in developing functional textiles. Hence, nanotechnology plays a major role in introducing unique and permanent functional fabrics.

2 Nanomaterials

The term 'nanomaterial' refers to the materials, whose size would be in the range of 1–100 nm at least in one physical dimension. The branch of studying materials and its properties in nano scale is commonly referred to as nanoscience. Nano-scaled materials are either naturally available (e.g. abalone shell of mollusca, volcanic ashes) or man-made (e.g. gold, silver nanoparticles) [17]. Synthesis route for the nanomaterials comes under two broad categories: (i) Top-Down approach—scaling down of bulk materials to desired smaller size and (ii) Bottom-Up approach—grouping atoms/molecules to form a nanoscale particle [18]. The implications of these evolved nanomaterials are evident in many fields, few to be mentioned are miniaturization of electronic devices, targeted drug delivery [19], multifunctional textiles [20] and effective catalysts owing to their unique electrical, optical, mechanical, thermal, magnetic properties and other material properties.

In bulk, much of the atoms are packed internally, where their interaction with the physical and chemical environments remain limited. The situation with particles in the nanoscale regime is different since in a given volume more atoms are involved in electrical and chemical reactions owing to increase in the surface area, in other words, at nano dimensions the surface to volume ratio is increased. As an illustration, let us calculate the surface area ($4\pi r^2$) to volume ($(4/3)\ \pi r^3$) ratio for the spherical particles of radius r,

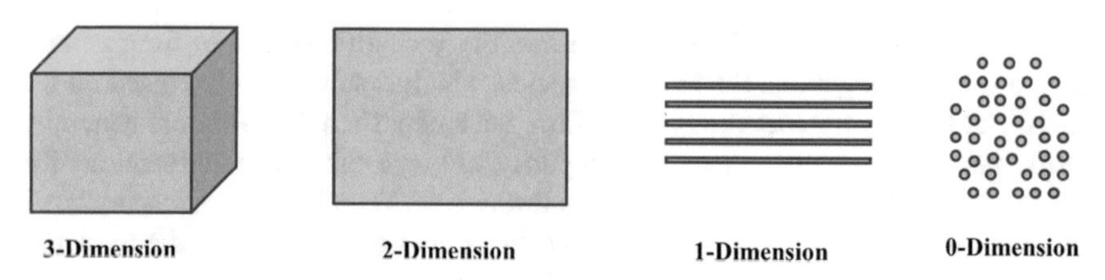

Fig. 1 Classification of materials among various dimensions

$$\text{Surface to Volume ratio} = \frac{4\pi r^2}{\frac{4}{3}\pi r^3} = \frac{3}{r} \tag{1}$$

From this relation, it is clear that size reduction of the system obviously increases the number of atoms on the surface that would be effectively involved in electrical and chemical reactions. To numerically validate the influence of the size, consider silicon samples with two different radii (a) 2 μm and (b) 2 nm and by substituting it in the Eq. (1), the obtained surface to volume ratio were 1.5×10^6 and 1.5×10^9 respectively. Hence it could be inferred that the number of idle atoms in bulk has been considerably reduced in nano regime, as 99.9% of atoms reside on its surface.

Reduction of material dimensions to nanoscale results in discretization of bands and their exposure compared to a bulk material due to limited number of atoms in the former. In addition, by reducing the dimensions of material to less than that of the de-Broglie wavelength, the propagation of electron in that dimension stands restricted resulting in its confinement within that particular dimension. This electron confinement ultimately gets reflected in the energy band structure: staircase function for 2-D materials, spike like function for 1-D materials and vertical line function for 0-D materials. Based on electron confinement, nanomaterials are classified as 3-D (Bulk), 2-D (Thin Films), 1-D (Nano Wires) and 0-D (Quantum Dots) materials. The representations of materials in various dimensions are shown in Fig. 1.

Irrespective of material dimensions, nanomaterials can also be categorized based on its properties. In this chapter, distinguished materials such as carbon derivatives, conducting polymers, metal-organic frameworks (MOFs) and metals and metal oxides towards textile functionalization for electronic applications are discussed elaborately. In each sub-set, the nature of the materials, its interaction mechanism with the physical environment and the outcome due to the incorporation of the materials on to the flexible substrate are discussed.

2.1 *Carbon Derivatives*

Carbon has received keen interest due to its unique bonding nature and different allotropic forms. Despite its bulk form, allotropes of carbon are found at nanoscale as well in three dimensions namely fullerenes, nanotubes (CNTs), graphene and

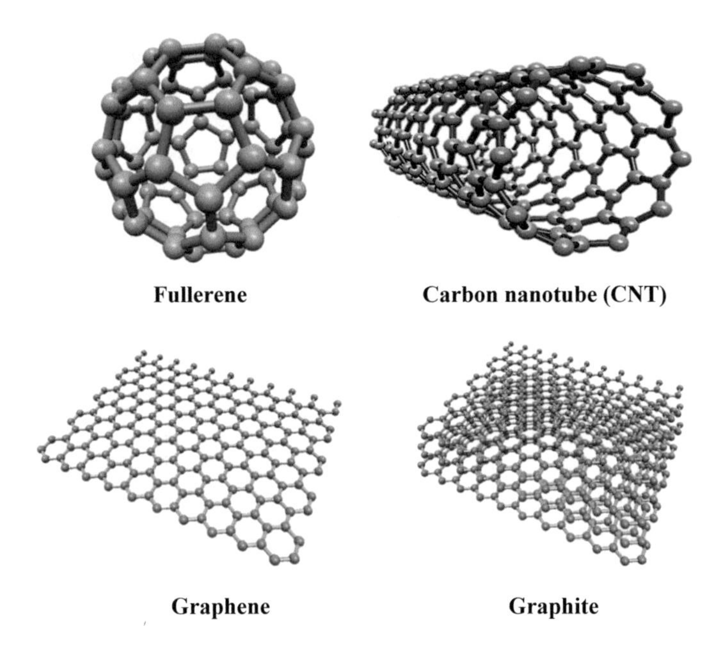

Fig. 2 Allotropes of carbon

graphite/diamond, as shown in Fig. 2. These allotropes possess sp^3, sp^2 and sp hybridization, which results in three different bond angles such as 109° 28′, 120° and 180° respectively. More than the elemental carbon, allotropes of carbon especially CNTs and graphene have found wide range of applications due to their versatile properties. Based on the structural orientation, graphene structure is elementary for graphite materials in all other dimensionalities. Graphene possesses a hexagonal arrangement of monolayer carbon atoms in a 2-D lattice [21].

Depending on the atom's crystallographic orientation, the formation of 2-D graphene structure could be categorized into armchair and zig-zag models. Each structure possesses unique electronic property owing to their different orientation; zig-zag structure behaves as metals while armchair acts as semiconductors. It was well known that an atomic layer of graphene is a zero-bandgap semiconductor and offers high mobility and conductivity [22], remarkable mechanical, thermal and electrical stability as a consequence of its π-conjugated electrons.

In practice, the separation of an atomic layer of graphene is a cumbersome process because of extensive van der Waals interactions between any two layers of graphite; hence graphite sheets are feasible materials for commercial applications. Thickness of layers plays a key role in describing and distinguishing the properties between graphene and graphite [23]. Hereby, focus has been shifted the scalable production of chemically modified graphene-based materials like graphene oxide (GO) and reduced graphene oxide (rGO) either by oxidation, reduction or exfoliation.

Unlike graphene, GO has polar and reactive groups that pose limitations in terms of thermal stability [24]. Although, presence of these functional groups promotes the interaction with polar solvents and play a major role in various fields such as supercapacitors, catalysis, batteries and it is expected to be an active material for next generation electronics. For example, negative charges of GO have the tendency to interact with functional groups of cotton fabrics and leads to better fixation with fabrics. This could be a prominent solution to fabricate a wearable and flexible e-textile sensor at least for continuous health monitoring [25]. In overall, the feasibility and potential of graphene-based materials make it a good candidate in the race of active material for flexible electronics.

Apart from graphene as a sheet, it is necessary to recognize the potential of another carbon allotrope, CNTs. When the graphene sheet is rolled along a particular axis, it forms 1-D carbon nano tubes (CNTs) and further processing of CNTs leads to the formation of carbon quantum dots (CQDs). It is classified into single walled CNTs and multi-walled CNTs depending on the layer of carbon atoms rolled over, with the diameter in the range of <1 nm and >100 nm respectively. CNTs offer high carrier mobility of more than 10,000 $cm^2 V^{-1} s^{-1}$ [26] and tensile strength of about 63 GPa [27], which is somewhat lesser than that of graphene. Employing these mobility characteristics of CNTs in transistor technology helps in designing a high electron mobility transistor [28]. In terms of sensing, large surface to volume ratio features enable highly sensitive response towards target analytes.

2.2 *Conducting Polymers*

Repetition of monomer units in all possible directions leads to the formation of polymers. In earlier stages, polymers were used as an electrically insulating material, whereas a series of conducting polymers (CPs) like polyacetylene, polyaniline (PANI), poly (3,4-ethylenedioxythiophene) (PEDOT), poly (styrenesulfonate) and polypyrrole (PPy) were discovered later. It can be synthesized either by the process of oxidative polymerization or by electrochemical polymerization. Basically, CPs exhibit one-dimensional (1-D) delocalized conjugated structure with remarkable electrical, optical properties and tunable bandgap nature. An interesting fact about them is that their conductivity could be tuned either by the process of doping or de-doping, which will influence its electrical and optical properties [29].

Based on the charge transport phenomena, conducting polymers have been classified based on the electron and proton conductivity. Further, the electron conducting polymers are subdivided into redox and intrinsically conducting polymers on the basis of electron transport via hopping mechanism and delocalized electrons through conjugated systems respectively. In summary, the conduction in CPs is mainly dictated by the presence of conjugated bonds [30].

With regard to applications, several studies have been reported on biomedical applications, where it was found that CPs response to the electric fields arise from biological samples like tissues, epithelium, muscles and retain its biocompatibility [31]. In addition, CPs of high electroactive nature have the tendency to catalyse the reduction/oxidation (redox) reaction-based applications like biosensing, fuel cells, and supercapacitors [32]. Based on the reports published, design of flexible energy storage/harvest device is possible once the CPs based textile supercapacitor is fabricated.

In terms of sensing application, the electrochromic behaviour of CPs helps to detect the analyte either in gas/aqueous phase, where change in colour (transparent to opaque) was visibly observed during interaction. For example, if conducting polymer PANI is exposed to ammonia (NH_3), it will undergo deprotonation and form NH_4^+, where distinct change in optical spectra was observed. Moreover, PANI was proven for humidity sensing application, where it was deposited on cantilever surface and shows ultra-sensitive deflection by varying the humidity conditions [33].

2.3 *Metal Organic Frameworks (MOFs)*

Metal Organic Frameworks, a highly porous and crystalline solid made of metal units and organic linkers, could be self-assembled together forming a highly ordered structure in different dimensions [34, 35]. In particular, the structure of MOFs can be intentionally designed by selecting the appropriate combination of metal clusters and organic linkers or by post modification treatment. Limitless choices as a result of such combinations provide us with diverse framework structures and different functionalities [36]. Hence, tunable characteristics of MOFs have received more interest in the contemporary research, which includes gas storage/separation [37], catalysis [38], drug delivery [39] and sensing [40]. The interesting feature of the interaction mechanism of MOFs with the approached guest molecules is mainly driven through: (i) uncoordinated metal sites [41], (ii) organic linker defects [42] and (iii) functional groups [43]. In specific, the Lewis acid/base nature between the MOFs and guest molecules [44] and its resultant performance decides the particular application. The schematic representation of MOF is represented in Fig. 3.

On other hand, porosity, another key parameter dominates the MOFs performance especially in gas adsorption. For example, tuning the MOFs pore size and shape selectively allows the analyte to diffuse into its bulk structure [45]. The size and

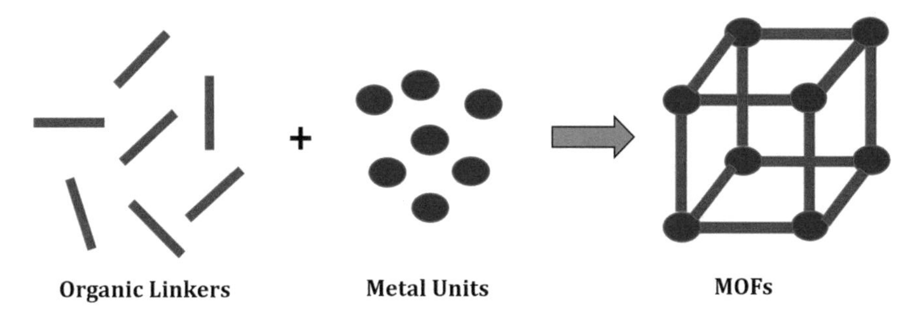

Fig. 3 Building blocks of MOFs

shape of pores could be estimated with the aid of N_2 adsorption isotherm, which reveals the MOFs reversible adsorption characteristics as well. Typical MOF exhibits micro and mesoporous nature. Moreover, the material parameters and properties of MOFs could be examined with the help of physical measurement techniques like X-ray Diffraction (XRD), Fourier Transform Infrared Spectroscopy (FTIR), Nuclear Magnetic Resonance (NMR) and X-ray Photoelectron Spectroscopy (XPS).

In addition, MOFs address the inadequacy of structural accessibility possessed by the conventional porous materials like activated carbon and mesoporous silica. Even though thousands of MOFs have been developed, only some have shown promising characteristics towards the real time applications. This could be attributed to the limitations of MOFs in terms of thermal, mechanical and chemical instability in different environment. In most cases, metal units of MOFs are divalent cations (e.g. Zn^{2+}, Cu^{2+}) and organic linkers are carboxylate linkers, which are susceptible to cleavage in the presence of water even at room temperature [46]. Other than that, σ-bonding interaction between the metal clusters and linkers make the MOFs less conductive.

In recent times, new materials (dopant, metal composite) [47], synthesis procedures and modification treatments have been tried upon to improve the stability and conductivity of MOFs [48]. Although the challenges concerning the material aspects were sorted out, the key research question of how it could be utilized effectively in today's scenario stands unaddressed. For example, most of the existing literature on MOFs as an active material in gas sensing predominantly, reports that isolated MOFs in powder form exhibited better performance than any of its functionalized forms (e.g. thin film) [49]. Surprisingly, recent reports state that MOFs modified cotton surface can act as a gas sensor that bridges the complete utilization of MOFs potential with the flexible substrate, which is a promising approach to meet the requirements of flexible electronics technology [50].

2.4 Metals and Metal Oxides

Metals are highly conductive in nature because of the overlapping of their conduction and valence band [51]. In metal oxides, the bands are separated with a forbidden energy gap termed as bandgap, which is an attribute for its semiconducting and insulating behaviour. Metal oxides are one of the interesting class of materials exhibiting unique electrical, optical and catalytic properties. Metal oxides having bandgap in the range of ~0.1–3.9 eV (~0.1–1.6 eV—narrow bandgap & ~1.6–3.9 eV—wide bandgap) are referred to as semiconductors and above 3.9 eV, as insulators. The differences in the bandgap of the above described materials is represented in Fig. 4. In this chapter, we have focussed our discussion on the metal oxide semiconductors alone with regard to application perspectives.

Metal oxides such as Zinc oxide (ZnO), Cerium oxide (CeO_2), Titanium oxide (TiO_2), Tungsten oxide (WO_3) and Copper oxide (CuO/Cu_2O), etc. have been reported for various applications so far [52]. It has been noted that the oxygen ratio

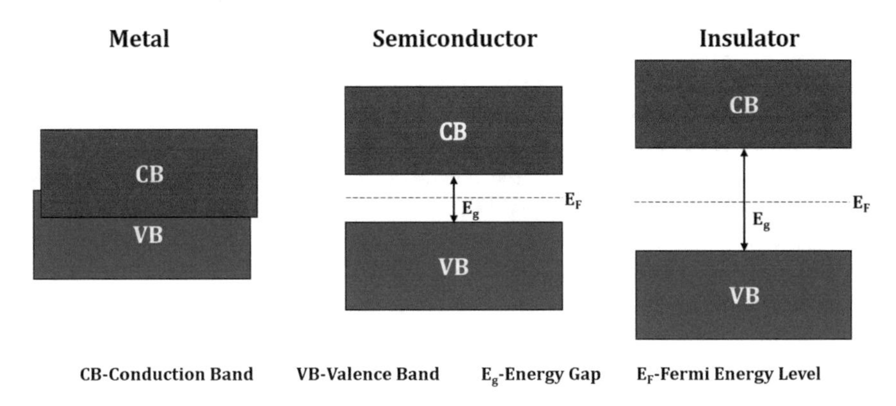

Fig. 4 Bandgap representation of metal, semiconductor and insulator

in metal oxides composition (e.g. Zn_xO_{1-x}) has an influence on the materials characteristics. Metal oxides get further oxidized (up to its solid solubility limit) if the oxide samples are kept in an ambient environment at elevated temperature. The possible adsorption of oxygen species (O^-, O_2^- & O^{2-}) on to the surface of metal oxide semiconductor [53] are as follows:

$$O_{2(gas)} \rightleftarrows O_{2(adsorbed)}$$

$$O_{2(adsorbed)} + e^- \rightleftarrows O^-_{2(adsorbed)}$$

$$O^-_{2(adsorbed)} + e^- \rightleftarrows 2O^-_{(lattice)}$$

Adsorption and desorption of oxygen on a surface plays a key role in gas sensing phenomenon, where the sensing response varies with respect to analytes/volatile organic compounds (VOCs). Initially, the atmospheric oxygen species get adsorbed onto the surface of metal oxides by capturing the conduction band electrons, when exposed to the ambient atmosphere. The evidence for this trapping of electrons could be seen as an increase in the resistance of sensing element (metal oxide) as a consequence of decrease in the charge carrier concentration and an increase in the space charge region width. After sometime, the system (surface resistance) attains equilibrium. When the same sensing element is exposed towards oxidizing/reducing gases, the adsorbed oxygen ions desorb from the surface due to their interaction with the surface, inducing a decrease in resistance and vice versa [53]. Such interactions highlight the versatile nature of metal oxides and opens up opportunities of their utility in various fields.

In addition, it is essential to address another described material, metals that hold a significant place in electronics. During device fabrication, it is required to pattern each individual electronics component to make it as a wholesome electronic device. As stated earlier, metals offer high electrical conductivity even in its bulk form and confining those metals to nano scale offers much more conductivity (e.g. gold, silver). In general, metals are defined as solid, hard and mechanically stable materials. In transistor technology, especially in metal oxide semiconductor field effect transistor (MOSFET), metals were used as the gate material. This was because the material that was employed as the gate in the transistor must have high conductivity, with the applied voltage between terminal (source) and metal (gate) regulates the channel voltage. Metals are also found in liquid form (e.g. mercury, gallium), offering both metallic and fluidic properties [54].

Moreover, metals provide appreciable deformable nature wherein it could retain their function under stretching, bending and twisting action. Such attractive properties facilitate the fabrication of flexible electronic devices. Deploying such electronic devices (e.g. sensor, antenna) interconnected by metal and deformable on a flexible substrate like textile, plastic would pave a way to attain a benchmark in wearable electronics.

2.5 *Multifunctional Textiles*

Textile is one of the basic commodities, which has been historically used for clothing over centuries. In this techno-era, attempts have been made to make it smart for multitasking. Improvisation of traditional fabric in properties such as hydrophobicity [55] (dirt-free), conductivity [56], antimicrobial activity [57], ultraviolet resistance [58], flame retardancy [59] and stimuli-responsiveness have been reported so far. Modified textile with super hydrophobicity results in self-cleaning [60] properties could be used to separate oil-water mixtures [61]. In addition, protective clothing against chemical warfare, trapping aerosols, windproof and wound dressing products for military purposes were other applications of functionalized textiles. At the onset of next generation smart textiles, the functionalities of these textile grade fibers have shifted focus towards smart fibers with the incorporation of wide panel of novel nanoparticles, few to be mentioned are ZnO [20], TiO_2 [62], SiO_2 [63], Au [64], Cu [65], nano-clay, Ag [66], polymers [67] and their composites. Owing to their stability, inorganic materials are well preferred than organic materials.

2.5.1 Functionalities Incorporated

When placed in the bacterial environment, nanoparticles (especially Ag, which exhibits anti-bacterial properties) infiltrate the bacterial cell resulting in cellular leakage, which eventually leads to the death of that particular cell [68]. To block the harmful UV rays, the choice of the nanoparticles should be such that its bandgap matches with the incident photons. When the textile is modified by this layer, the energy of the incident UV photons is absorbed, henceforth the penetration through textile stands prevented. Cellulose as such, is hydrophilic and easily flammable. But, when its surface is modified with nanoparticles, based on the morphology and porosity of the coated nanoparticle/nanocomposite, super hydrophobic nature can be incorporated onto the textile. The result of such modifications incorporates textile characteristics like dirt free, anti-icing and separation of mixtures of oil and water [69].

Antibacterial properties were incorporated into cotton fibres when treated with silver, gold nanoparticles and hexadecyltrimethoxysilane in addition to offering improved hydrophobicity and UV protective fabric [70]. Textiles finished with Ag-TiO_2 nanocomposites along with the polysilicane compounds/polymers exhibit enhanced aforementioned characteristics along with washing durability. The alkoxy group covalently bonds to polysiloxanes on the surface of the textile. Other compounds that effectively furnishes antibactericidal properties include triclosan, N-halamines, ammonium compounds and phosphonium salts [71].

Hence, the multi-faceted development in technology has modified the contemporary textile, an irreplaceable all-time commodity of human kind with the incorporation of multiple functionalities. The procedure involved and the imparted characteristics to be deployed in various fields are discussed in detail. Moreover, various techniques used for surface modification are also discussed in detail. Finally, surface modified textiles for applications like energy harvesting [72], UV filtering [73], electromagnetic interference shielding [74], healthcare and environmental monitoring [75] are highlighted.

3 Surface Modification Techniques

3.1 Sputtering

Sputtering is a widely used PVD technique that offers uniform coating and high purity thin films. Textiles can easily be sputter coated using radio frequency (RF) sputtering technique. There are several other types of sputtering such as DC sputtering, magnetron sputtering, and reactive sputtering, to name a few. But, the choice of the technique relies on the application and the type of thin film material to be coated [76]. Since textile is a dielectric medium, RF sputtering would be of more suitable option. It can also be combined with the reactive and magnetron sputter techniques depending on the applications like antimicrobial, UV protection and photocatalytic [77].

Sputtering involves the use of very high vacuum environments to ensure high purity thin films. The most commonly used pumps to generate vacuum include rotary vane pumps, molecular drag pumps, turbo molecular pumps, ion pumps and diffusion pumps. The selection of the pump depends on the operating pressure required for the particular deposition. Once the required pressure is reached, argon gas is purged into the vacuum chamber. This marks the initialization of the process. The argon gas reacts with the free electron in the chamber and forms Ar^{2+} and $2e^-$ [58]. The set-up also includes positive and negative terminals, the anode and the cathode, respectively. The target or the material to be sputter coated onto the textile is taken in the form of a solid and is generally referred to as the target. The target acts as the cathode, where the textile is generally stretched between two pulleys in front of a very powerful magnetron that helps in electron confinement thereby increasing the efficiency of the process. Once the potentials are applied, the Ar^{2+} ions are accelerated towards the cathode where the target is placed. Here, the ion loses its energy by ion bombardment with the target atoms. Because of the ion bombardment and the consequent momentum transfer, the surface atoms are knocked off the target (Fig. 5). These sputtered atoms reach the anode thereby forming a uniform coating across the exposed surface of the textile substrate. Once the exposed region is coated for the desired duration, the pulleys are rotated and the next region is exposed. This process is continued until a uniform coating is achieved for the entire length of the fabric.

The demand for silver deposition on fabrics keeps growing day by day for applications involving electrical conductivity, antimicrobial properties or a shiny metallic look for ornamental usage. This is made possible by producing fabrics with well-defined structures on its surface. Numerous techniques have been investigated with respect to silver thin films deposited on textile. These include electroless plating, electroplating, and vacuum deposition. Sputtering is highly advantageous owing to its simplicity, time consumption, eco-friendly, and superior adhesion of the film to substrate [77].

G. Segura et al. studied the antimicrobial actions of RF sputtered copper (Cu) on cotton fabrics against *Escherichia coli* (*E. coli*). In particular, a plasma technology

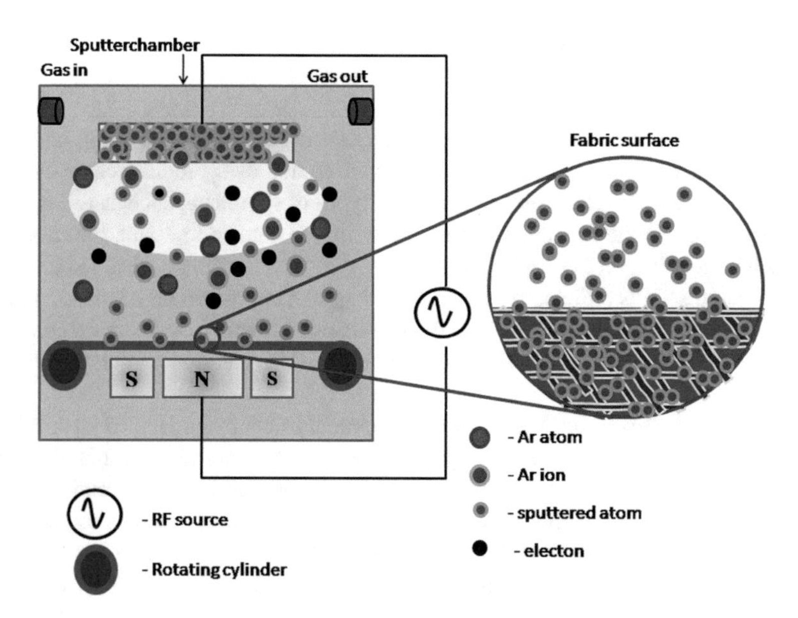

Fig. 5 Schematic of the textile coated using sputtering technique

RF magnetron sputtering was used to deposit the copper on fabrics primarily to confer antimicrobial properties. Moreover, use of argon plasma during deposition induced cotton cellulose fragmentation and free radicals generation by means of oxidation process. As a result, the amount of copper ions adhered to the surface of the fabric greatly improved. From the obtained results, it was revealed that bacterial growth drastically reduced for the highly Cu sputtered fabric. The mechanism behind bacteria inhibition by Cu was claimed to be the generation of hydroxyl radicals and oxygen molecules during redox reactions of Cu ions [78].

Instead of RF sputtering as discussed above, Laura Rio et al. came with Direct-Current magnetron sputtering (DCMS)/Direct-Current pulsed magnetron sputtering (DCPMS) for Cu deposition against *Staphylococcus aureus*. Bacterial viability was assessed through four different techniques such as "mechanical detachment, microcalorimetry, direct transfer onto plates and stereomicroscopy" [79]. Here, polyester was used as the substrate where the surface was sputtered with Cu by DC magnetron sputtering. Time for Cu deposition was varied between 90, 120 and 160 s in both the sputtering types. Akin to the above scenario, sputtered Cu by DCMS technique with greater time (160 s) showed efficient bactericidal activity which can be evaluated from the combined results of direct transfer onto plates and stereochemistry [79].

Even though cotton fabrics are promising to make wearable and flexible electronic devices, there are challenges to form continuous films on a cotton fabric. The existing literature on surface modified conductive fabrics addressed this limitation through different modification techniques. Chuanmei Liu et al. sputtered an electrically conducting material like Ti and Ag on polyethylene terephthalate (PET) fabrics using multi-target magnetron sputtering system. Extensive studies on electrical conductivity of the sputtered samples (Ti–Ag-Ti trilayer & Ti–Ag alloy) were carried out. Out of those, deposition of Ti on a trilayer structure was deposited by RF sputtering, whereas Ag was deposited by means of DC sputtering. Deposition of Ti–Ag alloy films was also carried out using the same deposition parameters. In order to ensure uniformity in deposition, the sample holder was rotated at a speed of 10 rpm. The increase in Ag content in both the films resulted in an increase in the electrical conductivity. In particular, Ti–Ag alloy composition showed lesser electrical resistivity (3.4×10^{-7} Ω m) than the trilayer configuration (5.1×10^{-7} Ω m) [80].

Potential of sputtering technique towards surface modification were utilized in various applications. Carneiro et al. reported the photocatalytic and UV protection activities of TiO_2 by pulsed magnetron sputtering on poly (lactic acid) fibres. High pure titanium target of 99.9% was used for sputtering, where the oxygen flow was allowed to the chamber to deposit TiO_2 on the fibre surface. The reason for employing pulsed magnetron sputtering was to suppress the arc occurrence during deposition, as arcs influence the stoichiometry in film growth ultimately influencing the optical and electrical properties. For photocatalytic and UV protection applications, performance metrices is mainly driven by its optical properties. In summary, it was concluded that deposited TiO_2 films on fibre source exhibits an ultraviolet protection factor (UPF) of 81.3% after washing compared to an unwashed fibre of 88.8% [81].

3.2 *Pulsed LASER Deposition (PLD)*

Pulsed Laser Deposition is yet another type of PVD technique, which uses electromagnetic radiation from a high intensity pulsed laser as its source. The high-power pulsed laser source is used to vaporize the material that is to be coated onto the textile surface. In the case of solid substrates, sufficient thermal energy can be supplied to the substrate in order to orient or align the surface deposited atoms as desired. This additional energy is used in phonon creation, which in-turn helps in the arrangement of atoms. The atoms that are vaporized tend to form a plasma plume that is then focused onto the surface of the substrate. The entire procedure is preferably carried out in vacuum. But in order to suit the desired application, the chamber can also be filled with a suitable reactive species such as oxygen and nitrogen [82].

In the process of PLD, the incident laser on the source materials results in the excitation of electrons in-turn lead to thermal energy, which favours the evaporation of elements from the source materials and the ejected elements form the nanostructured thin films of desired properties. Thin films deposited with this technique are observed to have desired composition, structure, thickness, crystallinity, conductivity, optical properties, mechanical properties, biocompatibility and roughness. Parameters of deposition, like the target material, temperature, gas/vacuum pressure, target-sample distance, laser wavelength, pulse width, pulse rate and radiation intensity can be controlled and optimized to achieve the desired film properties [83]. PVD techniques such as sputtering and PLD generally favour the deposition of seed layers which can then be grown to produce different morphologies with the help of post thermal treatments. A schematic of the surface modification of the fabric substrate via PLD technique is illustrated in Fig. 6.

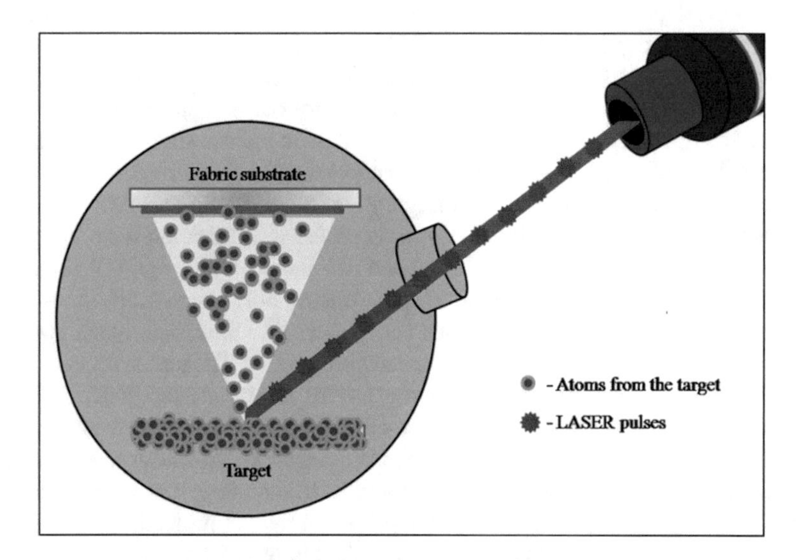

Fig. 6 Schematic representation of the PLD technique

PLD has been used to deposit superhydrophobic polytetrafluoroethylene (PTFE) thin films. By analysing the surface morphology, it was identified that surface roughness has a major part in enhancing hydrophobicity [84]. A similar study if conducted on textile substrates could offer several insights in the present-day demands related to industries and smart homes. Yet another study on ZnO was conducted to develop antifungal and UV protective clothing for biomedical applications [85]. Adhesion can be achieved when the substrate is pre-treated with glow discharge [86].

3.2.1 Thermal Evaporation

This technique involves vaporization of the material by resistive heating. The vaporised material is subsequently deposited on the desired target as a thin film. This process relies on the presence of a finite vapour pressure over the material surface when the material is heated particularly under reduced pressure. Thermal energy is supplied by passing a large current through a crucible/basket (a thermally tolerant material such as tungsten) holding the source material. The resistivity of the crucible generates sufficient heat such that the source material is vaporized as atom. The vaporized source material traverses a low-pressure space, to become incident on the textile, where it condenses to form a coating [87].

The source either sublimes (a solid to vapour transition), or evaporates (liquid to vapour transition via intermediate melting from solid to liquid). The process is achieved in a vacuum chamber as shown in the schematic in Fig. 7. In most cases,

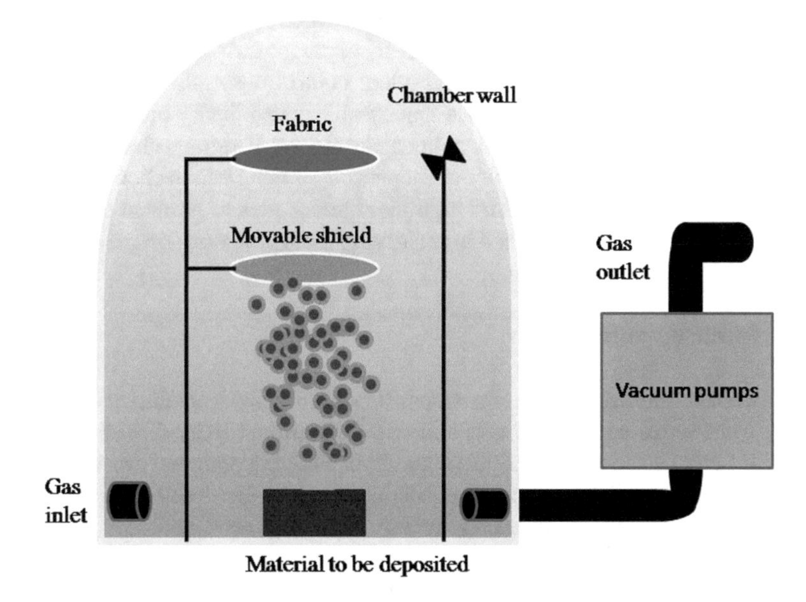

Fig. 7 Schematic representation of thermal evaporation technique

the source material is heated from its solid to liquid state before attaining the vapour phase. Hence, the source material is supported from beneath (bottom) and the targeted substrate where the material going to deposit will be placed horizontally at the top. A quartz crystal sensor enables control of the source material vaporization rate. The rate is coupled with a movable shield, to allow specific film thickness deposition on the sample fabric. Evacuation of the chamber gases is required before the source material can be deposited onto the textile. The first step is to achieve the required vacuum environment. Like all other PVD techniques, thermal evaporation makes use of different types of vacuum pumps to reach the desired operating pressure. The rotary pump generally helps to achieve 10^{-2} Pa. The turbo molecular pump is used to reach even lower pressures.

Resistive heating vaporizes the source material. Once the desired rate of vaporization is reached, the shield is moved to expose the textile surface. Vaporized source material impinges on the substrate surface where it settles and condenses eventually. Once a film of desired thickness is obtained, the shield is returned back to its original position to cover the sample and resistive heating slowly is ceased [88]. This method is directional due to a larger mean free path of the vaporized source material. One of the major draws of this method, however, is that the film adhesion can be poor in the case of fabric substrates.

3.3 Chemical Techniques

Surface modified textile is a growing field of research in nanotechnology in recent times. These modified textile materials are used predominantly in applications like UV-blocking, super-hydrophobicity and odour control materials. The textile materials are coated with the nanoparticles that exhibit extraordinary properties thereby changing the behaviour of the conventional materials. Various techniques are followed to coat the textile material with nanoparticles among which chemical solution deposition is one of the easiest and high yield techniques to produce the modified textiles. Some of the methods used in modifying the textiles are discussed below:

3.3.1 Electrospinning

Nanomaterial comprises zero-dimensional (Quantum dots), one-dimensional (nanotubes, rods, wires and nanofibers) and two-dimensional (nanosheets) structures. Among all other structures, nanofibers stand out for great potential due to high surface to volume ratio, higher aspect ratio, high porosity, etc., which are favourable for wide range of applications such as energy storage devices, batteries, solar cells, generators, fuel cells and supercapacitors [89]. Electrospinning is highly established and widely used techniques to prepare nanofibrous structures due to its merits of low cost, high efficiency, simple and reproducibility, which is working based on the principle of electrostatic interaction [90]. It consists of a syringe pump, electric filed

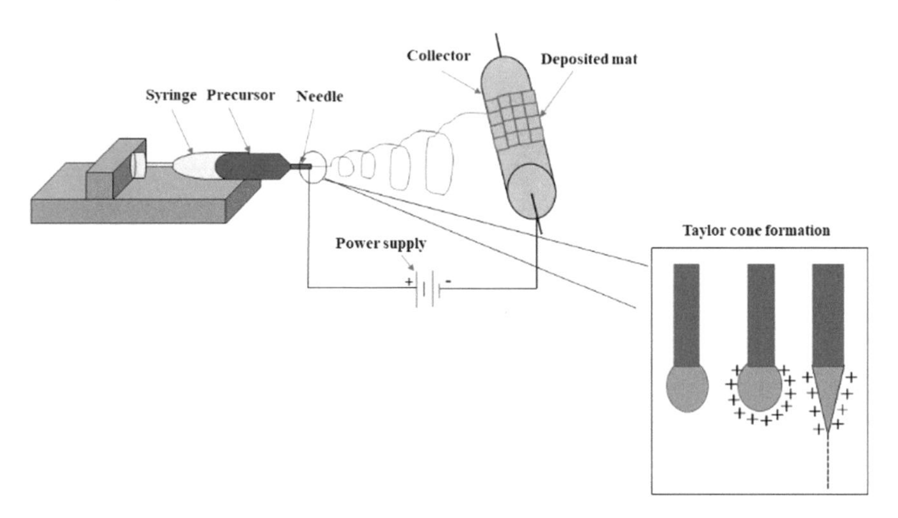

Fig. 8 Schematic of the textile coated using electrospinning technique

source and various collectors (called as target or counter electrode) such as different size rotating mandrel, stationery and X-Y direction movable collector (Fig. 8) [91]. The solution is held in the syringe nozzle and it is loaded to the syringe pump. The syringe pump and a counter electrode were maintained at certain distance to generate large electric filed. The huge potential difference between the nozzle and collector leads to the formation of charged jets, which are then accelerated towards the collector. During this process, the solvent gets evaporate in-turn leads to the formation of nanofibers. The physical properties of the nanofibers depend on viscosity, conductivity, surface tension of the liquids, humidity, temperature, flow rate, applied filed and nozzle-collector distance [92].

3.3.2 Sol-Gel Synthesis

Nanoparticles ranging from 1 to 100 nm in size can be effectively obtained through sol-gel synthesis. This method proves to be advantageous in terms of being simple, inexpensive, offering greater tailor-ability of microstructure of material and allows easy introduction of functional groups [93]. Here, the metal precursor solution is treated with a complexing agent, which in turn acts as a stabilizing agent or as catalyst. This complexing agent improves the adherent property of the film. The method involves hydrolysis of metal precursor solution to yield corresponding metal hydroxides, which subsequently forms the sol. The formed sol is then subjected to condensation to aid in metal-oxygen bond formation [93]; this is followed by drying procedures to yield the metal oxide [94]. The schematic representation of different stages of sol-gel technique is shown in Fig. 9.

$$-\mathrm{M}-\mathrm{OR} + \mathrm{H_2O} \rightarrow -\mathrm{MOH} + \mathrm{ROH} \quad (2)$$

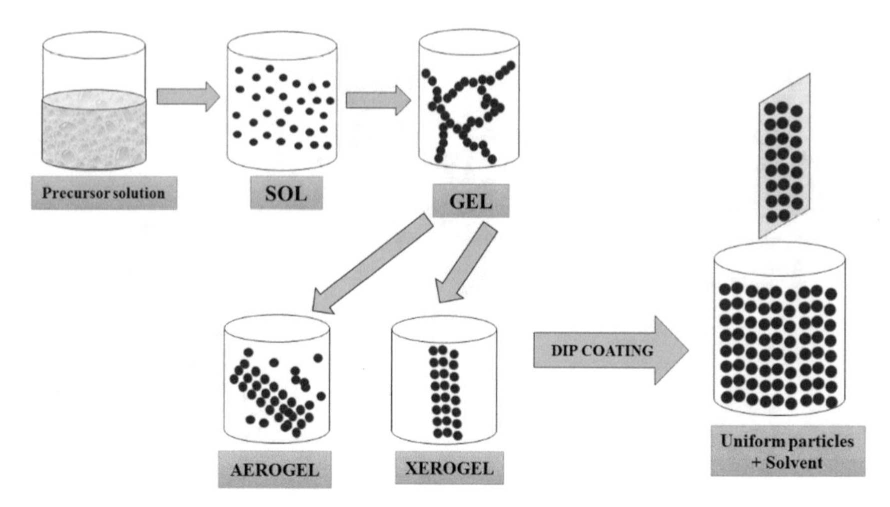

Fig. 9 Schematic representation of sol-gel technique

$$-\text{M}-\text{OH} + \text{XO}-\text{M}- \rightarrow -\text{MO}-\text{M}- + \text{XOH} \tag{3}$$

The sol-gel procedures can be categorised into two methods, Aqueous and Non-aqueous based; In case of the former method, metal alkoxides are predominantly used as precursor solution due to their greater affinity with water. However, this method is not suitable for synthesis of metal nano-oxides due to limitations associated with controlling of particle size and reproducibility owing to the occurrence of all the three process (hydrolysis, condensation and drying) simultaneously. This limitation is overcome in non-aqueous based sol-gel synthesis method rendering it much suitable for metal nano-oxide synthesis which improves the tunability of morphology, particle size, surface properties and composition of metal-oxide. The oxygen required for metal oxide formation, is obtained from organic solvents such as alcohols, aldehydes or ketones. The latter method is further categorised into surfactant based and solvent based; in case of the surfactant-based method, direct conversion of metal to metal oxide is achieved at higher temperature, whereas with the latter method interaction with metal halides and alcohol results in metal oxide formation.

The process of drying has a greater say in the microstructure of the final material formed; the conversion of wet gel to solid gel involves drying, based on the methodology of drying the obtained gels can be classified as "Aerogels" and "Xerogels". Aerogels are obtained when the wet gels are converted under supercritical conditions, subsequently yielding highly porous and low-density material through this process [93]. With respect to surface modification of cotton textile through sol gel synthesis, the substrate cotton textile is immersed on the sol precursor solution and by optimization of process parameters, the desired active materials will be deposited on the cotton textiles.

Flame retardancy is also one of the promising applications of surface modified textiles. Active research on this aspect is going on to find an active material that could withstand higher temperature and find a suitable technique to impregnate such a material onto the cotton textile. In this aspect, Bentis et al. [95] investigated the different ionic liquids (ILs) modified textile surface through sol-gel synthesis and pad-dry process. The sol-gel synthesis of organosilanes based ILs were deposited by means of padding method onto the cotton surface. From the obtained results, it was noted that incorporation of ILs onto cotton fabrics enhanced its flame-retardant properties and thermal stability without affecting its mechanical properties when compared to control fabric [95].

3.3.3 Spray Pyrolysis

Spray pyrolysis is a thermal stimulated chemical process of surface modification wherein the precursor solution of desired compound is sprayed on top of the substrate (Fig. 10). It involves atomization of precursor solution followed by evaporation and decomposition leads to formation of particles and thin films. The precursor might be sprayed through different sources such as compressed air, ultrasound and electric field. At initial, atomizing nozzle in pyrolysis unit generate droplets from a precursor

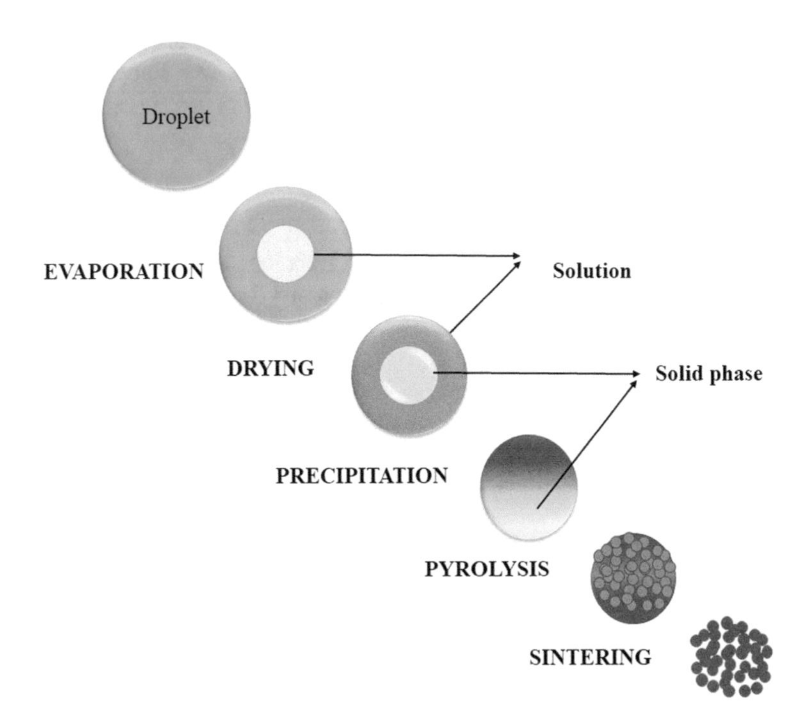

Fig. 10 Stages of spray pyrolysis technique

solution and the generated droplets will be sprayed over the surface of substrate that maintaining at certain temperature [96].

The sprayed droplets will undergo for an evaporation of solvents before reaching the surface of substrate. Then, the remaining residues will get impinge where further decomposition occurs and leads to deposition of desired compound as particles or thin films. The reactants in a precursor solution for spray pyrolysis are selected on the basis that other than desired compound, remaining products should be volatile at the temperature of deposition [97]. The substrate temperature is one of the main factors in determining the nucleation sites, which contributes to the morphology of the film. Apart from substrate temperature, the quality, properties, thickness of desired thin film over substrate will also be dictated by some other parameters start from anion-cation ratio in precursor, spray rate, distance between spray nozzle-substrate, droplet size and solvent evaporation rate. Hence, optimization of all these components enables to form a desired film for wide range of potential applications [98].

In transformation era of rigid (glass thin film) to flexible (cotton textile) device, spray pyrolysis still withstands in a row of surface modification techniques due to its versatility. In recent studies, significant number of research articles were published on surface modification of textiles by spray pyrolysis to various applications like photocatalytic [99], antimicrobial [100] and flame retardancy [101]. The principle and mechanism of deposition on cotton textile substrate are all same like deposition on glass substrate. It is noted to be that substrate temperature is an important concern. The given substrate temperature should not affect the nature of the substrate; withstand temperature of cotton textile is comparatively lower compared to glass substrate.

3.3.4 Pad-Dry-Cure Method

Pad-Dry technique is a well-known commercial technique to modify the surface of the textiles/cotton fabrics. In process, textile is passed (i.e. dipped) to the solution containing the active materials to be deposited followed by drying/curing using heat or pressure. The number of dipping cycles may increase or decrease that depends on need of material finishing towards end applications like anti-bacterial, water proofing, softening, etc. It offers low-cost, continuous and large-scale production of surface modified textiles for wearable applications. Karim et al. reported the scalable production of graphene modified textile using pad-dry technique for human activity monitoring application where change in resistance is calculated with respect to bending actions [25]. Initially, GO was synthesized using modified Hummers method and chemically reduced to rGO using the reducing agent, sodium hydrosulphite ($Na_2S_2O_4$). Dispersed rGO solution was then taken into the pad-dry unit where the unmodified fabric was passed through the padding bath containing rGO. The rGO coated fabric was then dried at 100 °C for 5 min to ensure the graphene deposition on the fabric surfaces. From the results, it was shown that rGO was uniformly deposited on the surface of textile and provided good electrical conductivity [25].

Xili Hu et al. studied the multifunctional properties of graphene/polyurethane modified textile for different applications such as UV blocking and far-infrared emission [102]. Here, the prepared graphene and polyurethane composites were deposited on the textile surface by means of pad-dry method. Initially, the control fabric was dipped into the graphene/polyurethane (GNP/WPU) solution and padded with certain pressure to obtain 100% wet pickup. Then, the modified fabrics were cured at different temperature in vacuum oven and calculated the weight of surface modified fabric with control fabric). At last, it was concluded that GNP/WPU modified fabric showed enhancement in far infrared emissivity of 0.911 and UV blocking with UPF factor of 500 [102].

Even though reports on the surface modified textile for antibacterial actions are available, they lack washing durability. On this basis, QingBo Xu et al. addressed antibacterial actions of carboxymethyl chitosan (CMC)/silver (Ag) modified fabric against different consecutive washing tests. The control fabric was initially immersed in CMC solution, underwent the padding procedure to maximize wet pickup and was dried to obtain the CMC modified fabric. Next, the padded fabrics were taken for Ag nanoparticles deposition through chemical reduction method. Uniform adherence between the materials deposited on the fabric was observed because of covalent linkage and coordination bonds. CMC contain amine and carboxylic acid groups where it forms coordination bonds with Ag nanoparticles and tend to react with the hydroxyl group of cotton cellulose. It showed bacterial reduction rate of 94% even after 50 consecutive washings [103].

Other than synthetic materials, naturally available materials also show antibacterial action. Studies by Mondal et al. suggested that *Aloe vera* (*A. vera*) gel has antibacterial characteristics towards gram positive *Staphylococcus aureus* [104]. Moreover, demonstration on *A. vera* and chitosan modified cotton fabric by pad-dry process against bacterial growth was also carried out. Antibacterial studies of modified cotton surface for the combination of chitosan/*A. vera* showed greater activity than individual components and paved the way to fabricate eco-friendly textile [104].

4 Technical Textiles for Flexible Wearable Applications

Wearable Technologies

Wearable technology refers to the utility of electronics, mechanical technologies and functional materials integrated in the wearable form. This has been achieved through textile surface modification or electronics skin in the form of tattoos [105]. These devices can be operated by the various forms of physical energies available in the environment as well as the activities made by human beings. Wearable electronics on textiles platform can be categorized into the following three types based on the changes in mechanical, chemical, electrical, magnetic and optical functions of the materials [106].

Active smart textiles: These materials react upon sensing a change in the surroundings or a perturbation. They are equipped with actuators in conjunction with sensors and take pre-determined actions.
Passive smart textiles: These are materials that can only sense a change in the surrounding environment or any stimuli given to them.
Ultra-smart textiles: These materials possess the ability to adapt themselves to an environmental perturbation or any variation in the signals that they receive. Mainly, they consist of a central processing unit that acts as a brain having the requisite predefined programs in place.

Nanoengineered textiles have been designed with a specific modality such as UV blocking, conductivity, antibacterial properties, hydrophobicity, antistatic behaviour, energy storage, chromic and sensing properties. This chapter highlights the field of wearable electronics applications on textiles platform in great focus.

4.1 Smart Textiles for Energy Harvesting and Storage

Energy has been essential to human life and the energy requirement issues in present times have been a major concern. Smart fabrics have emerged as one of the most promising platforms to address energy harvesting and storage applications. Based on this motivation, Wang et al. invented the piezoelectric [107] and triboelectric [108] nanogenerators in 2006 and 2012 respectively. Our surrounding environment has abundance in energies including wave, wind, droplet, thermal, chemical and other mechanical forms. By utilizing the most abundant energies, triboelectric nanogenerators (TENGs) can effectively convert mechanical energy into electrical energy [109], which facilitates contact electrification and electrostatic induction [109]. Generally, daily activities of human beings, such as arm movement or footstep of a 65 kg adult can generate approximately ~67 W of kinetic power [110]. In this context, nanostructured textiles provide a more effective way of approach to scavenge the energy from human motion. Zhong et al. [111] introduced the first textile based triboelectric nanogenerators. The procedure proposed and developed by Zhong et al. [112] has been described below: Generally, TENG comprises of four parts that includes an electrode, positive and negative triboelectric layers, and supporting structures (substrates, wires, spacers and power management systems, to name a few). The two dissimilar triboelectric surfaces consist high electron affinity, so that it becomes the negatively charged triboelectric layer and another layer comprises of positively charged triboelectric layer. As a result of mechanical friction between the two dissimilar triboelectric layers, electrostatic charges were created and these charges lead to the development of potential when they are separated by mechanical forces.

The working mechanism behind textile based TENGs is as follows:

1. During the contact establishment between two dissimilar materials under the tension of mechanical force results in the production of electrical energy due to the

equal and opposite charge density present in the inner surface of the triboelectric layers

2. This separation of triboelectric layers results in the electrons being driven to the electrode to equalize the potential drop created

Commonly used dielectric materials to attain higher tendency to gain or lose electrons for triboelectric applications are given in Table 1.

TENGs have four different types of operational modes as proposed by Zhong et al. [113] as given below:

- "*Vertical contact separation mode*
- *Lateral sliding mode*
- *Single electrode mode*
- *Free standing triboelectric layer mode*"

I. **Vertical Contact Separation Mode** [113]

The vertical contact separation mode of a triboelectric nanogenerator reported by Wang and Zhu [113] is shown in Fig. 11a. Here, two dissimilar dielectric materials are made to face each other, where the surfaces (top and bottom) of dielectric materials were deposited with copper for contact establishment by physical vapour deposition techniques like sputtering, thermal vapour deposition, pulsed laser deposition and so on [114]. The physical contact between the two dielectric layers promotes the development of oppositely charged surfaces on the materials. Potential drop is created when external kinetic energy is supplied to separate the two thin sheets. If the two electrodes are electrically coupled with an external load, the generated free electrons in one electrode terminal will try to flow to another electrode, to equalize this potential drop. Once the separation between the dielectric layers is padlocked, the created charges get vanished, and electrons flow back to the same electrode [115].

II. **Lateral Sliding Mode** [113]

Wang and Zhu [113] developed this mode and is shown in Fig. 11b. In this mode, relative sliding of the two dissimilar dielectric layers leads to polarization of the surface in-turn charge transport happens towards the oppositely charged electrodes to equalize the created electrostatic field by the triboelectric charges. In addition, periodic sliding can produce AC power as output. This mode of operation is called a sliding mode. TENG. Jing [116] and Lin [117] stated that the sliding can be carried out either in cylindrical rotation, disk rotation or planar motion.

III. **Single-electrode Mode** [113]

From the above two modes, physical contacts were established by connecting an intermediate load, which promoted the TENGs to move freely and have mobile portability. Though the TENGs have excellent portability, it cannot be electrically connected to the load due to its mobile object (i.e., man lifting an object). To overcome the issues from the previous modes and to harvest energy, a single-electrode mode TENG was proposed. In this mode, an electrode on the bottom side of the dielectric layer of the TENG is grounded and it is shown in Fig. 11c. In particular, the size of the

Table 1 Commonly used dielectric materials for triboelectric applications (Reused with permission from Ref. [112] Copyright 2013 American Chemical Society)

Positive	Polyformaldehyde 1.3-1.4	(continued)	Negative
	Ethylcellulose	Polyester (Dacron)	
	Polyamide 11	Polyisobutylene	
	Polyamide 6-6	Polyuretane flexible sponge	
	Melanimeformol	Polyethylene Terephthalate	
	Wool, Knitted	Polyvinyl butyral	
	Silk, Woven	Polychlorobutadiene	
	Aluminum	Natural rubber	
	Paper	Polyacrilonitrile	
	Cotton, Woven	Acrylonitrile-vinyl chloride	
	Steel	Polybisphenol carbonate	
	Wood	Polychloroether	
	Hard rubber	Polyvinylidine chloride (Saran)	
	Nickel, Copper	Polystyrene	
	Sulfur	Polyethylene	
	Brass, Silver	Polypropylene	
	Acetate, Rayon	Polyimide (Kapton)	
	Polymethyl methacrylate (Lucite)	Polyvinyl Chloride (PVC)	
	Polyvinyl alcohol	Polydimethylsiloxane (PDMS)	
	(Continued)	Polytetrafluoroethylene (Teflon)	
Positive			Negative
	Aniline-formal resin	Polyvinyl alcohol	
	Polyformaldehyde 1.3-1.4	Polyester (Dacron) (PET)	
	Ethylcellulose	Polyisobutylene	
	Polyamide 11	Polyurethane flexible storage	
	Polyamide 6-6	Polyethylene terephthalate	
	Melamine formol	Polyvinyl butyral	
	Wool, Knitted	Formo-phenolique, hardened	
	Silk, Woven	Polychlorobutadiene	
	Polyethylene glycol succinate	Butadiene-acrylonitrile copolymer	
	Cellulose	Nature rubber	
	Cellulose acetate	Polyacrylonitrile	
	Polyethylene glycol adipate	Acrylonitrile-vinyl chloride	
	Polydiallyl phthalate	Polybisphenol carbonate	
	Cellulose (regenerated) sponge	Polychloroether	
	Cotton, Woven	Polyvinylidine chloride (Saran)	
	Polyurethane elastomer	Poly (2,6-dimethyl polyphonyleneoxide)	
	Styrene-acrylonitrile copolymer	Polystyrene	
	Styrene-butadiene copolymer	Polyethylene	
	Wood	Polypropylene	
	Hard rubber	Polydiphenyl propane carbonate	
	Acetate, Rayon	Polyamide (Kapton)	
	Polymethyl methacrylate (Lucite)	Polyethylene terephthalate	
	Polyvinyl alcohol	Polyvinyl chloride (PVC)	
	(Continued)	Polytriflurochloroethylene	
		Polytetrafluroethylene (Teflon)	

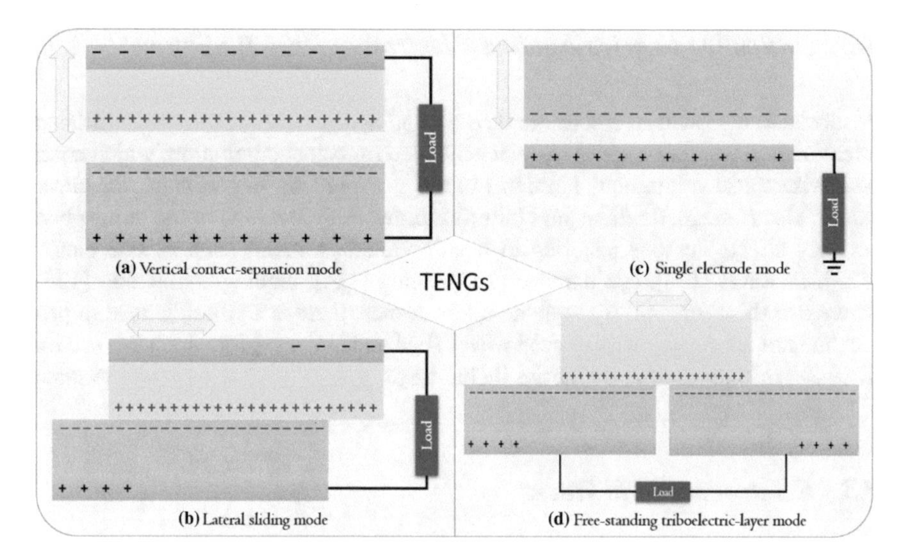

Fig. 11 Fundamental operation modes of TENGs. **a** Vertical contact-separation mode, **b** lateral sliding mode, **c** single-electrode mode and **d** free-standing triboelectric layer mode. Redrawn with permission from Ref. [110]. Copyright 2019 Elsevier

TENGs plays a crucial role in wearable electronics applications. If the size is finite, the movement between the top and bottom electrodes will change the generated electric field distribution [118–120]. In such a case, electron exchanges between electrode and the ground occurs to balance the potential change in the electrode. Single electrode mode can be adaptable for both contact-separation mode and sliding mode towards energy harvesting applications.

IV. **Freestanding Triboelectric-layer Mode** [112]

Naturally, a moving object can generate charge due to its physical contact with air or any another object. For instance, our sandals get charged while we walk on the ground. These charges remain for hours on the surface under necessary conditions such as contact or friction within the particular period and the charge density attains its maximum capacity during this period. To harness these charges, a pair of symmetric electrodes could be constructed under the dielectric layers. Electrodes' size and the intermediate gap between the two should be of the same order without affecting the flexible movement. This mode has been preferred for wearable devices due to the absence of mechanical contacts between the electrodes (Fig. 11d). Therefore, this mode can be used to extend the durability TENGs.

4.2 Protective Fabrics Against Electromagnetic Radiation

A substantive growth in electronic devices and equipment has raddled significant attention because they expose undetectable electromagnetic pollution, which could otherwise cause voluminous problems to the environment. In particular, the elevation of electromagnetic fields and harmful ultraviolet radiations in the atmosphere severely affects humans resulting in health-related problems such as skin cancer [121], cataracts [122], eye damage [123], aging [124], tanning of the skin [125], damage to the cornea [126], to name a few. Hence, there is a pressing need to protect humans from radiation-induced waves (EM and UV rays) for which this section discusses the nanostructured fabrics including protective finishing against the same.

4.3 Electromagnetic Waves

"Electromagnetic radiation is a form of energy that propagates as both electric and magnetic waves and travelling in packets of energy called photons" [127]. Electromagnetic emission comprises of a wide spectrum of wavelengths that includes infrared, ultraviolet, radio waves, microwaves, X-rays and gamma rays, which in turn impacts severe physical damage, changes in the nerve cell performance and stimulating nerves and muscles. In this context, we are in a situation to protect ourselves from these hazards and the same can be achieved by "shielding". Through shielding, one can reduce the exposure of high intensity electromagnetic waves by blocking them with using conductive materials, magnetic materials and nanocomposites as barriers.

Electromagnetic Waves Shielding Mechanism

When an electromagnetic wave propagates through a body, it could interact in three different phenomena based on its strength.

1. Attenuated absorption
2. Attenuation due to reflection
3. Attenuation due to successive internal reflections.

Usually, the blocking mechanism in electromagnetic shielding is based on two significant functions, namely, reflection phenomena by the conductive medium and absorption phenomena in the volume of conductive surface.

When an electromagnetic wave strikes the conductive surface, it encounters two types of losses, namely reflection and absorption [128, 129]. A few portions of the waves get reflected while the remaining waves get transmitted. These waves also gets attenuated when they pass through the conductive surface, which is depicted in Fig. 12 [130]. This combined loss will determine the shielding effectiveness. Though the barrier causes the absorption phenomena, the absorbed electromagnetic energy can be converted into thermal energy. In general, shielding of EM waves can be classified into two groups based on its material classification:

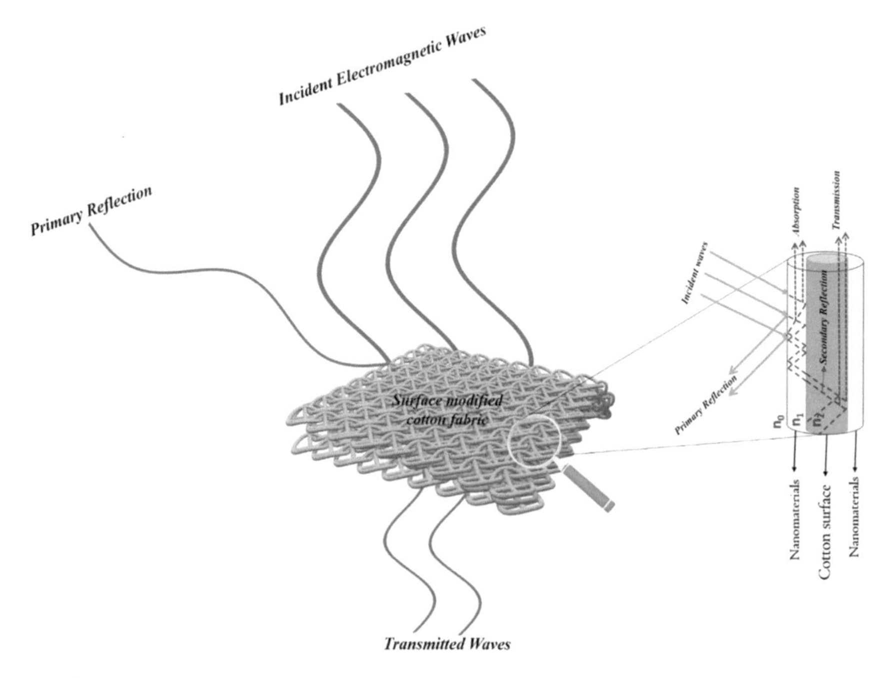

Fig. 12 Schematic representation of possible mechanism of EMI shielding by nanomaterials

1. High permeability materials [131], such as nickel, copper and ferrite derivatives
2. High dielectric constant materials [132], such as SiO_2 and carbon derivatives

High permeability materials can convert the absorbed magnetic energy into thermal energy [130]. Similarly, materials with a high dielectric constant converts the absorbed electric energy into thermal energy [133]. In the past few decades, metallic materials have received maximum attention due to its excellent shielding effectiveness. Functionalization of these materials to the surface of fabrics could be employed by various techniques such as dip-pad-dry-cure technique, sputtering, chemical bath deposition, electroless plating, spray coating and electrospinning deposition. Nowadays, the combination of metallic materials and conductive polymers has been extensively employed as EMI shielding materials to prevent the health-related risks associated with the harmful electromagnetic waves.

Generally, a fabric should have isolation properties with a surface resistance of ~$10^{15}\ \Omega\ cm^{-2}$ [130], which is sufficient enough to resist the electromagnetic waves. In this context, conductive fabrics have gained significant attention to improve the shielding effectiveness. The electromagnetic protection mechanism is based on the model of charge transfer dynamics by utilizing metallic materials on different substrates. Also, the protective barrier must contain higher electric and magnetic dipoles [134] as these dipoles will actively interact with electromagnetic waves and act as a shielding layer for defence applications. Such a barrier should possess high dielectric constant and high magnetic permeability [134–136]. Thus, metallic materials are one of the promising candidates for the protection of electromagnetic waves owing to its higher stability and conductivity. However, these materials possess few drawbacks such as high cost and their immediate ability to get oxidized or undergo corrosion under ambient conditions. In this context, development of wearable EM protective suits by modifying the surface of fabrics with nanomaterials has gained significant momentum to produce protective textiles against EM waves.

4.4 Ultraviolet Protection Fabrics

Prolonged exposure of ultraviolet (UV) radiation from the sun has been identified as the cause for several adverse human related effects such as skin cancer, eye damage, cataracts, wrinkles, sunburn, tanning, suppression of immune system and genetic damage to cells [137–139]. The Occupational Safety and Health Administration (OSHA) suggested the prevention of UV rays by covering up the skin with woven fabrics & hat, sunscreen lotions and UV absorbent materials [20]. Though, the sunscreen lotions and absorbents [140] play a crucial role in blocking the skin from these harmful rays, many of the sunscreen lotions are toxic, unstable and are moderately effective against UV-A region rays [141]. In the last few decades, literature relating to blocking of UV radiations by altering the physical parameters of the fabrics, i.e., fiber type, dyes, fabric porosity between warp and weft count and thickness has been increasing. As a consequence, researchers and scientists have started exploring multifunctional structures by modifying the surface of fabrics with such UV absorbing materials. These materials could be prepared by different modification techniques such as chemical bath technique, padding, sputtering, printing and solution growth process to satisfy the UV protection property of the material.

In this context, inorganic materials have been utilized for the development of multifunctional fabrics owing to the ease at which their properties could be tuned. Among the materials used in UV blocking fabrics, TiO_2 [142] and ZnO [20] were found to possess excellent UV blocking characteristics along with antibacterial property due to their photocatalytic behaviour. Generally, nanoparticles exhibit unique physical and chemical characteristics due to their large surface to volume ratio, which can absorb and scatter UV light. More recently, researchers have started focusing on exploring metal oxide materials, which could perform as a UV blocker even in the visible region by altering the bandgap. Also, metal oxide materials possess the combination of electronic structure alteration, charge transport behaviour, light absorption properties and excited lifetimes. From this background, incorporation of metal oxide materials on the surface of fabrics for UV protection fabrics is a viable and effective alternative. For instance, ZnO modified fabrics are kept in ambient atmosphere, oxygen from the atmosphere adsorb onto the surface where the electrons from the conduction band of the material forms a depletion layer of O_2^- ions on the surface of the modified fabrics. But under UV illumination, these fabrics could excite the photogenerated electrons from the valence band (VB) to the conduction band (CB) and result in

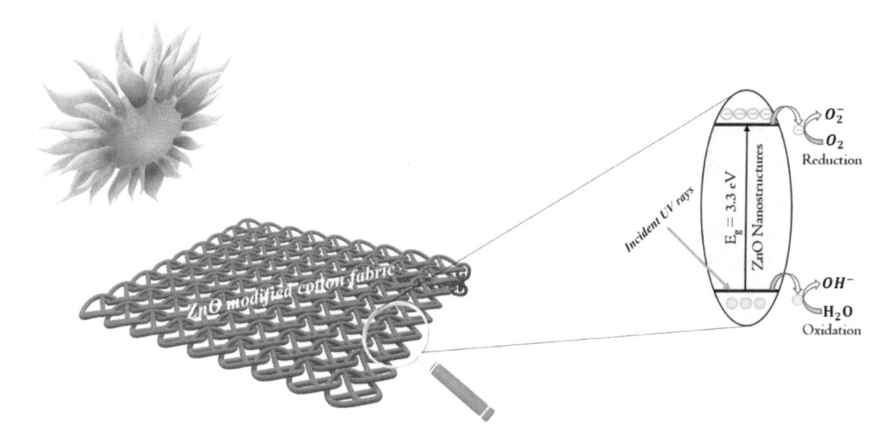

Fig. 13 Schematic representation of possible mechanism of UV blocking by ZnO modified cotton fabric. Redrawn with permission from Ref. [20] Copyright 2018 Elsevier

electron/hole pair generation [20, 142]. Subsequently, the photo-generated electron-hole pairs readily undergo redox reactions with chemisorbed oxygen molecules. This chemisorption process of oxygen molecules and interaction with photogenerated electron/hole pair progresses continuously, which effectively absorb/scatter the ultraviolet rays (Fig. 13). On the other hand, suitable materials such as carbon derivatives with self-assembled nanostructured layer will enhance the UV blocking characteristics, which plays an important factor for the development of novel nanostructured multifunctional fabrics for different applications [143].

4.5 Wearable Sensors

Wearable devices have received maximum attention due to their ability to perform continuous monitoring and stay non-invasive. In the late 1990s, Wearable Health Devices (WHDs) concept was introduced to monitor the health status and improve the quality of care for individual people [144, 145]. In this context, few traditional non-invasive techniques [146] such as nuclear magnetic resonance (NMR) imaging, X-ray imaging, Endotracheal cardiac output monitor (ECOM) and Impedance Cardiography (ICG) were employed by wearable health care devices to create greater opportunities for continuous and remote health care monitoring. According to Market Research Future (MRFR) [147], wearable electronic textiles is an emerging technology that consists of a miniaturized resistor, capacitor and optical sensors for continuous monitoring of health and environmental vital signs.

Due to the rapid industrialization, air pollution serves as a severe threat to living organisms. More specifically, indoor pollution due to the emission of various volatile organic compounds need to be monitored [148]. Generally, inhalation of various volatile organic compounds can cause several health problems like headache, nausea, irritation of eyes, respiratory and serious health disorders [148]. With this background, wearable devices are anticipated to make a mark in healthcare and environmental monitoring applications.

The basic criteria for the usage of smart materials should be compliant with the following [149–151]: (i) compatible with human skin, (ii) adequate detection limit and (iii) high sensitivity to subtle changes. These devices are a result of multiple scientific disciplines such as micro/nanotechnology, material science and engineering, information and communication engineering, biomedical technologies and electronic engineering so as to create a synergy, as shown in Fig. 14. These structured devices can be in different forms; they could be smart watches, shoes, socks, tattoo, wristbands and shirts.

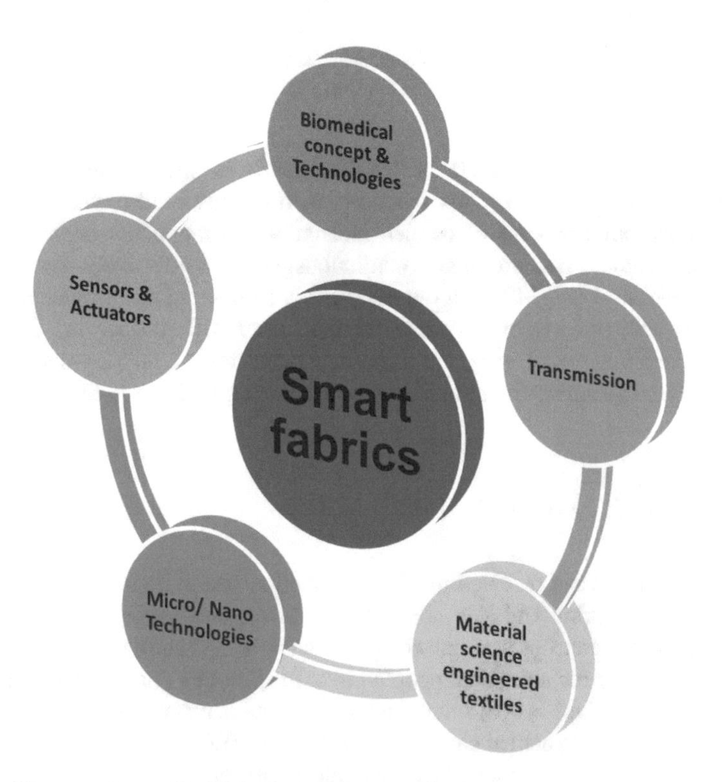

Fig. 14 Elements of wearable fabrics for healthcare applications

4.5.1 Fabric-Based Temperature Sensor

To achieve the requirement of self-monitoring of body temperature, wearable health care devices plays an important role with the superimposed advantages of non-invasive methodologies and quantitative monitoring. These sensors are biocompatible, light-weight, flexible and highly sensitive towards temperature (35–42 °C). The principle of detection is based on resistometric mechanism. Initially, electron hopping was utilized at the interfaces to enhance the sensitivity. For instance, carbon derivatives and poly(3,4-ethylenedioxythiophene) polystyrene sulfonate (PEDOT: PSS) mixed in water towards the fabrication of hypersensitive temperature sensor. The mixed solution was imprinted on the surface of fabrics via pattering technique. Finally, the surface modified fabric is cured at 70 °C for 1 h. Due to electron hopping between the interface [152], change in resistance was observed as a function of temperature in the developed sensing element.

4.5.2 Pressure Monitoring Fabrics

Integration of wearable sensor network with electrodes on life jacket/belt for remote health care monitoring could be employed (e.g., ECG, EG and EMG). As a primary platform, incorporation of nanostructured materials on fabrics can support a variety of sensing applications. For instance, flexible e-Nanoflex band-aid based sensors have been utilized for monitoring respiratory function, blood pressure, pulse rate and other parameters. Also, conductive carbon nanotubes (CNT) based fabrics have been developed as a strain and temperature-based sensor [152]. Piezo-resistive conductive polymers with their composites have also been receiving significant importance due to their excellent flexibility and biocompatibility for the development of strain sensors. The working mechanism behind pressure-based sensors could be realized through the following scheme:

The flexible pressor sensor consists of two different substrates, namely, flexible conductive material medium and suitable substrates. Under external pressure, changes in contact resistance between the substrate and the printed electrodes are observed [153],

(i) Under a constant voltage of 0.01 V, an increase in the current was observed. When an external pressure was applied, it caused a tiny deformation of the porous substrate (fabric) and established a closer contact between the substrate and the interdigitated electrodes, thus increased the conductivity.
(ii) Under unloading of external pressure, the deformed porous substrate returned back to its native state thereby increasing the contact area between the substrate and interdigitated electrodes, thus reducing the current flow.

These materials are robust for longer duration under temperatures ranging from −50 to +200 °C, which is also dependent on conductive phase content. The data obtained from these wearable fabrics is communicated to a nearby storage device, which is then examined by the medical team. By utilizing the wearable fabrics, blood pressure could be measured in a cuff-less method; by using a life belt, the mother and the unborn baby (fetus) could be supervised in the transabdominal condition for a longer duration. Such a design would be essential and beneficial for pregnant women staying in remote areas who face specific fitness issues.

4.6 Conductive Fabrics

In the past few decades, most of the developed surface modified fabrics showed great promise as wearable e-textiles owing to their high flexibility [154], high conductivity [155], low cost [156] and light weighted electronic platform [157]. Integration of several flexible electronic devices on fabrics, textiles, and fibers need electrically conducting and semi-conducting platforms such as polymers, metals, metal oxides and its composites. Thus, materials play an important role in realising wearable, light-weight, easy to process, tough, flexible and portable e-textiles for various versatile interactive electronic applications. The chemical, electrical and mechanical properties of the conductive fabrics are the crucial parameters. In this section, we briefly have discussed the conductive materials that could be integrated on textiles through surface modification with special focus on electrical properties of conductive fabrics, fibers and textiles (Fig. 15).

Surface Modified Fabrics with Metals, Metal Oxides and its Composites
Conductive textiles can be developed by surface modification with metals, metal oxides and composite materials. Especially, conductive metals can be imprinted on textiles/fibers/fabrics broadly through two methods, which are commercially been used for surface modification.

1. Chemical plating
2. Chemical bath deposition

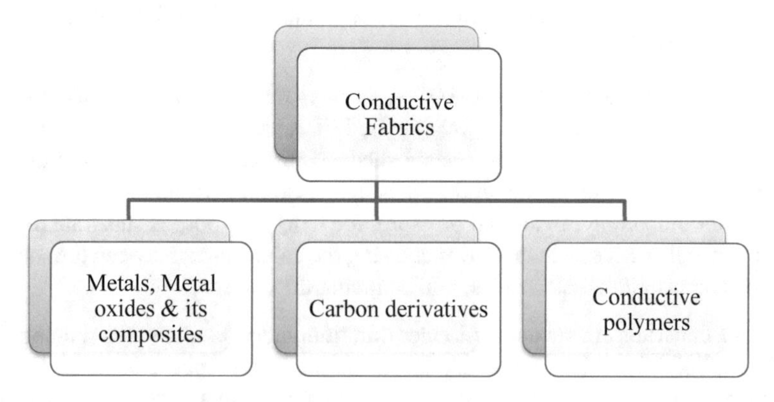

Fig. 15 Source of conductive fabrics

I. Metal and Metal Oxide Fabrics

Development of metal fabrics will be prepared from conductive elements [129] such as nickel, ferrous materials, titanium, zinc, aluminium and copper. These elements possess thin layered structures, with a diameter of 1–80 μm [129]. Although these materials have a high electrical conductivity, heavy weight, massive cost for manufacturing, and inability to produce uniform fibre bends than other surface modification materials. Metal oxide materials are often nearly colourless; so, that the usage of conducting materials onto the fabrics has been considered to overcome fewer problems associated with conducting carbon derivatives. Nanostructured metal oxide materials can be incorporated on the surface of the fabric, or into the sheath–core of the fibre, or chemically reacted with the source material present on the surface of fabric. Conductive fibers can also be modified by surface coatings techniques with pristine metal precursors such as copper iodide, copper sulfide and zinc acetate. Notwithstanding the fact that metallic materials can promote the conductivity of fibers present in the fabric with enhanced adhesion, these materials are limited by their ability to undergo corrosion.

II. Fabrics Containing Conductive Carbon

For the development of conductive fabrics using carbon derivatives, several methods such as chemical bath deposition, electrospinning, electroplating, evaporation deposition and sputtering deposition techniques have been adapted [129]. In these surface modification techniques, the source material is loaded with the higher concentrations of carbon content. Also, carbon could be incorporated into the core of a sheath–core bicomponent fabric via electrospinning technique. Few surface modifying techniques, such as sputtering, evaporation techniques can be utilized for the incorporation of source materials to the side/side fabric.

III. Conductive Polymers

Conductive fabrics can be prepared with spinning the fabric fibers with metallic materials such as zinc, copper, aluminium, silver or gold foil and can be used to produce electrically conductive fibers [56]. Conductive threads prepared using electrospinning technique possess inherent tensile and stiffness properties than conductive yarns [158]. These prepared conductive fibers can be stitched together to develop e-textiles. Similar surface modification processes [129] such as electrode-less plating, sputtering and chemical bath deposition with a conductive polymer by loading fibres and carbonising could be used to furnish conductive coatings to the surface of the fabrics.

Electrode-less plating can produce a uniform surface deposition using conductive materials, but this technique is expensive. Physical vapour deposition techniques such as RF/DC magnetron sputtering and thermal evaporation techniques could promote a uniform deposition with enhanced adhesion as well. Fabrics modified with conductive polymers such as polyacrylonitrile and polypyrrole showed better properties than metals with regards to adhesion. Though it possesses good adhesion characteristics, the growth/modification are difficult to process using conventional surface modification techniques.

5 Conclusion

Development of new gadgets at different times is to design the same by incorporating the modifications like portability, user friendliness, lightweight, durability on suitable substrates. Presently, polymer-based substrates are used to fabricate flexible devices and a competent alternate would be textiles. Nanomaterials have greatly enhanced the functional properties of matter, making them the most sought-after candidate. With triumphant evolution of nanotechnology, nanoparticle functionalized multifunctional textiles are gaining commercial momentum. Nanomaterials offer enhanced functionalities to textiles such as enhancement of oil/water repellence, UV blocking ability, reduction of wrinkles, elimination of static charge build up, continuous monitoring of bodily functions and metabolism, rehabilitation, toxicity reduction, long term durability, and environmental impact without compromising their flexibility or comfort. This new approach has opened up several windows in the wearable and flexible technologies including garments, which is capable of sensing and responding to environmental stimuli including mechanical, chemical, electrical, thermal, optical, or magnetic sources. Enormous research has been carried out by different groups in improving and acquiring the desired properties by selecting a proper surface modification technique, where we can achieve the desired functional group. Each PVD/CVD technique has its own impact and importance in modifying the surface property based on its outer end application. Extensive research works have been carried out to study the UV blocking properties of certain nanoparticles like ZnO, which can then be coated over the fabrics to protect the wearer from the harmful UV rays. The antibacterial properties of silver nanoparticles and ZnO nano particles have been investigated so that it can be coated on socks and help in reducing bacterial growth and odour. The multi versatility of the nanoparticles offer several functionalities to garments and when the knowledge of internet of things (IOT) is put to implementation, the scope of such garments modified with the suitable nanoparticles will be widened. There is an increasing demand for such technology in the present-day society owing to its cost-effective nature, affordability, versatility, high durability and bio compatibility. Surface modified textiles are a promising technology that is yet to be explored in several other aspects of applying science and technology to everyday life.

Acknowledgements The authors are grateful to Nano Mission, Department of Science & Technology, New Delhi for the funding support (SR/NM/NT- 1039/2015) and one of the authors Mr. Dinesh Kumar Subbiah thanks Council of Scientific and Industrial Research, New Delhi for Senior Research Fellowship (09/1095(0035)/18-EMR-1). We also acknowledge SASTRA Deemed University, Thanjavur and Indian Institute of Technology Delhi, New Delhi, for extending infrastructure support to carry out this work.

References

1. I.B.T.-S. in Capek IS (ed) (2006) Chapter 1 Nanotechnology and nanomaterials. Stud. Interface Sci 23:1–69
2. Khan I, Saeed K, Khan I (2019) Nanoparticles: properties, applications and toxicities. Arab J Chem 12:908–931
3. Hubbell JA, Chilkoti A (2012) Chemistry. Nanomaterials for drug delivery. Science 337:303–5
4. Dahman Y (2017) Nanosensors. In: Y.B.T.-N. F.M. for Dahman E (ed), Nanotechnology function Materials Engineering. Elsevier, pp 67–91
5. Yetisen AK, Qu H, Manbachi A, Butt H, Dokmeci MR, Hinestroza JP, Skorobogatiy M, Khademhosseini A, Yun SH (2016) Nanotechnology in textiles. ACS Nano 10:3042–3068
6. Joshi M, Bhattacharyya A, Agarwal N, Parmar S (2012) Nanostructured coatings for super hydrophobic textiles. Bull Mater Sci 35:933–938
7. Dhineshbabu NR, Bose S (2018) Smart textiles coated with eco-friendly UV-blocking nanoparticles derived from natural resources. ACS Omega 3:7454–7465
8. Ghafooripour A (2014) Application of nano technology in civil engineering. J Eng Technol Res 5:104–111
9. Xu P, Wang W, Chen SL (2005) Application of nanosol on the antistatic property of polyester. Melliand Int 11:56–59
10. Faisal AI, Majumder S, Mondal T, Cowan D, Naseh S, Deen MJ (2019) Monitoring methods of human body joints: state-of-the-art and research challenges. Sensors (Basel). 19
11. Mitrano DM, Limpiteeprakan P, Babel S, Nowack B (2016) Durability of nano-enhanced textiles through the life cycle: releases from landfilling after washing. Environ Sci Nano 3:375–387
12. Quandt BM, Braun F, Ferrario D, Rossi RM, Scheel-Sailer A, Wolf M, Bona GL, Hufenus R, Scherer LJ, Boesel LF (2017) Body-monitoring with photonic textiles: A reflective heartbeat sensor based on polymer optical fibres. J R Soc Interface 14:20170060
13. Wang J, Jákli A, Guan Y, Fu S, West J (2017) Developing liquid-crystal functionalized fabrics for wearable sensors. Inf Disp 33:16–20
14. Dhiman G, Chakraborty JN (2015) Antimicrobial performance of cotton finished with triclosan, silver and chitosan. Fash Text 2:13
15. Wei Q (2009) Surface modification of textiles. Woodhead Publishing in Textiles, Elsevier
16. Baptista A, Silva F, Porteiro J, Míguez J, Pinto G (2018) Sputtering physical vapour deposition (PVD) coatings: a critical review on process improvement and market trend demands. Coatings 8:402
17. Ashby MF, Ferreira PJ, Schodek DL (2009) Nanomaterials and Nanotechnologies: an Overview. In: Ashby MF, Ferreira PJ, Schodek DLBT-N (eds) Nanotechnologies and design nanomaterials nanotechnologies design. Butterworth-Heinemann, Boston pp 1–16
18. Ramachandra Rao MS, Singh S (2013) Nanoscience and nanotechnology: fundamentals to frontiers. Wiley
19. Baig B, Halim SA, Farrukh A, Greish Y, Amin A (2019) Current status of nanomaterial-based treatment for hepatocellular carcinoma. Biomed Pharmacother 116:108852
20. Subbiah DK, Mani GK, Babu KJ, Das A, Balaguru Rayappan JB (2018) Nanostructured ZnO on cotton fabrics—a novel flexible gas sensor & UV filter. J Clean Prod 194:372–382
21. Garg R, Dutta N, Choudhury N (2014) Work function engineering of graphene. Nanomaterials 4:267–300
22. Geim AK, Novoselov KS (2009) The rise of graphene, In: Nanoscience and technology a collection review from nature journals. Nature Publishing Group, pp 11–19
23. Partoens B, Peeters FM (2006) From graphene to graphite: electronic structure around the K point. Phys Rev B—Condens Matter Mater Phys 74:75404
24. Pinto AM, Gonçalves IC, Magalhães FD (2013) Graphene-based materials biocompatibility: a review. Colloids Surf B Biointerfaces 111:188–202
25. Karim N, Afroj S, Tan S, He P, Fernando A, Carr C, Novoselov KS (2017) Scalable production of graphene-based wearable E-textiles. ACS Nano 11:12266–12275

26. Zhao Y, Liao A, Pop E (2009) Multiband mobility in semiconducting carbon nanotubes. IEEE Electron Dev Lett 30:1078–1081
27. Yu MF, Lourie O, Dyer MJ, Moloni K, Kelly TF, Ruoff RS (2000) Strength and breaking mechanism of multiwalled carbon nanotubes under tensile load. Science 287:637–640
28. Dürkop T, Getty SA, Cobas E, Fuhrer MS (2004) Extraordinary mobility in semiconducting carbon nanotubes. Nano Lett 4:35–39
29. Han Y, Dai L (2019) Conducting polymers for flexible supercapacitors. Macromol Chem Phys 220:1800355
30. Inzelt G (2017) Synthesis; redox behaviour; composites; sensors; biosensors; supercapacitors; electrocatalysis. J Electrochem Sci Eng 8:3–37
31. Nezakati T, Seifalian A, Tan A, Seifalian AM (2018) Conductive polymers: opportunities and challenges in biomedical applications. Chem Rev 118:6766–6843
32. Zhou Q, Shi G (2016) Conducting polymer-based catalysts. J Am Chem Soc 138:2868–2876
33. Celiesiute R, Ramanaviciene A, Gicevicius M, Ramanavicius A (2019) Electrochromic sensors based on conducting polymers, metal oxides, and coordination complexes. Crit Rev Anal Chem 49:195–208
34. Hoskins BF, Robson R (1990) Design and construction of a new class of scaffolding-like materials comprising infinite polymeric frameworks of 3D-linked molecular rods. A reappraisal of the $Zn(CN)_2$ and $Cd(CN)_2$ structures and the synthesis and structure of the diamond-related framework. J Am Chem Soc 112:1546–1554
35. Furukawa H, Cordova KE, O'Keeffe M, Yaghi OM (2013) The chemistry and applications of metal-organic frameworks. Science 341:1230444
36. Li B, Wen HM, Cui Y, Zhou W, Qian G, Chen B (2016) Emerging multifunctional metal–organic framework materials. Adv Mater 28:8819–8860
37. Li B, Wen HM, Zhou W, Chen B (2014) Porous metal-organic frameworks for gas storage and separation: what, how, and why? J Phys Chem Lett 5:3468–3479
38. Liu Y, Howarth AJ, Vermeulen NA, Moon SY, Hupp JT, Farha OK (2017) Catalytic degradation of chemical warfare agents and their stimulants by metal-organic frameworks. Coord Chem Rev 346:101–111
39. Wu MX, Yang YW (2017) Metal–organic framework (MOF)-based drug/cargo delivery and cancer therapy. Adv Mater 29:1606134
40. Kumar P, Deep A, Kim KH (2015) Metal organic frameworks for sensing applications. TrAC—Trends Anal Chem 73:39–53
41. Caratelli C, Hajek J, Cirujano FG, Waroquier M, Llabrés i Xamena FX, Van Speybroeck V (2017) Nature of active sites on UiO-66 and beneficial influence of water in the catalysis of Fischer esterification. J Catal 352:401–414
42. Gil-San-Millan R, López-Maya E, Hall M, Padial NM, Peterson GW, DeCoste JB, Rodríguez-Albelo LM, Oltra JE, Barea E, Navarro JAR (2017) Chemical warfare agents detoxification properties of zirconium metal-organic frameworks by synergistic incorporation of nucleophilic and basic sites. ACS Appl Mater Interfaces 9:23967–23973
43. Cmarik GE, Kim M, Cohen SM, Walton KS (2012) Tuning the adsorption properties of UiO-66 via ligand functionalization. Langmuir 28:15606–15613
44. Ploskonka AM, Decoste JB (2017) Tailoring the adsorption and reaction chemistry of the metal-organic frameworks UiO-66, UiO-66-NH_2, and HKUST-1 via the incorporation of molecular guests. ACS Appl Mater Interfaces 9:21579–21585
45. Zhou T, Sang Y, Wang X, Wu C, Zeng D, Xie C (2018) Pore size dependent gas-sensing selectivity based on ZnO@ZIF nanorod arrays. Sens Actuators B Chem 258:1099–1106
46. Devic T, Serre C (2014) High valence 3p and transition metal based MOFs. Chem Soc Rev 43:6097–6115
47. Cao Y, Zhang H, Song F, Huang T, Ji J, Zhong Q, Chu W, Xu Q (2018) UiO-66-NH_2/GO composite: synthesis, characterization and CO_2 adsorption performance. Materials (Basel) 11:589
48. Chowdhury P, Bikkina C, Meister D, Dreisbach F, Gumma S (2009) Comparison of adsorption isotherms on Cu-BTC metal organic frameworks synthesized from different routes. Microporous Mesoporous Mater 117:406–413

49. Peterson GW, Lu AX, Epps TH (2017) Tuning the morphology and activity of electrospun polystyrene/UiO-66-NH_2 metal-organic framework composites to enhance chemical warfare agent removal. ACS Appl Mater Interfaces 9:32248–32254
50. Bunge MA, Davis AB, West KN, West CW, Glover TG (2018) Synthesis and characterization of UiO-66-NH_2 metal-organic framework cotton composite textiles. Ind Eng Chem Res 57:9151–9161
51. Mizutani U (2001) Introduction to the electron theory of metals. Cambridge University Press, Cambridge
52. Alim MA, Batra AK, Aggarwal MD, Currie JR (2011) Immittance response of the $SnO_2Bi_2O_3$ based thick-films. Phys B Condens Matter 406:1445–1452
53. Sivalingam D, Gopalakrishnan JB, Rayappan JBB (2012) Structural, morphological, electrical and vapour sensing properties of Mn doped nanostructured ZnO thin films. Sens Actuators B Chem 166–167:624–631
54. Dickey MD (2017) Stretchable and soft electronics using liquid metals. Adv Mater 29:1606425
55. Qiu Q, Zhu M, Li Z, Qiu K, Liu X, Yu J, Ding B (2019) Highly flexible, breathable, tailorable and washable power generation fabrics for wearable electronics. Nano Energy 58:750–758
56. Ding Y, Invernale MA, Sotzing GA (2010) Conductivity trends of pedot-pss impregnated fabric and the effect of conductivity on electrochromic textile. ACS Appl Mater Interfaces 2:1588–1593
57. Gao A, Zhang H, Sun G, Xie K, Hou A (2017) Light-induced antibacterial and UV-protective properties of polyamide 56 biomaterial modified with anthraquinone and benzophenone derivatives. Mater Des 130:215–222
58. Subbiah DK, Babu KJ, Das A, Rayappan JBB (2019) NiO_x nanoflower modified cotton fabric for UV filter and gas sensing applications. ACS Appl Mater Interfaces 11(22):20045–20055
59. Lee K, Lee S (2012) Multi functionality of poly(vinyl alcohol) nanofiber webs containing titanium dioxide. J Appl Polym Sci 124:4038–4046
60. Khandual A, Luximon A, Sachdeva A, Rout N, Sahoo PK (2015) Enhancement of functional properties of cotton by conventional dyeing with TiO_2 nanoparticles. Mater Today Proc 2:3674–3683
61. Liu Y, Xin JH, Choi CH (2012) Cotton fabrics with single-faced superhydrophobicity. Langmuir 28:17426–17434
62. Xu ZJ, Tian YL, Liu HL, Du ZQ (2015) Cotton fabric finishing with TiO_2/SiO_2 composite hydrosol based on ionic cross-linking method. Appl Surf Sci 324:68–75
63. Wang L, Zhang X, Li B, Sun P, Yang J, Xu H, Liu Y (2011) Superhydrophobic and ultraviolet-blocking cotton textiles. ACS Appl Mater Interfaces 3:1277–1281
64. Ganesan RM, Gurumallesh Prabu H (2019) Synthesis of gold nanoparticles using herbal *Acorus calamus* rhizome extract and coating on cotton fabric for antibacterial and UV blocking applications. Arab J Chem 12:2166–2174
65. Vihodceva S, Barloti J, Kukle S, Zommere G (2015) Natural fibre textile nano-level surface modification, environ. In: Technology resource proceedings of international science practical conference, vol 2, 113
66. Ma R, Lee J, Choi D, Moon H, Baik S (2014) Knitted fabrics made from highly conductive stretchable fibers. Nano Lett 14:1944–1951
67. Hu X, Ren N, Chao Y, Lan H, Yan X, Sha Y, Sha X, Bai Y (2017) Highly aligned graphene oxide/poly(vinyl alcohol) nanocomposite fibers with high-strength, antiultraviolet and antibacterial properties. Compos Part A Appl Sci Manuf 102:297–304
68. Qing Y, Cheng L, Li R, Liu G, Zhang Y, Tang X, Wang J, Liu H, Qin Y (2018) Potential antibacterial mechanism of silver nanoparticles and the optimization of orthopedic implants by advanced modification technologies. Int J Nanomed 13:3311–3327
69. Gupta RK, Dunderdale GJ, England MW, Hozumi A (2017) Oil/water separation techniques: a review of recent progresses and future directions. J Mater Chem A 5:16025–16058
70. Xue CH, Chen J, Yin W, Jia ST, Ma JZ (2012) Superhydrophobic conductive textiles with antibacterial property by coating fibers with silver nanoparticles. Appl Surf Sci 258:2468–2472

71. Morais DS, Guedes RM, Lopes MA (2016) Antimicrobial approaches for textiles: from research to market. Materials (Basel) 9:498
72. Goswami S, dos Santos MA, Nandy S, Igreja R, Barquinha P, Martins R, Fortunato E (2019) Human-motion interactive energy harvester based on polyaniline functionalized textile fibers following metal/polymer mechano-responsive charge transfer mechanism. Nano Energy 60:794–801
73. Mondal S, Hu JL (2007) A novel approach to excellent UV protecting cotton fabric with functionalized MWNT containing water vapor permeable PU coating. J Appl Polym Sci 103:3370–3376
74. Tian M, Du M, Qu L, Chen S, Zhu S, Han G (2017) Electromagnetic interference shielding cotton fabrics with high electrical conductivity and electrical heating behavior: Via layer-by-layer self-assembly route. RSC Adv 7:42641–42652
75. Kinkeldei T, Zysset C, Cherenack KH, Troster G (2011) A textile integrated sensor system for monitoring humidity and temperature. In: 16th international solid-state sensors, actuators microsystems conference, pp 1156–1159
76. Wei Q, Xu Y, Wang Y (2009) Textile surface functionalization by physical vapor deposition (PVD) In: Q.B.T.-S.M. of Wei T (ed) Surface modified text. Woodhead Publishing, pp 58–90
77. Jiang SX, Qin WF, Tao XM, Zhang ZM, Yuen CWM, Xiong J, Kan CW, Zhang L, Guo RH, Shang SM (2011) Surface characterization of sputter silver-coated polyester fiber. Fibers Polym 12:616–619
78. Segura G, Guzmán P, Zuñiga P, Chaves S, Barrantes Y, Navarro G, Asenjo J, Guadamuz S, Vargas VI, Chaves J (2015) Copper deposition on fabrics by rf plasma sputtering for medical applications. In: Journal of Physics: Conference Series. IOP Publishing, p 12046
79. Rio L, Kusiak-Nejman E, Kiwi J, Bétrisey B, Pulgarin C, Trampuz A, Bizzini A (2012) Comparison of methods for evaluation of the bactericidal activity of copper-sputtered surfaces against methicillin-resistant *Staphylococcus aureus*. Appl Environ Microbiol 78:8176–8182
80. Liu C, Xu J, Liu Z, Ning X, Jiang S, Miao D (2018) Fabrication of highly electrically conductive Ti/Ag/Ti tri-layer and Ti–Ag alloy thin films on PET fabrics by multi-target magnetron sputtering. J Mater Sci Mater Electron 29:19578–19587
81. Carneiro JO, Teixeira V, Nascimento JHO, Neves J, Tavares PB (2011) Photocatalytic activity and UV-protection of TiO_2 nanocoatings on poly(lactic acid) fibres deposited by pulsed magnetron sputtering. J Nanosci Nanotechnol 8979–8985
82. Krämer A, Kunz C, Gräf S, Müller FA (2015) Pulsed laser deposition of anatase thin films on textile substrates. Appl Surf Sci 353:1046–1051
83. Jelínek M, Trtík V, Jastrabík L (1997) Pulsed laser deposition of thin films BT—physics and materials science of high temperature superconductors, IV. In: Kossowsky R, Jelinek M, Novak J (eds) Springer Netherlands, Dordrecht, pp 215–231
84. Kwong HY, Wong MH, Wong YW, Wong KH (2007) Superhydrophobicity of polytetrafluoroethylene thin film fabricated by pulsed laser deposition. Appl Surf Sci 253:8841–8845
85. Duta L, Popescu AC, Dorcioman G, Mihailescu IN, Stan GE, Zgura I, Enculescu I, Dumitrescu I (2012) ZnO thin films deposited on textile material substrates for biomedical applications: ZnO thin films deposited on textiles. NATO Sci Peace Secur Ser A Chem Biol, 207–210
86. Depla D, Segers S, Leroy W, Van Hove T, Van Parys M (2011) Smart textiles: an explorative study of the use of magnetron sputter deposition. Text Res J 81:1808–1817
87. Kern W, Schuegraf KK (2001) Deposition technologies and applications. In: KBT-H of TFDP, T. (Second Seshan E (ed) Handbook thin film deposition process. Tech. William Andrew Publishing, Norwich, NY, p 11–43
88. Ohring M (2002) Thin-film evaporation processes. In: MBT-MS of TF (Second Ohring E (ed) Material science Thin Film. Academic Press, San Diego, pp 95–144
89. Kenry, Lim CT (2017) Nanofiber technology: current status and emerging developments. Prog Polym Sci 70:1–17
90. Pillay V, Dott C, Choonara YE, Tyagi C, Tomar L, Kumar P, Du Toit LC, Ndesendo VMK (2013) A review of the effect of processing variables on the fabrication of electrospun nanofibers for drug delivery applications. J Nanomater 789289

91. Shi X, Zhou W, Ma D, Ma Q, Bridges D, Ma Y, Hu A (2015) Electrospinning of nanofibers and their applications for energy devices. J Nanomater 140716
92. Xue J, Wu T, Dai Y, Xia Y (2019) Electrospinning and electrospun nanofibers: methods, materials, and applications. Chem Rev 119:5298–5415
93. Rao BG, Mukherjee D, Reddy BM (2017) Novel approaches for preparation of nanoparticles. In: Ficai D, AMBT-N for Grumezescu NT (eds) Nanostructures Nov. Ther. synthesis character application. Elsevier, pp 1–36
94. De M (2014) Solid catalysts. NPTEL. 1–53
95. Bentis A, Boukhriss A, Boyer D, Gmouh S (2017) Development of flame retardant cotton fabric based on ionic liquids via sol-gel technique. In: IOP conference series materials science engineering, p 122001
96. Jung DS, Ko YN, Kang YC, Bin Park S (2014) Recent progress in electrode materials produced by spray pyrolysis for next-generation lithium ion batteries. Adv Powder Technol 25:18–31
97. Mooney JB, Radding SB (1982) Spray pyrolysis processing. Annu Rev Mater Sci, 81–101
98. Vinay Kumar Singh HNS (2017) Versatility of spray pyrolysis technique for the deposition of carbon thin films. J Emerg Technol Innov Res 4
99. Bourfaa F, Lamri Zeggar M, Adjimi A, Aida MS, Attaf N (2016) Investigation of photocatalytic activity of ZnO prepared by spray pyrolis with various precursors. In: IOP conference series materials science engineering. IOP Publishing, p 12049
100. Manoharan C, Pavithra G, Bououdina M, Dhanapandian S, Dhamodharan P (2016) Characterization and study of antibacterial activity of spray pyrolysed ZnO:Al thin films. Appl Nanosci 6:815–825
101. Rajendran V, Dhineshbabu NR, Kanna RR, Kaler KVISIS (2014) Enhancement of thermal stability, flame retardancy, and antimicrobial properties of cotton fabrics functionalized by inorganic nanocomposites. Ind Eng Chem Res 53:19512–19524
102. Hu X, Tian M, Qu L, Zhu S, Han G (2015) Multifunctional cotton fabrics with graphene/polyurethane coatings with far-infrared emission, electrical conductivity, and ultraviolet-blocking properties. Carbon NY 95:625–633
103. Xu QB, Ke XT, Shen LW, Ge NQ, Zhang YY, Fu FY, Liu XD (2018) Surface modification by carboxymethy chitosan via pad-dry-cure method for binding Ag NPs onto cotton fabric. Int J Biol Macromol 111:796–803
104. Mondal MIH, Saha J (2019) Antimicrobial, UV resistant and thermal comfort properties of chitosan- and *Aloe vera*-modified cotton woven fabric. J Polym Environ 27:405–420
105. Gong S, Lai DTH, Wang Y, Yap LW, Si KJ, Shi Q, Jason NN, Sridhar T, Uddin H, Cheng W (2015) Tattoo like polyaniline microparticle-doped gold nanowire patches as highly durable wearable sensors. ACS Appl Mater Interfaces 7:19700–19708
106. Chittenden T (2017) Skin in the game: the use of sensing smart fabrics in tennis costume as a means of analyzing performance. Fash Text 4:1–21
107. Wang ZL, Song J (2006) Piezoelectric nanogenerators based on zinc oxide nanowire arrays. Science 312(5771):242–246
108. Zhu Y, Murali S, Stoller MD, Ganesh KJ, Cai W, Ferreira PJ, Pirkle A, Wallace RM, Cychosz KA, Thommes M, Su D, Stach EA, Ruoff RS (2011) Carbon-based supercapacitors produced by activation of graphene. Science 332(6037):1537–1541
109. Yi F, Wang J, Wang X, Niu S, Li S, Liao Q, Xu Y, You Z, Zhang Y, Wang ZL (2016) Stretchable and waterproof self-charging power system for harvesting energy from diverse deformation and powering wearable electronics. ACS Nano 10:6519–6525
110. Paosangthong W, Torah R, Beeby S (2019) Recent progress on textile-based triboelectric nanogenerators. Nano Energy 55:401–423
111. Zhong J, Zhang Y, Zhong Q, Hu Q, Hu B, Wang ZL, Zhou J (2014) Fiber-based generator for wearable electronics and mobile medication. ACS Nano 8:6273–6280
112. Wang ZL (2013) Triboelectric nanogenerators as new energy technology for self-powered systems and as active mechanical and chemical sensors. ACS Nano 7:9533–9557
113. Zhang C, Wang ZL (2018) Triboelectric nanogenerators. Springer Nature, Singapore, pp 1335–1376

114. Wang Y, Yang Y, Wang ZL (2017) Triboelectric nanogenerators as flexible power sources. Npj Flex Electron 1:1–9
115. Lin L, Wang S, Xie Y, Jing Q, Niu S, Hu Y, Wang ZL (2013) Segmentally structured disk triboelectric nanogenerator for harvesting rotational mechanical energy. Nano Lett 13:2916–2923
116. Jing Q, Zhu G, Bai P, Xie Y, Chen J, Han RPS, Wang ZL (2014) Case-encapsulated triboelectric nanogenerator for harvesting energy from reciprocating sliding motion. ACS Nano 8:3836–3842
117. Wang S, Lin L, Xie Y, Jing Q, Niu S, Lin Wang Z (2013) Sliding-triboelectric nanogenerators based on in-plane charge-separation mechanism. Nano Lett 13:2226–2233
118. Yang Y, Zhou YS, Zhang H, Liu Y, Lee S, Wang ZL (2013) A single-electrode based triboelectric nanogenerator as self-powered tracking system. Adv Mater 25:6594–6601
119. Zhang H, Yang Y, Zhong X, Su Y, Zhou Y, Hu C, Wang ZL (2014) Single-electrode-based rotating triboelectric nanogenerator for harvesting energy from tires. ACS Nano 8:680–689
120. Niu S, Liu Y, Wang S, Lin L, Zhou YS, Hu Y, Wang ZL (2014) Theoretical investigation and structural optimization of single-electrode triboelectric nanogenerators. Adv Funct Mater 24:3332–3340
121. Hati S, Das BR, Ishtiaque SM, Rengasamy RS, Kumar A (2010) Ultraviolet absorbers for textiles. Rjta 14(1)
122. Abney JR, Scalettar BA (1998) Saving your students' skin. Undergraduate experiments that probe UV protection by sunscreens and sunglasses. J Chem Educ 75:757
123. Marrett LD, Chu MBH, Atkinson J, Nuttall R, Bromfield G, Hershfield L, Rosen CF (2016) An update to the recommended core content for sun safety messages for public education in Canada: a consensus report. Can J Public Heal 107:e473–e479
124. Ibrahim NA, Allam EA, El-Hossamy MB, El-Zairy WM (2007) UV-protective finishing of cellulose/wool blended fabrics. Polym—Plast Technol Eng 46:905–911
125. Morganroth PA, Lim HW, Burnett CT (2013) Ultraviolet radiation and the skin: an in-depth review. Am J Lifestyle Med 7:168–181
126. https://www.epa.gov/sites/production/files/2016-09/documents/triethylamine.pdflast visited 1 March 2019, n.d. https://www.epa.gov/sites/production/files/2016-09/documents/triethylamine.pdf. Accessed 25 Nov 2018
127. Dean JA (1940) Handbook of chemistry. McGraw-Hill, Inc., United States of America, pp 1–1561 http://journals.lww.com/soilsci/Abstract/1944/07000/Handbook_of_Chemistry.13.aspx
128. Zhao H, Hou L, Bi S, Lu Y (2017) Enhanced X-band electromagnetic-interference shielding performance of layer-structured fabric-supported polyaniline/cobalt-nickel coatings. ACS Appl Mater Interfaces 9:33059–33070
129. Pandey DN, Basu A, Kumar P (2018) Study of electro conductive textiles : a review. Int J Sci Res Eng Technol 7:744–749
130. Bashari A, Shakeri M, Shirvan AR, Najafabadi SAN (2018) Functional finishing of textiles via nanomaterials. Nanomater Wet Process Text, pp 1–70
131. Wan YJ, Zhu PL, Yu SH, Sun R, Wong CP, Liao WH (2017) Ultralight, super-elastic and volume-preserving cellulose fiber/graphene aerogel for high-performance electromagnetic interference shielding. Carbon NY 115:629–639
132. Liu P, Ng VMH, Yao Z, Zhou J, Lei Y, Yang Z, H. Lv., L.B. Kong (2017) Facile synthesis and hierarchical assembly of flowerlike NiO structures with enhanced dielectric and microwave absorption properties. ACS Appl Mater Interfaces 9:16404–16416
133. Lee J, Liu Y, Liu Y, Park SJ, Park M, Kim HY (2017) Ultrahigh electromagnetic interference shielding performance of lightweight, flexible, and highly conductive copper-clad carbon fiber nonwoven fabrics. J Mater Chem C 5:7853–7861
134. Kolanowska A, Janas D, Herman AP, Jędrysiak RG, Giżewski T, Boncel S (2018) From blackness to invisibility—carbon nanotubes role in the attenuation of and shielding from radio waves for stealth technology. Carbon N Y 126:31–52

135. Raagulan K, Braveenth R, Jang HJ, Lee YS, Yang CM, Kim BM, Moon JJ, Chai KY (2018) Electromagnetic shielding by MXene-Graphene-PVDF composite with hydrophobic, lightweight and flexible graphene coated fabric. Materials (Basel) 11:1–19
136. Tang W, Lu L, Xing D, Fang H, Liu Q, Teh KS (2018) A carbon-fabric/polycarbonate sandwiched film with high tensile and EMI shielding comprehensive properties: an experimental study. Compos Part B Eng 152:8–16
137. Sivakumar A, Murugan R, Sundaresan K, Periyasamy S (2013) UV protection and self-cleaning finish for cotton fabric using metal oxide nanoparticles. Indian J Fibre Text Res 38:285–292
138. Feng XX, Zhang LL, Chen JY, Zhang JC (2006) New insights into solar UV-protective properties of natural dye. J Clean Prod 15:366–372
139. Wang RH, Xin JH, Tao XM (2005) UV-blocking property of dumbbell-shaped ZnO crystallites on cotton fabrics. Inorg Chem 44:3926–3930
140. Subramaniyan G, Sundaramoorthy S, Andiappan M (2013) Ultraviolet protection property of mulberry fruit extract on cotton fabrics. Indian J Fibre Text Res 38:420–423
141. Sharma A, Bányiová K, Vrana B, Justan I, Čupr P (2017) Investigation of cis-trans isomer dependent dermatotoxicokinetics of UV filter ethylhexyl methoxycinnamate through stratum corneum in vivo. Environ Sci Pollut Res 24:25061–25070
142. Mahmoudifard M, Safi M (2012) Novel study of carbon nanotubes as UV absorbers for the modification of cotton fabric. J Text Inst 103:893–899
143. Bharathi Yazhini K, Gurumallesh Prabu H (2015) Study on flame-retardant and UV-protection properties of cotton fabric functionalized with ppy-ZnO-CNT nanocomposite. RSC Adv 5:49062–49069
144. Paradiso R, Loriga G, Taccini N (2005) A wearable health care system based on knitted. IEEE Trans Inf Technol Biomed 9:337–344
145. Lymberis A, Gatzoulis L (2006) Wearable health systems: From smart technologies to real applications. In: Annual international conference of the IEEE engineering medical biology—proceedings Supplement, pp 6789–6792
146. Yang G, Pang G, Pang Z, Gu Y, Mantysalo M, Yang H (2019) Non-invasive flexible and stretchable wearable sensors with nano-based enhancement for chronic disease care. IEEE Rev Biomed Eng 12:34–71
147. Technology C, Services C, Consulting S, Acquisition C, Publications O, Reports S, Reports S, Reports U, Releases P, Studies C, Us A, Market A, Team O, Us C, Device M, Medical smart textile market research report- forecast To 2027, Half-Cooked Res. Reports. (n.d.), pp 1–13. https://www.marketresearchfuture.com/reports/medical-smart-textile-market-1123. Accessed 7 Aug 2019
148. Subbiah DK, Kulandaisamy AJ, George RB, Shankar P, Mani GK, Jayanth Babu K, Rayappan JBB (2018) Nano ceria as xylene sensor—role of cerium precursor. J Alloys Compd 753:771–780
149. Saetiaw C (2017) Design of textile capsule-shaped patch antenna for WBAN applications. In: 9th international conference on information technology and electrical engineering ICITEE, 2017. 978-1-5090-6477-9/17
150. Dittmar A, Lymberis A (2005) Smart clothes and associated wearable devices for biomedical ambulatory monitoring BT. In: 13th international conference on solid-state sensors and actuators and microsystems. TRANSDUCERS '05, 1 221–780389948
151. Bakker E, Gooding JJ (2016) Wearable sensors—an exciting area of research for sensor scientists. ACS Sens 1(7):834–834
152. Blasdel NJ, Wujcik EK, Carletta JE, Lee KS, Monty CN (2015) Fabric nanocomposite resistance temperature detector. IEEE Sens J 15(1):300–306
153. Takamatsu S, Lonjaret T, Crisp D, Badier JM, Malliaras GG, Ismailova E (2015) Direct patterning of organic conductors on knitted textiles for long-term electrocardiography. Sci Rep 5:15003
154. Afroj S, Karim N, Wang Z, Tan S, He P, Holwill M, Ghazaryan D, Fernando A, Novoselov KS (2019) Engineering graphene flakes for wearable textile sensors via highly scalable and ultrafast yarn dyeing technique. ACS Nano 13(4):3847–3857

155. Wu YZ, Sun JX, Li LF, Ding YS, Xu HA (2010) Performance evaluation of a novel cloth electrode. In: 2010 4th international conference bioinformatics biomedical engineering ICBBE 2010, pp 1–5
156. Yu A, Pu X, Wen R, Liu M, Zhou T, Zhang K, Zhang Y, Zhai J, Hu W, Wang ZL (2017) Core-shell-yarn-based triboelectric nanogenerator textiles as power cloths. ACS Nano 11:12764–12771
157. Cheng Q, Ye D, Yang W, Zhang S, Chen H, Chang C, Zhang L (2018) Construction of transparent cellulose-based nanocomposite papers and potential application in flexible solar cells. ACS Sustain Chem Eng 6:8040–8047
158. Dadvar S, Tavanai H, Morshed M (2011) UV-protection properties of electrospun polyacrylonitrile nanofibrous mats embedded with MgO and Al_2O_3 nanoparticles. J Nanoparticle Res 13:5163–5169

Functional Finishing of Cotton Textiles Using Nanomaterials

N. Vigneshwaran and A. Arputharaj

Abstract Cotton textiles are conventionally used by human beings due to their comfort properties and well-established production and processing technologies. Mainly suitable for tropical conditions and other hot and humid environments, cotton performs better in terms of soft-feeling and eco-friendliness. But, cotton could not compete for use in sportswear, medical textiles, filtration, agro-textiles and other technical textiles due to its inherent weakness in terms of absorbency, susceptibility to microbial attack and poor strength properties. In order to impart or improve the required functional properties, various nanomaterials made from metals, metal oxides, ceramics, polymers and carbon are being used as finishing agents. The methods of application vary widely, starting from the traditional pad-dry-cure process and cross-linking to electrospraying, electrospinning, in situ synthesis and layer-by-layer deposition. Novel and un-conventional properties like superhydrophobicity, electrical conductivity, photocatalytic/self-cleaning activity, antimicrobial, UV-protective and flame retarding properties could be achieved in cotton using the nanomaterials. Even, energy production using nanostructures on the surface of cotton fibres is being evolved. This chapter describes the basic mechanisms involved in the use of nanomaterials to impart functional properties in cotton textiles, relevant issues and the future scope for commercial exploitation.

Keywords Antimicrobial finish · Cotton textiles · E-textiles · Flame retardancy · Nanofinishing · UV protection

1 Introduction

Cotton, the King of Fibres, is conventionally used for apparels, medical and technical textiles due to their inherent comfort properties and established production and processing technologies. But, with the stiff competition from synthetic fibres, cotton has to improve its performance in various other aspects so that it can retain its marketability. Multiple functionalities need to be incorporated into cotton textiles

N. Vigneshwaran (✉) · A. Arputharaj
ICAR-Central Institute for Research on Cotton Technology, Mumbai 400019, India
e-mail: vigneshwaran.n@icar.gov.in

M. Shahid and R. Adivarekar (eds.), *Advances in Functional Finishing of Textiles*, Textile Science and Clothing Technology, https://doi.org/10.1007/978-981-15-3669-4_2

to enhance its scope of application. Various chemical and polymeric treatments are traditionally used to impart functionalities like durable press finish [1, 2], wrinkle resistant and flame retardant finish [3], water and oil repellent finish [4], perfumed finish [5], better dyeability [6] and abrasion resistance finish [7].

Recently, various natural biomaterials are being used to impart diversified functionalities to cotton textiles. Cotton textiles finished with Aloe vera gel having 1,2,3,4-butanetetracarboxlic acid as crosslinking agent is reported to show excellent antibacterial property [8] due to destruction of bacterial cell wall. Similarly, neem seed extract treated fabric could impart antibacterial activity in cotton textiles having fastness up to five machine washes [9]. The use of combination of aloe vera, chitosan and curcumin on cotton substrate imparts antibacterial property having fastness up to twenty five washing cycles [10]. The Gallnut extract is used to impart antioxidant property in cotton textiles [11]. In many cases, the natural dyes could add multifunctional properties in addition to its primary aim of colouring the cotton fabrics [12–15]. In spite of numerous research work being carried out to use the natural materials for functional finishing of cotton textiles, the issues like affinity to cotton textiles, uniformity and stability of natural materials/dyes, less wash/light fastness and lack of adequate availability limits their use on an industrial scale.

Nanotechnology during the last two decades revolutionized every fields of science and technology with the first ever industrial application in textiles. Nanotex® is a leading fabric innovation company, started in 1998, provides nanotechnology-based textile enhancements to the apparel, home and commercial/residential interiors markets. This company replicated the natural Lotus leaf phenomenon in the textiles using nanomaterial finish. This was followed by numerous research and development activities in the field of nanofinish in cotton and synthetic textiles. This chapter focuses on the methods of nanofinish and various multi functionalities that could be achieved using nanofinish in cotton textiles. Figure 1 shows the possible advantages of nanofinish as compared to conventional finishes using chemicals and polymers.

2 Nanomaterials Finishing Techniques

Most of the research work has been reported to establish the UV absorption properties metal oxide nano particles such as TiO_2, ZnO, Al_2O_3, CeO_2 on textile materials. The functionalization of textile fabrics with NPs to improve the ability of the fabric to block UVR can be achieved by ex situ and in situ methods. Ex situ methods consist of two steps: (1) Top down or bottom up approach to synthesize nanoparticles, (2) Application of nanoparticles to textiles by means of wet-chemical procedures. For fixing the nanoparticles to fabrics, the steps include impregnating with a wet pick-up, drying followed by curing process. Wherever required, binders or cross-linkers are also added to fix the nanoparticles on the surface of textiles materials.

Wang et al. [16] synthesized dumb-bell shaped nano-ZnO and applied on cotton fabric by pad-dry-cure method. They observed that the higher the curing temperature employed better the UPF was achieved. The UPF of more than 400 has been achieved

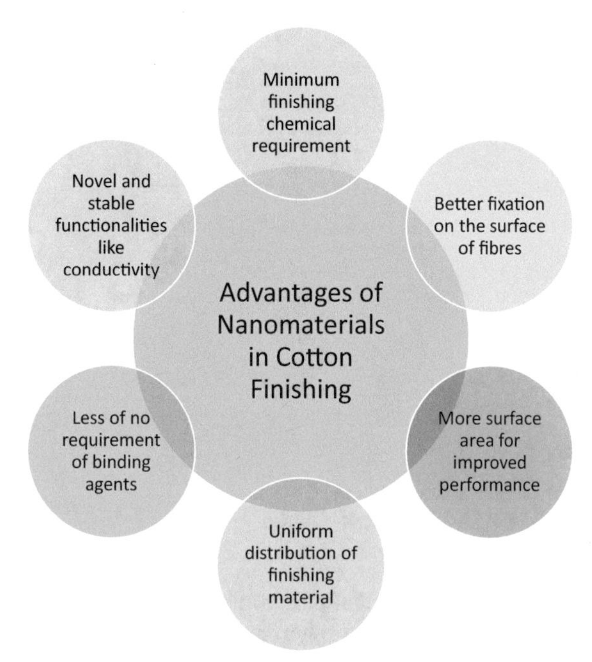

Fig. 1 Advantages of using nanomaterials in finishing of cotton textiles

when a curing temperature of at least 150 °C was employed at the UV-blocking range was 352–280 nm. Similarly, the pad-dry-cure process to fix the nano-ZnO (Starch stabilized) on to the surface of cotton fabrics to impart antibacterial and UV-absorbing properties was successfully evaluated in our lab [17, 18]. Ex situ methods involve time consuming synthetic protocols and reduced functional properties after laundering process. The process sequence in such cases either involves several steps or is chemical intensive. Recent research efforts indicate that in situ techniques will be remedy for such problems.

In situ or one-step methods are characterized by the synthesis of nanoparticles in the presence of the fabric/substrate. Mao et al. [19] reported in situ growth of ZnO on SiO_2 sol coated cotton fabrics via hydrothermal method using relatively higher reagent concentrations. The cotton fabric was then treated in hot water to obtain needle-shaped nano-ZnO crystallites. As a result, ZnO coated cotton fabric had better UV-blocking property. After treating in boiling water for 3 h, ZnO coated cotton fabric had excellent UV protection (UPF > 50). After 20 washes, cotton fabric still had good UV protection. Arputharaj et al. [20] synthesized nano-ZnO on cotton fabric using the precursors of $Zn(NO_3)_2{\cdot}6H_2O$ and NaOH. Plain cotton woven fabric was treated with methanolic solution of precursors and nano-ZnO was developed as given in the following schematic diagram. UPF of in situ synthesized nano-ZnO treated fabric (42) was obtained as compared to control fabric UPF mean (5). The UV protection functionality was found to be unaltered even after 30 laundering. The schematic of in situ process is shown in Fig. 2.

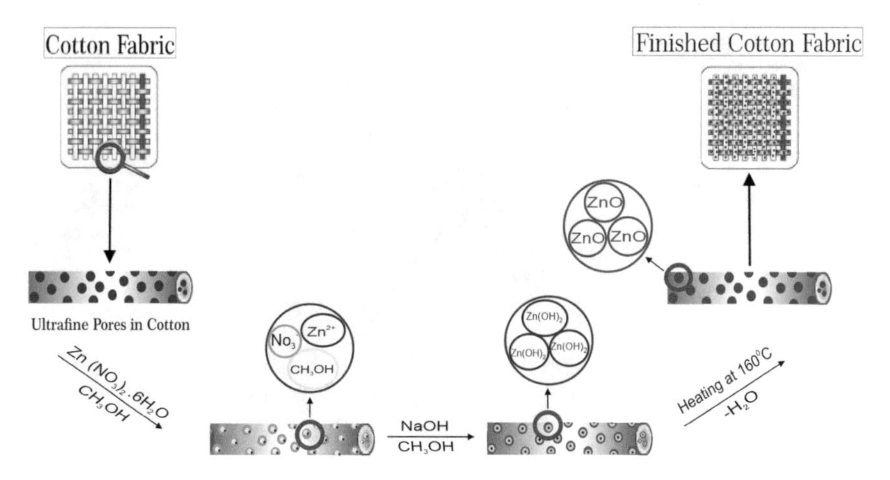

Fig. 2 Schematic diagram of the in situ synthesis. Reprinted with permission from Ref. [20], Copyright (2017) Springer Nature

Our lab has also earlier reported the in situ synthesis of nano silver in the cotton fabrics that resulted in brown coloured fabric due to the surface plasmon resonance of silver nanoparticles [21]. This process resulted in antibacterial property of the fabric in addition to providing the colour to the fabric. Similar in situ functionalization of the cotton fabric with ZnO NPs was carried out by Román et al. [22] via exhaust dyeing method using same precursors in aqueous medium. Cotton fabrics exhibited excellent UV protection, even after 20 washing cycles. Prasad et al. [23] synthesized nano-ZnO onto 100% cotton fabrics (terry or woven) by spraying or dipping process to impart UV protection property. It was reported that spraying process resulted in 3 times less uptake of nano-ZnO than that of dipping process, without significant reduction in UPF properties.

Akhavan and Montazer [24] simultaneously synthesized and loaded nanotitania onto the cotton fabric by hydrolysis of titanium tetra isopropoxide (TIP) using ultrasonic irradiation (50 kHz, 50 W) using acetic acid as a dispersant. They reported that due to the formation of covalent bonding between the OH groups of cotton and the OH groups of TiO_2 excellent UV protection property of the fabric resulted even after 20 launderings. El-Naggar et al. [25] treated the cotton fabric with aqueous solution of TIP and dried. After that the samples were treated with urea nitrate solution. The treated cotton samples dried then cured at 130 °C. It is reported that increasing the amount of TiO_2 NPs deposited on the surface of cotton increased the UV protection properties.

A novel method to express the durability of nano-finish (Nano-ZnO) was reported as "Half-life time (t1/2) of zinc concentration on the fabric" for the durability to sweat [26]. The results indicated that nanofinished antibacterial cotton fabric was relatively sensitive to the acid artificial sweat while that was durable in saline or alkaline solution, and the t1/2 exceeded 3000 min in them.

3 Antimicrobial Finish

Our earliest work [18] reported the use of zinc oxide—soluble starch nanocomposites to impart antibacterial activity on the surface of cotton fabrics and it demonstrated excellent antibacterial activity against the two representative bacteria, *Staphylococcus aureus* (Gram positive) and *Klebsiella pneumoniae* (Gram negative). Later, nano-silver was also demonstrated to impart antibacterial activity on the surface of cotton [21]. Due to the ease of synthesis, nano-silver is the widely used material to impart antibacterial activity in cotton textiles [27–31]. Though the nano-silver is a potent nanomaterial for antibacterial activity, the challenge is to avoid the oxidation of nano-silver during its exposure and uneven visible deposition on the surface of cotton fabrics that is quickly shown due to its colour. Hence, the nano-ZnO and nano-titania are also the viable alternative materials to impart antimicrobial finish on cotton textiles.

ZnO is a bio-safe material that possesses photo-oxidizing and photocatalysis impacts on chemical and biological species. The bactericidal and bacteriostatic mechanisms of metal oxide nanoparticles are due to the generation of reactive oxygen species (ROS) like hydrogen peroxide, hydroxyl radicals and peroxide radicals. ROS has been a major factor for several mechanisms including cell wall damage, enhanced membrane permeability, and uptake of toxic dissolved metal ions. In some cases, the antibacterial activity is also attributed to abrasive surface texture of nanomaterials. The mechanisms of antibacterial activity by nano-ZnO is discussed in detail in a recent review [32]. The antibacterial activity of nano silver is mainly due to its binding effect with the proteins/enzymes of the microbes and their inactivation. The chitosan, a known antibacterial biopolymer, when converted to nano-form and applied on the surface of cotton fabrics, exhibited excellent antibacterial property [33]. Various mechanisms involved in the antibacterial activity of nanomaterials are summarized in Fig. 3.

4 UV Protection Finish

Now a day, the level of ultraviolet (UV) radiation reaching the Earth's surface has increased due to the ozone depletion and other environmental related issues. It is well known that small doses of UV rays are beneficial for the body as this is essential for the synthesis of Vitamin D. However, an overdose of it can be highly detrimental to the skin. UV rays react with the compound in the human skin which is called as melanin (Fig. 4). As a protective mechanism, the precursors of melanin absorb UV rays and getting converted into a black coloured polymer. But some of human population doesn't have much of melanin in their skin due to their genetic nature. This can results in harmful health conditions including skin cancer, cataracts, premature ageing and also sunburns. The use of textiles as a means of sun protection is getting popular apart from the conventional use of sunscreen and lotions. UV protective

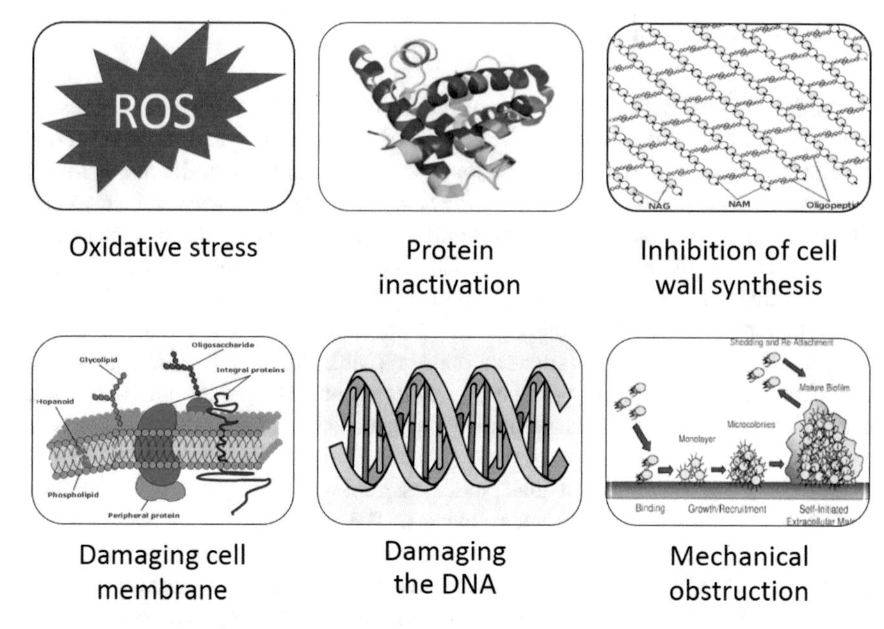

Fig. 3 Mechanisms of antimicrobial properties of nanomaterials

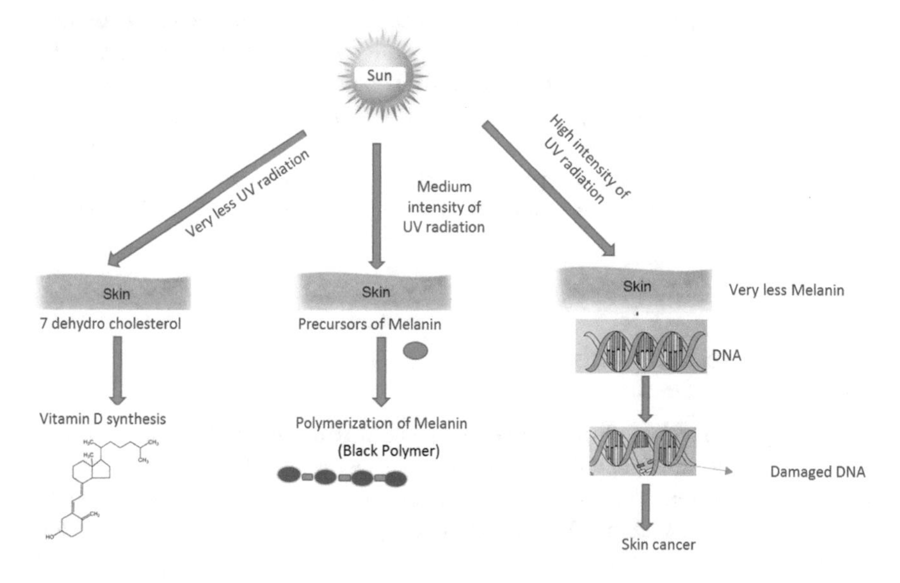

Fig. 4 Positive and negative effects of UV radiations

clothing represents the most convenient and reliable method of protecting the skin against the harmful effects of the Sun. The demand for UV protective clothing has been growing significantly as consumers have become more aware of the dangers of excessive exposure to the sun.

Fabric parameters such as the porosity, type, color, weight and thickness have direct influence on the UV protection of a finished garment. The application of UV absorbers into the surface of the fabric plays a vital role in the improvement of UPF of apparel. Stretching and laundering during use can alter the UV-protective properties of a textile. The use of UV-blocking cloths can provide excellent protection against the hazards of sunlight; this is especially true for garments manufactured as UV-protective clothing.

UV-protective finishing agents are the chemicals which are used to absorb UV radiations between 290 and 350 nm. UV protective chemicals can be majorly classified into organic and inorganic absorbers (Fig. 5). Organic UV absorbers are colourless organic aromatic molecules with conjugated double bonds having high absorption coefficients. After absorption, they transform UV radiation energy into vibration energy. If the molecules of UV absorbers are permanently transformed into their non-absorbing isomers, their UV absorbing proper-ties are destroyed. Compounds with phenolic group which form intra molecular O–H–O bridges, such as salicylates, 2-hydroxybenzophenones, 2,2′-dihydroxy benzophenones, are used as UV absorbers. The growing use of synthetic organic UV absorbers in recent years have caused environmental concerns since different toxic degradation products of UV absorbers can bio accumulate and will result environmental problems.

Inorganic oxides such as TiO_2, CeO_2, and ZnO are used as UV-protective agents. As semiconductors, metal oxides are characterized by an electron band structure that includes bands with orbitals and gaps in the UV spectral region. The band gap energies are corresponding to their absorption spectra and refractive index. UV light is absorbed by excitation of electrons from the valance band to the conduction band.

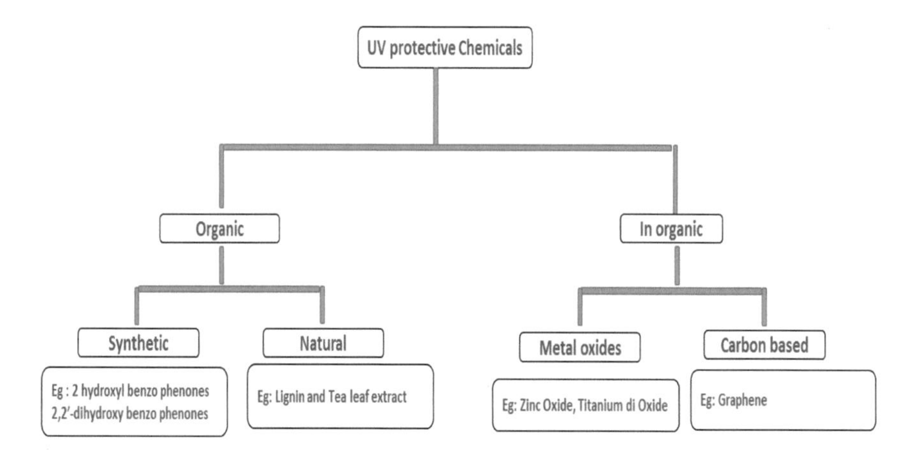

Fig. 5 Classification of UV protective chemicals

Upon the absorption of UV radiation with energy that matches or exceeds the band gap energy, the electrons from the valence band are excited to the conduction band, thus leaving a positively charged hole in the valence band. Therefore, light below these wavelengths has enough energy to excite electrons and is absorbed by metal oxides. Since energy always has to go somewhere, and UV is quite strong, metal oxides absorb UV and turn it into comparably harmless infrared radiation, which they dispose of as heat.

$$I_s \propto \frac{Nd^6}{\lambda^4}\left|\frac{m^2 - I}{m^2 + 2}\right|^2 I_i \quad (1)$$

Where, I_s is the intensity of scattered light, N is the number of particles, d is the diameter of the particle, λ is the wavelength of incident light, m is the relative refractive index and I_i is the intensity of incident light.

If we reduce the particle size of the metal oxides they become very good UV absorption materials. Mie theory explains that the intensity of scattered light is a function of the size of a single particle. Equation (1) gives the relationship between the particle size and the intensity of scattered light (I_s). ZnO has a band gap energy of ~3.3 eV that corresponds to the wavelengths of ~375 nm. If we reduce the size of the ZnO the wave length of the absorption is also reduced. Reducing the size of inorganic UV absorbers to lower than 50 nm results in higher transparency of the UV-blocking agents. The advantage of nano inorganic UV absorbers than organic UV absorbers are

- Chemical stability under both high temperature and UV-ray exposure
- Broader and tailor made spectrum of UV absorption
- Less toxic to the environment due to non-bio accumulation
- Comparatively cheaper
- Multi-functional property like antimicrobial, self-cleaning etc.

Graphene has recently gained keen interest owing to its outstanding electronic properties. It has a UV absorption peak around 281 nm, hence it can absorb UV radiation with a wavelength comprised in the 100–281 nm range Graphene derivatives have been widely employed as UV-blocking materials for coating the fabrics and obtain functional fabrics. Hu et al. [34] coated cotton fabrics with GNPs to get UV protection functionality. The UPF value of the treated fabric was found to be 500 (0.8% wt GNPs), which is a 60-fold increase when compared to control cotton fabrics. Kale et al. used titanium chloride as a reducing agent to convert graphene into graphene oxide on polyester fabric. Both graphene as well as titanium dioxide was formed due to self-oxidation of titanium chloride into titanium dioxide. Electro conductive, antistatic, UV protected and mechanically strong polyester fabric was produced by using this technique.

5 Flame Retardant Finish

Natural cellulosic textile substrates like cotton, linen are made-up of the natural polymer i.e. cellulose. These materials are being used in the production of value-added home textiles and other upholstery fabrics. All these products exhibit high flammability and combustibility. Due to its chemical composition, cotton is highly prone to flammability that can cause immediate combustion leading to the flames and fire. To prevent fire accidents and loss of human life, quality and safety of textile materials is to be assured. Due to the upcoming government regulations, flame retardant property of textiles become very important attribute for the materials that are used as fabrics in airplanes, trains, buses, hotels, restaurants, and other public places. Flame retardant finish describes a finish that imparts slow burning capacity of self-extinguishing property to the fabrics. The flame retardant chemicals that are used in textile finishing are classified as given in Fig. 6.

The health issues associated with the halogenated flame retardant chemicals are discussed in detail by Shaw in 2011 [35]. It was suggested for a more systematic study to understand the actual impact of such chemicals during human exposure. Though the degradation of brominated polymeric flame retardants cause no acute toxicity, chronic toxicity might relevant [36]. To overcome the various problems associated with the flame retardant toxicity, nanomaterials might be an alternative for finishing of cotton textiles. Arputharaj et al. [20] reported that the presence of nano-ZnO which was synthesized in situ in cotton fabric improved the thermal stability of cotton fabric. Similar results were reported by Samanta et al. [37] for the flame retardant property of nano-ZnO treated jute fabric. Sheshama et al. [38] reported that 1% nano ZnO treated sisal yarn showed more LOI (Limiting Oxygen Index) and less burning rate compared to the 12% bulk ZnO treated yarn. These results indicate that

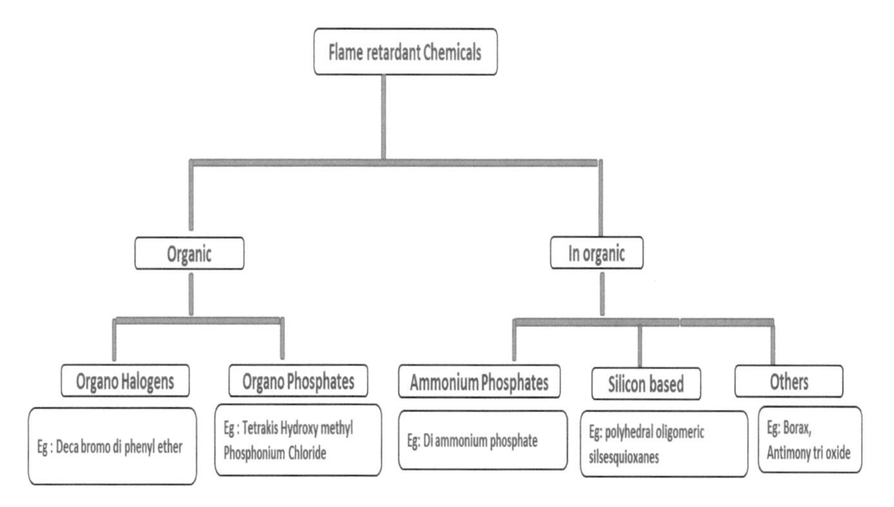

Fig. 6 Classification of flame retardants

the consumption of the traditional chemicals can be reduced by using flame retardant nanoparticles.

FRs made with naturally occurring clay called montmorillonite (MMT) are poised to have a huge influence on future fire safety due to their great potential for being applied to many different fields in textile industry. Chang et al. used continuous layer by layer deposition process to apply 50 bilayers of flame-retardant clay nanoparticles using a modified pad-steam unit. They found that FR properties of coated fabric was significantly greater than the uncoated fabrics [39].

6 Superhydrophobic Finish

Superhydrophobic finish of textile materials, also called as lotus effect, is carried out by imparting roughness on the surface in combination with hydrophobicity. The water contact angle above 150° is generally accepted as a superhydrophobic finished surface. While the nanomaterials could be used to impart roughness on the surface, the hydrophobic chemistry is required to make it to superhydrophobicity. By sol-gel process, superhydrophobic cotton fabrics was demonstrated using silica nanoparticles and perfluorooctylated quaternary ammonium silane coupling agent [40]. Another work reported the formation of superhydrophobic nature by electrostatic layer-by-layer assembly of polyelectrolyte/silica nanoparticle multilayers on cotton fibers, followed with a fluoroalkylsilane treatment [41]. In this case, the hydrophobicity was tailored by controlling the number of layers used for the assembly. For imparting durability to the finish, a robust and self-healing superhydrophobic cotton fabric was fabricated by facile dip coating and UV curing [42]. The fabrics were dip-coated with tri-functionality vinyl perfluorodecanol, vinyl-terminated polydimethylsiloxane and octavinyl-polyhedral oligomeric silsesquioxane followed by UV curing. These obtained cotton fabrics exhibited superior resistance to various liquid pollutants, and had excellent resistance to the acid and alkali liquid. Furthermore, they were durable to withstand 10,000 cycles of abrasion, 120 h of accelerated weathering test and heating or freezing test. Thus, the superhydrophobic treatment helps in self-cleaning also, wherein the dust particles/pollutants are not allowed to settle on the surface of the finished fabrics.

7 Electronics Textiles

Electronic textiles (E-textiles) represent the textile materials having the electronic components impregnated or interwoven in them. Though synthetic textile materials have a lead in the e-textiles, cotton textiles are also being considered in diversified applications due to their comfort and eco-friendliness. Electro-conductive cotton textiles are under development using simple dip-coating process for depositing functionalized carbon nanotubes [43] to develop a flexible electro thermal heating element. Graphene is being routinely evaluated for coating on the surface of cotton fabrics to make it electro conductive [44]. Another research team [45] deposited graphene oxide on the surface of cotton to form the conductive cotton fabric that viability as a strain sensor even after 400 bending cycles. The deposited graphene oxide was reduced by hot press method at 180 °C for 60 min without the use of chemical reducing agent. In yet another work, the cotton T-shirt was converted into highly conductive and flexible activated carbon textiles by a chemical activation route, resulting in an ideal electrical double-layer capacitive behavior [46]. A conductive activated cotton textile with porous tubular structure was prepared from cotton textile to load sulfur that was further wrapped with partially reduced graphene oxide to immobilize the lithium polysulfides [47]. This was used to prepare lithium-sulfur battery that exceptional capacity and rate performance.

8 Safety Aspects

The use of nanomaterials for nanofinishing of cotton textiles is on the rise and, it simultaneously indicates the necessity to look into the safety aspects of nanofinishing. Figure 7 shows the diversified safety aspects to be considered while using nanofinishing in cotton textiles. The areas required attention is the site of production of nanomaterials, location of nanofinishing of cotton textiles, effluent management and disposal of cotton textiles treated with nanomaterials. Industrial personnel and customers' awareness are required for handling the nanofinished textile materials.

Fig. 7 Safety aspects to be considered while using nanomaterials for cotton finishing

The toxicity of nanomaterials is mostly of chemical-specific apart from the size-related issues. It is assumed that the organic nanoparticles are less toxic than their inorganic counterparts. But, a systematic study is required to prove this hypothesis. Also, a complete life cycle analysis needs to be done to weigh the overall impact of the use of nanomaterials in finishing of cotton textiles.

9 Conclusion

The use of nanomaterials might impart novel functionalities to the cotton textiles and has the potential to extend the use of cotton textiles beyond apparel segments. It can diversify the applications in medical textiles, agro-textiles, home furnishings and other technical textiles. But, care is required to avoid the excess usage of nanomaterials, restriction in use of inorganic nanoparticles and creation of awareness among various stakeholders. By following stringent norms and responsible behavior in the work place, the use of nanomaterials for cotton textiles could be used for their fullest potential.

References

1. Cashen NA (1979) Reduced formaldehyde release in durable-press finishing of cotton textiles. Text Res J 49:480–484
2. Raheel M, Guo C (1998) Single-step dyeing and formaldehyde-free durable press finishing of cotton fabric. Text Res J 68:571–577
3. Reinhardt RM, Harris JA (1980) Ultraviolet radiation in treatments for imparting functional properties to cotton textiles 1. Text Res J 50:139–147
4. Simončič B, Tomšič B, Černe L, Orel B, Jerman I, Kovač J, Žerjav M, Simončič A (2012) Multifunctional water and oil repellent and antimicrobial properties of finished cotton: influence of sol–gel finishing procedure. J Sol-Gel Sci Technol 61:340–354
5. Fouda MMG, Fahmy HM (2011) Multifunctional finish and cotton cellulose fabric. Carbohyd Polym 86:625–629
6. Gupta D, Haile A (2007) Multifunctional properties of cotton fabric treated with chitosan and carboxymethyl chitosan. Carbohyd Polym 69:164–171
7. Zhou W, Yang CQ, Lickfield GC (2004) Mechanical strength of durable press finished cotton fabric part V: poly(vinyl alcohol) as an additive to improve fabric abrasion resistance. J Appl Polym Sci 91:3940–3946
8. Ali SW, Purwar R, Joshi M, Rajendran S (2014) Antibacterial properties of aloe vera gel-finished cotton fabric. Cellulose 21:2063–2072
9. Purwar R, Mishra P, Joshi M (2008) Antibacterial finishing of cotton textiles using neem extract. AATCC Rev 8:36–43
10. Ammayappan L, Jeyakodi Moses J (2009) Study of antimicrobial activity of aloevera, chitosan, and curcumin on cotton, wool, and rabbit hair. Fibers Polym 10:161–166
11. Koh E, Hong KH (2014) Gallnut extract-treated wool and cotton for developing green functional textiles. Dyes Pigm 103:222–227
12. Ibrahim NA, El-Gamal AR, Gouda M, Mahrous F (2010) A new approach for natural dyeing and functional finishing of cotton cellulose. Carbohyd Polym 82:1205–1211

13. Davulcu A, Benli H, Şen Y, Bahtiyari Mİ (2014) Dyeing of cotton with thyme and pomegranate peel. Cellulose 21:4671–4680
14. da Silva MG, de Barros MASD, de Almeida RTR, Pilau EJ, Pinto E, Soares G, Santos JG (2018) Cleaner production of antimicrobial and anti-UV cotton materials through dyeing with eucalyptus leaves extract. J Clean Prod 199:807–816
15. Grifoni D, Bacci L, Di Lonardo S, Pinelli P, Scardigli A, Camilli F, Sabatini F, Zipoli G, Romani A (2014) UV protective properties of cotton and flax fabrics dyed with multifunctional plant extracts. Dyes Pigm 105:89–96
16. Wang RH, Xin JH, Tao XM (2005) UV-blocking property of dumbbell-shaped ZnO crystallites on cotton fabrics. Inorg Chem 44:3926–3930
17. Yadav A, Prasad V, Kathe AA, Raj S, Yadav D, Sundaramoorthy C, Vigneshwaran N (2006) Functional finishing in cotton fabrics using zinc oxide nanoparticles. Bull Mater Sci 29:641–645
18. Vigneshwaran N, Kumar S, Kathe AA, Varadarajan PV, Prasad V (2006) Functional finishing of cotton fabrics using zinc oxide–soluble starch nanocomposites. Nanotechnology 17:5087–5095
19. Mao Z, Shi Q, Zhang L, Cao H (2009) The formation and UV-blocking property of needle-shaped ZnO nanorod on cotton fabric. Thin Solid Films 517:2681–2686
20. Arputharaj A, Vigneshwaran N, Shukla SR (2017) A simple and efficient protocol to develop durable multifunctional property to cellulosic materials using in situ generated nano-ZnO. Cellulose 24:3399–3410
21. Vigneshwaran N, Kathe AA, Varadarajan PV, Nachane R, Balasubramanya R (2007) Functional finishing of cotton fabrics using silver nanoparticles. J Nanosci Nanotechnol 7:1893–1897
22. Román LE, Huachani J, Uribe C, Solís JL, Gómez MM, Costa S, Costa S (2019) Blocking erythemally weighted UV radiation using cotton fabrics functionalized with ZnO nanoparticles in situ. Appl Surf Sci 469:204–212
23. Prasad V, Arputharaj A, Bharimalla A, Patil P, Vigneshwaran N (2016) Durable multifunctional finishing of cotton fabrics by in situ synthesis of nano-ZnO. Appl Surf Sci 390:936–940
24. Akhavan Sadr F, Montazer M (2014) In situ sonosynthesis of nano TiO_2 on cotton fabric. Ultrason Sonochem 21:681–691
25. El-Naggar ME, Shaheen TI, Zaghloul S, El-Rafie MH, Hebeish A (2016) Antibacterial activities and UV protection of the in situ synthesized titanium oxide nanoparticles on cotton fabrics. Ind Eng Chem Res 55:2661–2668
26. Li Q, Chen S-L, Jiang W-C (2007) Durability of nano ZnO antibacterial cotton fabric to sweat. J Appl Polym Sci 103:412–416
27. Ghosh S, Yadav S, Reynolds N (2010) Antibacterial properties of cotton fabric treated with silver nanoparticles. J Text Inst 101:917–924
28. Sadanand V, Tian H, Rajulu AV, Satyanarayana B (2017) Antibacterial cotton fabric with in situ generated silver nanoparticles by one-step hydrothermal method. Int J Polym Anal Charact 22:275–279
29. Zhang D, Zhang G, Chen L, Liao Y, Chen Y, Lin H (2013) Multifunctional finishing of cotton fabric based on in situ fabrication of polymer-hybrid nanoparticles. J Appl Polym Sci 130:3778–3784
30. Nateri AS, Oroumei A, Dadvar S, Fallah-Shojaie A, Khayati G, Emamgholipur O (2011) Antibacterial nanofinishing of cotton fabrics using silver nanoparticles via simultaneous synthesizing and coating process. Synth React Inorg, Met-Org, Nano-Met Chem 41:1263–1267
31. Vankar PS, Shukla D (2012) Biosynthesis of silver nanoparticles using lemon leaves extract and its application for antimicrobial finish on fabric. Appl Nanosci 2:163–168
32. Kumar R, Umar A, Kumar G, Nalwa HS (2017) Antimicrobial properties of ZnO nanomaterials: a review. Ceram Int 43:3940–3961
33. Hebeish A, Sharaf S, Farouk A (2013) Utilization of chitosan nanoparticles as a green finish in multifunctionalization of cotton textile. Int J Biol Macromol 60:10–17
34. Hu X, Tian M, Qu L, Zhu S, Han G (2015) Multifunctional cotton fabrics with graphene/polyurethane coatings with far-infrared emission, electrical conductivity, and ultraviolet-blocking properties. Carbon 95:625–633

35. Shaw S (2010) Halogenated flame retardants: do the fire safety benefits justify the risks? Rev Environ Health 261
36. Koch C, Sures B (2019) Degradation of brominated polymeric flame retardants and effects of generated decomposition products. Chemosphere 227:329–333
37. Samanta AK, Bhattacharyya R, Jose S, Basu G, Chowdhury R (2017) Fire retardant finish of jute fabric with nano zinc oxide. Cellulose 24:1143–1157
38. Sheshama M, Khatri H, Suthar M, Basak S, Ali W (2017) Bulk vs. nano ZnO: influence of fire retardant behavior on sisal fibre yarn. Carbohyd Polym 175:257–264
39. Chang S, Slopek RP, Condon B, Grunlan JC (2014) Surface coating for flame-retardant behavior of cotton fabric using a continuous layer-by-layer process. Ind Eng Chem Res 53:3805–3812
40. Yu M, Gu G, Meng W-D, Qing F-L (2007) Superhydrophobic cotton fabric coating based on a complex layer of silica nanoparticles and perfluorooctylated quaternary ammonium silane coupling agent. Appl Surf Sci 253:3669–3673
41. Zhao Y, Tang Y, Wang X, Lin T (2010) Superhydrophobic cotton fabric fabricated by electrostatic assembly of silica nanoparticles and its remarkable buoyancy. Appl Surf Sci 256:6736–6742
42. Qiang S, Chen K, Yin Y, Wang C (2017) Robust UV-cured superhydrophobic cotton fabric surfaces with self-healing ability. Mater Des 116:395–402
43. Rahman MJ, Mieno T (2015) Conductive cotton textile from safely functionalized carbon nanotubes. J Nanomater 2015:10
44. Shateri-Khalilabad M, Yazdanshenas ME (2013) Fabricating electroconductive cotton textiles using graphene. Carbohyd Polym 96:190–195
45. Ren J, Wang C, Zhang X, Carey T, Chen K, Yin Y, Torrisi F (2017) Environmentally-friendly conductive cotton fabric as flexible strain sensor based on hot press reduced graphene oxide. Carbon 111:622–630
46. Bao L, Li X (2012) Towards textile energy storage from cotton T-shirts. Adv Mater 24:3246–3252
47. Gao Z, Zhang Y, Song N, Li X (2017) Towards flexible lithium-sulfur battery from natural cotton textile. Electrochim Acta 246:507–516

Environmental Profile of Nano-finished Textile Materials: Implications on Public Health, Risk Assessment, and Public Perception

Luqman Jameel Rather, Qi Zhou, Showkat Ali Ganie and Qing Li

Abstract Antimicrobial modification via the use of nanomaterials or nanocomposites have emerged very strongly from recent past because of increasing public concerns related to the health hygiene. However, negative environmental implications, concerns related to human health and possible harmful effects on aquatic life after the left over/unexhausted baths are released to wastewaters have restricted the use of many antimicrobial agents more likely the synthetic ones. The increased use of nanomaterials necessitates assessing of the potential negative impacts of this novel technology on humans and the environment. Application of nano-biocomposites could be a good alternative to most of the synthetic antibacterial agents due to their high environmental compatibility, biodegradable, and non-toxic nature. This chapter has focused on the characteristics and achieved functionalities of nanofinished textile materials with their environmental health profile, implications on human health and possible risk assessments.

Keywords Nanoparticles · Nanocomposites · Environmental assessment · Health profile

1 Introduction

Textile materials have played an important and major role in the development and industrialization of earlier civilizations. However, there is a growing demand for modern functional textiles which has forced scientists to synthesize new functional materials and develop new technologies. Therefore, high-tech materials and new strategies of fabric constructions can improve the wearing comfort and provide unique functional value-additions. Imparting different functional values into textiles can increase

L. J. Rather · Q. Zhou · S. A. Ganie · Q. Li (✉)
College of Textile and Garments, Southwest University, Chongqing 400715, People's Republic of China
e-mail: qingli@swu.edu.cn

Chongqing Engineering Research Center for Biomaterial Fibres and Modern Textile, Chongqing 400715, China

M. Shahid and R. Adivarekar (eds.), *Advances in Functional Finishing of Textiles*, Textile Science and Clothing Technology, https://doi.org/10.1007/978-981-15-3669-4_3

their potential for different applications since textiles are now broadly used in different application sectors such as clothing, pharmaceutical, medical, engineering, agricultural, and food industries. For that scientists have explored different synthetic, semi-synthetic, and natural products to add values to textile surfaces in terms of antibacterial, UV protective, antifungal, and fluorescent properties in addition to coloration of textile materials [1]. However, a new revolution in the textile industry is going on with the apparition of new technologies, which could add special functions and prominent finishes to the fabrics surfaces without decreasing comfort properties. For instance, there has been distinguished enhancement in technologies for synthetic and natural smart fabrics, textile finishing and high-performance functional textiles [2].

Nanomaterials play a vital role because of the interesting surface properties that allow increasing their effect in comparison with bulky traditional additives and materials in technological evolution. Conventional nanomaterials such as metal oxide agents, carbon-based materials, host-guest compounds, and so on are examples of nano-structured materials used in antimicrobial, deodorant, UV-protection, self-cleaning, and other common finishing methods [3]. In non-commercial research of sunscreens that contain TiO_2 and ZnO nanoparticles and their composites, the subject of safety mainly concerns the penetration of nanocomposites into the skin. Also, the safety nanofinished sunscreens are also determined by physicochemical properties of the nanoparticles, coatings, formulations, and the interaction of these components with UV radiation [4]. In the finishing process it is important to use environment-friendly methods and materials. So, substitute ingredients with great environmental safety are preferred. This chapter reviews the most relevant contributions of the use of common nanoparticles and their composites to functionalize textile materials with an environmental assessment and impact on human health.

2 Nanomaterials for Textile Finishing

Increasing customer demand from last few decades for robust and practical clothing produced in environmentally friendly and sustainable means has shaped a prospect for nanomaterials to be integrated into textile substrates as active functionalization agents to alter their surface properties [4]. Nanomoieties/nanoclusters can induce static elimination stain repellence, electrical conductivity and wrinkle-freeness to fibers without conceding their flexibility and comfort. Textiles materials are universal interface/materials for the fabrication of nanomaterials, electronics, and optical devices. Such technologies and integrated materials offer a stage that retorts to electrical, thermal, optical, chemical, magnetic or mechanical stimuli [5, 6]. Nanomaterials similarly offer a broader application platform to produce functional garments that can intellect and reply to external stimuli via color, electrical or physiological signals [7].

This created the concept of nano-engineered fabrics/fibers which offers above mentioned functionalities deprived of the properties of textile material comfort properties of the textile material. These engineered substances should impeccably assimilate into garments and be stretchy and relaxed while having no allergic reaction to the body. Moreover, such substances need to satisfy performance, weight and appearance properties [4]. The conventional approaches used in the textile industry have the challenge not to functionalize fabrics that do not lead to permanent effects. Antibacterial modification of textiles with metal oxide nanoparticles and their composites can prevent the growth of fungi, alga, bacteria and other microorganisms on them. Finishing of textiles with metal nanoparticles such as silver, zinc, and titanium have been reported for antibacterial properties [8, 9]. Because of their multi-targeted mechanism of action, high surface area-to-volume ratio and unique properties of these nanoparticles, metal nanoparticles are more effective in comparison to native biopolymers [10]. An important advantage of nanomaterials is the increasing surface area which increases the contact of the antibacterial agent with infectious microorganisms. Among all the antibacterial nanomaterials, silver have proved to be the most effective against most common bacteria in antibacterial textile finishing [10]. Inorganic nanoparticles such as TiO_2, ZnO, SiO_2, Cu_2O, CuO, Al_2O_3, and reduced graphene oxide have been abundantly exploited for their high temperature thermal and chemical stability, permanent photostability, and non-toxicity compared to the organic ones [4, 7]. Metal oxides as UV protective agents can improve the absorbency percentage in the UV region. UV protection can be increased by modifying textile and polymers via increasing the total reflectance and absorbance properties which in turn reduces the amount of transferred UV light [4].

Another area where nanomaterials can be handy is to generate anti-odor textiles which is one of the main sectors of the textile industry, and has got considerable and increased attention over the last few years. In this section, most important textile materials are sportswear, underwear, socks, and shoes. Previously, physical (absorption by activated carbon) and chemicals (turning the odor into common smell) approaches have been used along with some aromatic compounds to eliminate unpleasant odors. Although, aroma textiles is not new concept but great progress has been made in the field of aroma textile production via surface finishing to produce effective and long-lasting bioactive textiles [7]. Earlier scientists used aromatic extracts via direct spraying on the surface of textiles but now-a-days new techniques have been developed and encapsulation is one such effective method to increase the lifetime of odors on textiles. Textile materials are finished with micro/nano fragrances via encapsulation method which are then released due to pressure or abrasion of garments [11]. Some applications of nanofinished textiles materials are summarized in graphical form (Fig. 1).

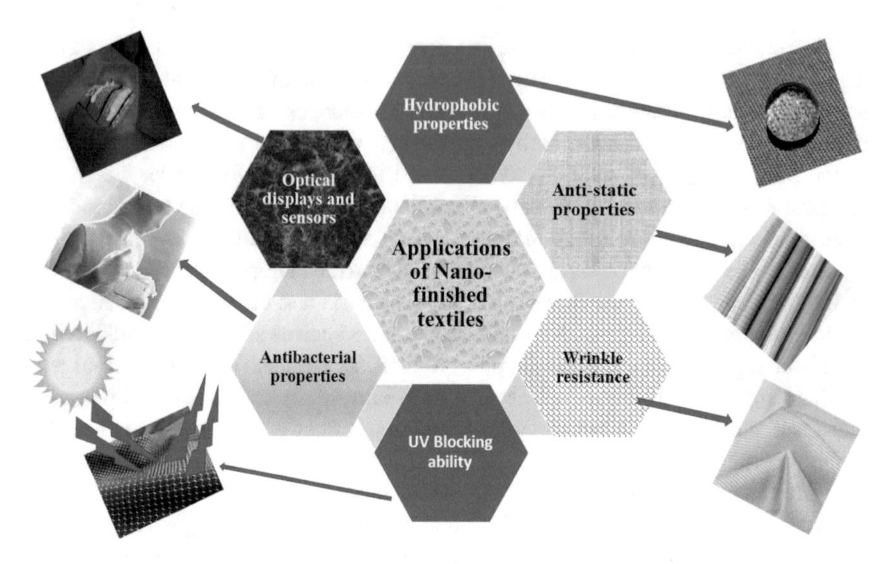

Fig. 1 Different applications of nanofinished textile materials

3 Classes of Nanomaterials

Most of the nanomaterials are based on the inorganic based active agents with good potential of antimicrobial action on textile materials. However, broadly they can be classified into two main categories: (a) Inorganic nanomaterials and their composites; (b) Inorganic nano-structured loaded organics (Nano-biocomposites) (Fig. 2).

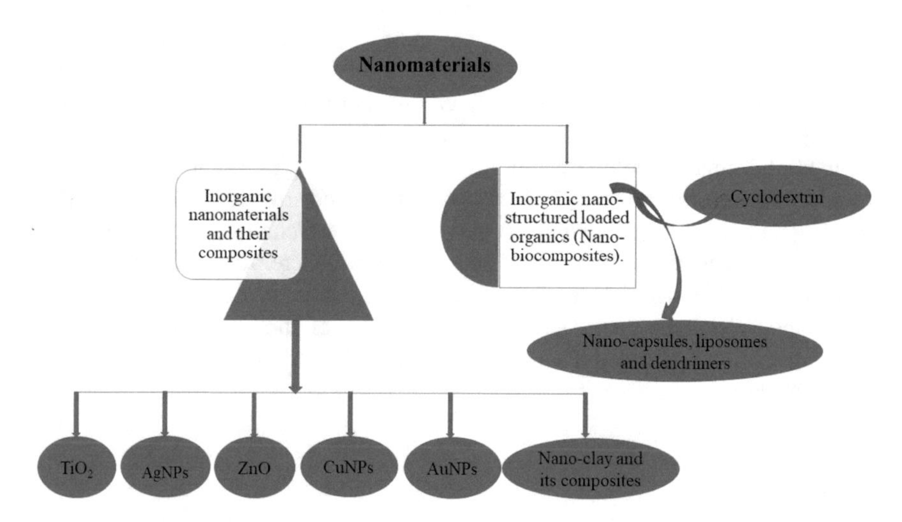

Fig. 2 Classification of nanomaterials

3.1 Inorganic Nanomaterials and Their Nanocomposites

The most common inorganic nanomaterials are titanium dioxide (metallic and non-metallic composites), silver (nano-structured materials based on silver and silver-modified montmorillonites), zinc oxide, copper, gallium, gold nano-particles, carbon nanotubes, nano-layered clay including its modified species (silver-nano-clay/chitosan and clay–polyvinyl pyridinium), and their respective nano-composites.

3.1.1 Titanium Dioxide (TiO_2) Nanoparticles

Currently, titanium dioxide in the form of nano-antibacterials has created remarkable additions in its application field for developing attractive multi-functional materials. It possesses high stability and broad spectrum activity [12]. Photo-catalytic activity of TiO_2 is a function of structure and powder purification which makes it one of the efficient candidates for anti-bacterial, UV protecting, self-cleaning, water and air purifier, environmental purification, in solar cells and gas sensors [13–17]. Among the various crystal structures of TiO_2, anatase due to its high surface area finds extensive use in photo-catalysis [18]. Redox reactions at the surface of TiO_2 are carried out via electron–hole pair generation by irradiation with UV light which reduces band gap. The free electrons created react with oxygen to form O_2^- which in presence of water and positive holes produces hydroxyl radicals. The highly reactive oxygen species oxidizes organic matter (odor molecules, bacteria and viruses) to CO_2 and H_2O (Fig. 3) [19].

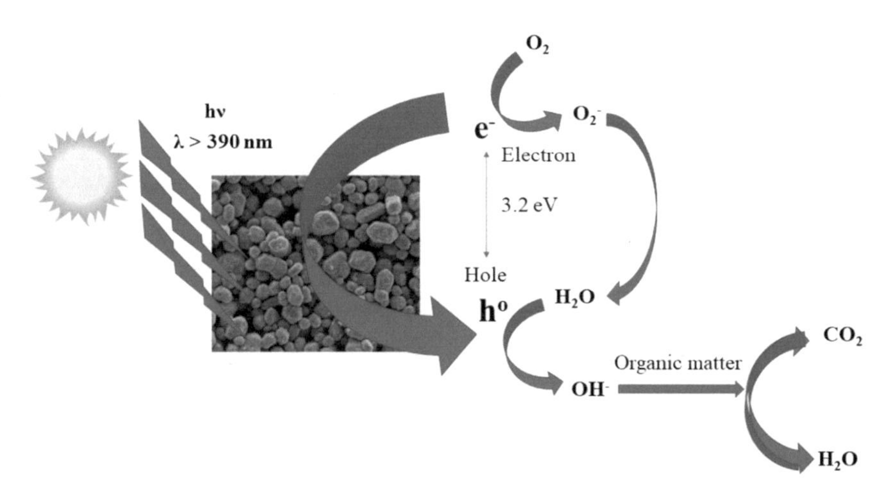

Fig. 3 Photocatalytic oxidation on the surface of TiO_2 nanoparticles [19]

3.1.2 Silver Nanoparticles and Its Composites

One of profound approaches to fabricate smart textiles is nanomaterials finishing onto their surfaces for enriching them with dirt-repellent, self-cleaning, water repellent, antistatic properties, decreased gas permeability, conductive, flammability and antimicrobial properties. Silver nanoparticles (AgNPs) have been enormously used for some of the outstanding mechanical properties along with above mentioned bioactivities. Natural and synthetic textiles have been employed to generate functional textiles treated with AgNPs. Several methods have been employed to enhance the durability of nano finish against washing processing. Therefore, chemical binders and cross-linkable polymers have been used for stabilizing AgNPs on fabric surfaces [20–24]. The deposition of AgNPs mainly remains bound to electrostatic interactions between AgNPs and textile constituents (functional groups). There are 3 stages of AgNPs finish in terms of successful chemical deposition: (a) adsorption of silver ions on textile surfaces, (b) diffusion of AgNPs into the interior of textile fabric/fiber, and (c) interaction of AgNPs with textile fibers. For instance, X-ray photoelectron spectroscopy observation have revealed the nature of possible interactions between AgNPs and sulfur atoms, which are likely to be interact via the cleavage of S–S bond in the wool fiber [4, 25, 26].

3.1.3 Zinc Oxide (ZnO) Nanoparticles

Recently, ZnO has got high attention from scientific community for its remarkable applications in photo-diodes, UV light emitting devices solar cells, sensors, electro-acoustic transducers, displays, gas sensors, varistors, piezoelectric devices, sun-screens, gas sensors, UV absorbers, anti-reflection coatings, catalyst due to low band gap (3.37 eV) and photo-catalysis The stability of ZnO has been acceptable with reducing option of electron-hole recombination (Fig. 2) [27–33]. ZnO nanoparticles have some additional advantages compared to Ag nanoparticles in terms of low cost, UV-blocking property and white appearance [28, 33]. ZnO enhances the properties of polymeric nano-composites [33]. It has been also proven that there appears an enhancement in wear resistant and anti-sliding phase in zinc oxide nanocomposites as a consequence of their high elastic modulus and strength [32]. Li et al. have investigated the antibacterial action of ZnO nanoparticles functionalized on cotton fabric and have evaluated the durability of antibacterial action under conditions of alkaline, acidic and inorganic salt artificial perspiration [34]. As compared to normal conditions a negative surface charge has been presumed for ZnO nanoparticles and illumination can increase anti-bacterial performance. ZnO nano-particle like tetrapod was also used for producing acrylic composite resin [32]. ZnO nanotubes, nanocages, nano-belts and nanowires have been also produced [30, 35–37].

3.1.4 Copper Nanoparticles

Copper nano-particles were found strongly active against various bacteria compared to triclosan when embedded into submicron particles of Sepiolite [38]. However, it was found that copper nanoparticles were less active compared to Ag nano-particles [39]. UV protection were imparted to polypropylene (PP) nonwoven by deposition of copper nanoparticles by magnetron sputter deposition method along with enhancement in electrical conductivity [40].

3.1.5 Gold Nanoparticles

Gold nanoparticles have been expensively used in commercial soap and cosmetics industries for its effective antibacterial properties against acne or scurf and have potential to eliminate waste substances from the skin and control sebum [41–43]. Gold nanoparticles are effective against extensive range of bacteria including gram positive and gram negative and fungi [44]. Grace and Pandian showed intense antibacterial efficiency of Au nanocomposites with antibiotics like streptomycin, gentamycin and neomycin against *E. coli*, *P. aeruginosa*, *S. aureus*, and *M. luteus* [41]. Park et al. filled Au nanoparticles inside the liposome structure leading to permeability of barrier of the lipid, an increase in the fluidity and provide thermally sensitive liposome for controlled delivery at particular temperatures [42].

3.1.6 Nano-clay and Its Modified Species

Recently, nanoclay has been found one of the interesting materials and has been investigated for its outstanding advantages like increase of tensile strength, gas barrier property, modulus, dimensional stability, HDT and flame retardancy transparency [45–49]. Sterilizing effect, antibacterial efficiency, membrane coating and adsorption of toxins are some other important applications related to the biomedical field. The first use of clay in medical field was carried out by Romanes in 60 BC as poultice in wound plaster. The healing properties were mainly due to the physical absorption of water, toxic bacteria, viruses and organic matter. The presence of inorganic metal and their oxides in clay accounts for its antibacterial properties without knowing exact mechanism of action. However, complete sterilization of *E. coli* was observed by Agricure clay [50].

Haydel et al. utilized clay of iron-rich to evaluate its use as a therapeutic agent against antibiotic-susceptible and antibiotic-resistant pathogenic bacteria [51]. Seckin et al. used clay-polyvinyl pyridinium matrix as an adsorbent against bacterial cells from water [52]. Ion exchanged montmorillonites and silver-chitosan/clay nanocomposites were investigated by Hu et al. and Zhou et al. for their antibacterial and bacteriostatic effect, respectively [53, 54]. Antibacterial poly propylene fiber/clay composite was successfully produced by Mlynarcikova et al. and Pavlikova et al. [47, 48].

3.2 *Inorganic Nano-structured Loaded Organics (Nano-biocomposites)*

The inorganic nano-structured loaded in organics usually known nano-biocomposites include cyclodextrin and modified cyclodextrin, nano and micro-capsules with inorganic nanoparticles, inorganic nanoparticles loaded in liposomes and dendrimer nano-composites.

3.2.1 Cyclodextrin

Cyclodextrins, composed of α-1,4-glucopyranose units are the family of cyclic oligosaccharides, produced during enzymatic degradation/cleavage of starch and are mainly useful in molecular complexation processes [55, 56]. Three are types of cyclodextrins: α-cyclodextrin (6 glycosyl units), β-cyclodextrin (7 glycosyl units), and γ–cyclodextrin (8 glycosyl units). Among the above three cyclodextrins, β-cyclodextrin having molecular weight of 1135 and height of 750–800 pm, is most commonly used in textile applications and is assessable at low price. The two sides of β-cyclodextrin have different affinities for water as the inner hole of average volume of 260–265 A^3 is hydrophobic and the external section is hydrophilic [57]. The most important property of cyclodextrins is their complexation capacity with a wide range of compound including metallic ions and gaseous compounds (host-guest relationship), which can deliver the perfumed substances and drug carriers in textile industry via controlled release mechanism [58, 59]. Recently, β-cyclodextrin has been grafted onto cotton and wool textiles with the help of cross linking agents such as epichlorohydrin, BTCA, etc. [60]. Moreover, β-cyclodextrin has the potential to be loaded with inorganic nano materials at different textile processing stages and making them stabilized by using different cross-linking agents on the fabric surfaces [61].

3.2.2 Nano-capsules, Liposomes, and Dendrimers

Nano, micro-capsules, liposomes and dendrimers have been used for the delivery of different active agents to target organs in drug delivery system [62]. Recently, scientists have developed nano-encapsulation technologies for everlasting anti-microbial efficacy on cotton fabrics and have been recognized on global and industrial levels by the Ciba Company [63]. In general, nano and micro-capsules are made of a shell of natural or synthetic polymers with a core which can host various active agents, vitamins catalysts, bioactive materials, drugs, proteins, phase change materials and deodorants. However, a releasable capsule is a concept where the trapped bioactive agents in the core of the capsule should be released through a continuous precise release mechanism. A delicate shell which can be excited by changing intensity of light, temperature and pressure is highly recommended [64, 65]. This can be achieved by decreasing the size of capsules to nano size which can be generated by high vigorous stirring or ultra-sonication of micro-capsules [66]. Zhang et al. revealed that the size of nano-capsules can be altered by increasing the stirring rate and by changing the concentration of emulsifier [67]. Shim et al. effectively used poly(methyl methacrylate) frame work for the encapsulation of zinc oxide nano-particles [68]. In another similar kind of work, Oku et al. encapsulated silver nano-particles of silver nitrate by the chemical reduction with an aim of improving the electronic properties of the nano-captured boron nitride nano-cages [69].

Liposomes are self-assembled amphiphile liquid dispersed in aqueous solvent. Due to their dual properties, they can hold and release both hydrophobic as well as hydrophilic materials. This particular property of liposomes makes their efficient use in drug delivery systems [62, 70]. Additionally, liposomes are being used in low temperature and energy saving dyeing processes of different types of textile fibers and fabrics [71–73]. The size, fluidity, and permeability of liposomes has been altered by loading nano sized silver and gold nanoparticles for better release of some biological agents under thermal stimulation [74]. However, it has been seen that the thermal and mechanical stability of liposomes should be determined for better controlled release. For increasing stability of liposomes on textile surfaces, use of functionalized stabilizer has been suggested. In case of antibacterial modifications, phospholipid layers act as fertilizing agents for increasing bacterial growth exerting a negative effect of antibacterial efficacy. Nano silver has been used to overcome this drawback [71, 73].

Dendrimers are synthetic and branched 3D molecules produced by including repeating diverging arrangements to produce a exceptionally new architecture with a rare and great level of consistency [75–77]. Unique physical and chemical properties have been created due to the molecular scattering of the guest molecules in the dendrimer host. Due to large number of active sites, dendrimer carboxylate salts can interact with large number of silver cations which will serve as nanoscopic transport carrier of different quantities of bioactive substances. In vitro and in vivo experiments have revealed the bio-compatibility of dendrimer conjugates [78, 79].

4 Textile Applications of Nanomaterials

4.1 Antimicrobial Applications

Textile materials are being treated with a wide range of synthetic and natural antimicrobial agents for various explanations liable on the market segment and application area. Antimicrobial agents are mostly functional on the textile surfaces to improve their resilience against mildew, fungi, and bacteria (e.g. discoloration, colonization of odor producing bacteria and preventing destruction of polymers) and increase the durability of materials leading to longer lifetime [80]. The antimicrobial agents in general are applied to play a role in enhancing the clinical hygiene and delicate surroundings by reducing the colonization of microbes and their communication through fabric surfaces [81]. Antimicrobial textiles requiring less care can deliver the goal of the reduction in the environmental footprints. Requirements for antimicrobial treatments for different textile materials vary according to the use of end product and the type of reactivity of antimicrobials towards textiles. The toxicity profile of human is vital for garments of skin-contact, whereas the photostability of the antimicrobial agents is vital for outdoor textiles. Zinc pyrithione (ZnPT), n-octyl-isothiazolinone (OIT), benz-isothiazolinone (BIT) or 10,10′-oxybispheloxarsine (OBPA)nanomaterials are favored for non-skin contact uses such as mattresses or bedding whereas silver (Ag), triclosan (TCS), silane quaternary ammonium (Si-QAC) compounds are mainly used for apparel textiles. Moreover, silver and silane quaternary ammonium compounds are also widely used in the medical field.

Application concentration is one of the most important parameters that influence the antimicrobial action and rates of biocidal agents [82]. The application rate further depends upon the type of antimicrobial, method of application (bulk incorporation vs. topical treatment), fabric construction, type of interactions (physisorption or chemisorption), and stability during use (durability). To verify the applications rates, several tests were conducted in USA on textiles treated with Ag, TCS, Si-QAC and ZnPT which were then given the product registration labels extracted from the National Pesticide Information Retrieval System. PAN Pesticides Database gives detailed categories and corresponding PC codes [83, 84].

Durability of the finished product is another fundamental requirement to ensure its significant role throughout its life cycle. Successive cycles of washing and wear of the textile product can contribute much to the progressive loss of the activity due to leaching process during washing treatments or mechanical impairment to the fabric or fiber. The leaching can occur until the antimicrobial activity reaches its limiting factor. In line with best industrial practices the finished textile fabric should service at least 50 machine washes [85]. The highest wash durability with lowest application rates is mostly favored in marketplace. Many studies reveal the leaching of different antimicrobials from textile fabric surfaces: Ag [86–89] the leaching of TCS [90, 91], the leaching of Si-QAC [92] or the leakage of ZnPT from textiles [93]. However, strong fixation of the antimicrobials is desirable to improve the durability of finished products. Treatments of antimicrobials characteristically have a choice of suggested application rates that reflect how much of the active is required to offer performances and durability aspects [94].

4.2 UV Protection Applications

Recently, consumers all over the world are looking for well-developed protective clothing materials which can provide dual purposes of coloration and simultaneous UV radiation protection. UV protective nature of food and packaging materials retard their chemical degradation and enhances their life span, whereas UV protective agents in sunscreens is imperious for the protection of human health as extreme UV contact leads to massive increase of skin cancers/diseases (Fig. 4).

In this regard, UV protective coatings on food, packaging materials and textile substrates must meet following two requirements and challenges:

(a) To offer extensive shield over the whole UVA/UVB spectrum, and
(b) To preserve photo-stability after prolonged phases of irradiation.

Presently, extensive spectrum UV protective nature of functional finishes has been achieved by the use of various organic UV-filters e.g. octinoxate, oxybenzone, avobenzone, octocrylene padimate-O. The main concern to the use of above-mentioned organic filters is the formation of carcinogenic reactive oxygen species (ROS) during photodegradation process [95, 96]. So, use of antioxidants in sunscreen formulations is highly advised as it scavenges generated ROS, thus improving photo-stability [97, 98]. Systemic absorption of UV-filters by skin can have adverse effects on human health. This led scientists to synthesize environmental and human compatible UV-filters by encapsulating nanoparticles to decrease their photo-toxicity [99, 100].

However, there is further necessity to fabricate nanoparticles which can successfully encapsulate various UV-filters and antioxidants together irrespective of their early physical states (liquid/solid). Additionally, the developed/synthesized nanoparticles-UV filters composite should be bio-based for maximum cosmetic and consumer's appeal [100]. Textile and polymer surfaces have been significantly investigated in terms of anti-UV properties coated with inorganic nanoparticles such as TiO_2, ZnO, SiO_2, Cu_2O, CuO, Al_2O_3, and modified graphene oxide, owing to their high temperature thermal and chemical stability, photo-stability and non-toxicity related to the organic finishing agents [101]. Among different inorganic

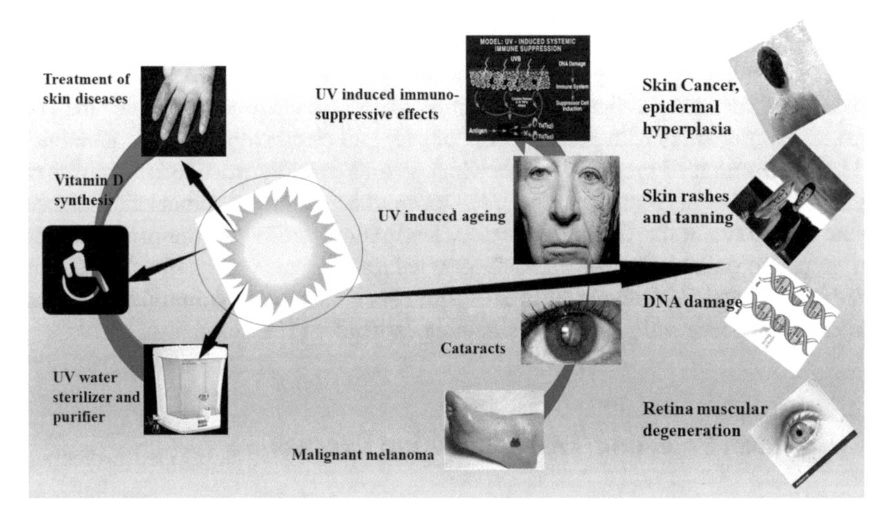

Fig. 4 Effects of UV radiations on human health

nano-materials, TiO_2 and ZnO have been successfully and widely used due to low price [102], white appearance, photo catalytic oxidation ability [103], UV-blocking and self-cleaning properties [104], and wide range of absorption capacity [103, 105]. The percentage UV absorbance of textile and polymer materials can be improved by the use of nano metal oxides. The extent and mechanism of protection is specialized by the nano metal oxides electronic structure and the surface area to volume ratio [105].

Moreover, use of nano-metals in conjunction with TiO_2 (Ag/TiO_2) produced profound effects on the UV-protection property, antibacterial activity, self-cleaning ability and durability to washing process of linen/cotton (50/50%) fabrics [106]. In another research, Cu_2O in conjunction with ZnO nanoparticles (Cu_2O/ZnO) were synthesized and the subsequent finishing onto the cotton fabric improved UV protection ability compared to the individual use of ZnO and Cu_2O nanoparticles [107, 108]. Prasad et al. synthesized nanoparticles of ZnO by dipping and spraying procedures for the modification of cotton with antibacterial and UV protection ability [109]. Under alkaline conditions the negative charge of cellulosic chains is the reason of fast nucleation of ZnO nanoparticles [110]. UV-Visible radiations cause photo-fading of dyed fabric/fibers, reduction in the mechanical properties, and sometimes photo yellowing in wool fabrics. In order to decrease or overcome these limitations wool textiles have treated with nanoparticles of inorganic metal oxides like TiO_2, ZnO, and Al_2O_3 which delay the photo yellowing [111, 112].

Use of nano layered double hydroxide of Mg/Al has improved the properties on textile fabrics from the UV shield and flame retardancy. Use of nano fibers in the form of protective mat onto the surface of fabric is also recommended to improve UV protection [113]. Lee et al. reported increasing UV shielding of electrospun PU/ZnO thin layer with increasing ZNO concentrations [114]. Dadvar et al. reported TiO_2 nanofilm coating on PAN nanofibrous mats possessing very high UV protection. Decreasing the nanofiber diameter increased UPF values of finished mats [115]. The comparative results of ZnO/PAN and MWCNT/PAN nanofibrous mats showed better UV shielding ability of composites comprising of MWCNT/PAN nanofiber mats of 3 g/m^2 area. Carbon nanostructures have also been used to enhance the UV protection properties with great hydrophobicity and electrical/thermal conductivity [116]. Utilization of graphene oxide which is a chemically modified derivative of graphene can simply bind to fibers [117]. Dry approach and conventional dip methods were used to synthesize GO nano-sheet finished cotton fabric under thermal reduction of graphene oxide with low transmittance values. Compared to unfinished cotton (UPF: 14.9) and GO/cotton (UPF: 20.4), graphene nano-sheet/cotton fabrics blend (UPF: 35.8) possessed greater UV shielding capability [118].

5 Release, Fate, and Transformation Process

After the successful finishing of textile and polymer materials with different nano-based active agents, it is important to check the environmental implications and possible effects on human health. The concept of green technology will become successful if the finished products are environmentally safe and possess very least threats to aquatic life after being released during the washing of textile and polymeric materials. The environmental threats and risks associated with the exposure of metal oxide nanomaterials or nano-biocomposites or engineered nanomaterials (ENM) will be examined by the environmental developments that govern the fate, transport, and transformation of these nanomaterials. Significant amounts of nano-silver from silver functionalized consumer products such as socks, underwear, slippers and shoe liners are freed into the aquatic environment through washing liquids and laundering of textiles [88, 119].

The release into the environment is expected to arise in agricultural fields as a consequence of the use of biosolids for soil amendment and during there leases of municipal wastewater. Recent research studies have shown that the fate of nanosilver under washing conditions such as agitation, using bleaching and detergents undergoes considerable alteration. Impellitteri et al. described more than 50% of nano-Ag from socks is released with detergents during washing. It is transformed into AgCl which is comparatively insoluble in water [120]. The toxicity and transformation of nano-Ag is linked to the physicochemical parameters such as concentration, pH, ionic strength, redox environment and presence or absence of inorganic ligands [88, 121]. Moreover, the nano-Ag biosolids may get dumped into landfills. So, suspension of nano-Ag, separating, and transformation processes may direct the chance for this particular system. Incinerated biosolids subjected to high temperature may result in the release (airborne) of carbon nanotubes that will be fully controlled by combustion processes.

Hypothetical case investigations have been used to demonstrate the providence and passage processes of ENM in recent past with examples of TiO_2 in paint and sunscreen, in textiles as nano silver, carbon nano tubes in composites, and cerium oxide nanoparticles in diesel engines [122]. Above mentioned nanomaterials were selected for their diversity in their fundamental properties which include their solubility, redox activity, and the end use of the product. Among several other chemical processes, the fate and transport processes of ENM in finished products comprise of reduction, dissolution, oxidation, desorption, precipitation, combustion, adsorption, biotransformation and photochemical transformation. The above-mentioned chemical processes occur in almost all finished product and ENM.

5.1 Photochemical Transformation

In this process incident light penetrates the photoreactive center of the product (ENM) inducing excitation of ENM [123, 124], generation of free radicals [125], or by direct interaction with other components of the product (Fig. 2) [126]. The level of reactivity of photoreactive site and the influence of on the creation of product weathered engineered nanomaterials (PW-ENM) depends upon the intensity of the incident radiation (penetration capacity), wavelength, and the nature of outer layers of the

ENM. The surface-modified particles have higher resistance to the light penetration efficiency which reduces the excitation or photodegradation. The rate of photochemical transformation depends upon the rate determining step which is the mass transfer from material surface to the exterior media. Photoactivation of TiO_2 has been found to have profound effects on its binding with the dissolved organic matter [127].

5.2 Oxidation and Reduction

The substances in a particular oxidation states are subjected to reduction or oxidation processes if they are thermodynamically favorable [128]. Redox reactions are influenced by various factors including the pH, occurrence of reducing or oxidizing agents and presence of stabilizers on the ENM surfaces which will decrease the rate of transformation [129].

5.3 Dissolution

Dissolution refers to the release of water-soluble separate ions/molecules [130–132]. The dissolution involves the reaction of molecules on the surface and the release of the ionic forms or direct solubilization of the component materials, tracked by diffusional passage [133].

5.4 Precipitation

Precipitation occurs under certain conditions in presence of precipitating agents in which new solid materials are formed from the solutions (suspended materials from solutions) [119, 134]. The precipitation is governed by the precipitation laws involving the solubility product (Ksp) and ionic product (K_w) under equilibrium conditions, which in turn depends upon the ionic strength, ligand availability, pH, and temperature of the media [128]. The existence of adsorbed materials may increase or decrease the rate of dissolution by surface protection phenomenon which can be further confirmed by thermodynamic calculations. However, slow diffusion rates may delay in the achievement of the final product [135–137].

5.5 *Adsorption and Desorption*

Adsorption processes are of two main types' physical adsorption (Physisorption) and chemical adsorption (Chemisorption) involving following forces of interactions: electrostatic interactions (ion exchange), chemical bonding or Van der Waals attractions, [138–140]. Physisorption means weakly and non-specific adsorption to the ENM surface, chemisorption and ion exchange involves chemical bonding or a charged communication on the surface's sites of available active sites. Under partially covered surfaces, dispersion may lead to the destabilization and aggregation through bridging effects among the surface and the non-adsorbed functional groups of the adsorbate. However, if the surface is fully occupied, stabilization of the dispersion may occur which may decrease the accumulation brought by both steric and electrochemical interactions [141]. The material may adsorb impurities and act as a route in the environment for their transport [142–144]. Sorption procedures may be principally significant with respect to altering the surface physiognomies of the product modified ENM to that of the product weathered ENM and environmentally transformed ENM.

6 Environmental and Health Profile

6.1 *Environmental Compatibility*

The quantities of raw materials used and the release of nanomaterials into the environment over the entire life cycle of the textile materials are the major aspects for the environmental compatibility of nanomaterials in textiles. On energy consumption and raw material quantities used there are no clear statements or evidences reported in literature. Walser et al. examined the greenhouse gas potential based on the CO_2 footprint using T-shirts as examples [145]. They concluded that the eco-toxicity of the T-shirts increases during washing if biocidal materials are added. The study shows that the use phase of the textile can make a significant influence to saving energy with active biocidal elements since these textiles need fewer washing cycles, which saves power and laundry detergents. The longer life cycle of the nanofinished textiles can be achieved through UV protection caused by nanomaterials to improve the durability of the textiles (like awnings). Saving laundry cycles and detergents can also achieve through self-cleaning surfaces [146]. Raw material and wastewater can be saved by improved dyeability of textiles because it reduces the number of dyeing and washing cycles. The nanomaterials via industrial and urban wastewater can enter wastewater treatment plants. For nanomaterials the level of their elimination from the water cycle with the sewage sludge must be explored in a broader way. Glover et al. showed that silver ions that are released from silver particles into the environment can form nanoscale silver particles under certain conditions and suggested that more than 90% of the nanomaterials investigated ware separated via the sewage sludge

[147–150]. Also no severe loss of the nitrification performance of microorganisms in the activated sludge of wastewater treatment plants was observed. Disposal in incinerators must be another good option for input path of nanomaterials from textiles. Walser et al. from their studies showed that the behavior of CeO_2 nanoparticles in a waste incineration plant has found that nanoparticles do not discharge into the atmosphere if the incineration plant is precisely well-equipped but that they attach themselves to residues and can then be found in recycled raw materials or landfills [151].

6.2 Effects on Human Health

An overall conclusion on possible risks with respect to human health cannot derived due to the diversity of nanomaterials-containing textiles with respect to manufacturing processes, the synthetic nanomaterials used, and their uses. Some of the nanomaterials used have the risks for toxic or eco-toxic effects but these nanomaterials must be pertinently absorbed into the system to act in this way. The possible gateways of entry into the human body for particles potentially released from textiles during wearing are skin and lung. Particles inhaled were transported out of the lung via natural clearance mechanisms such as the "mucus elevator", and thereafter were swallowed, so a contact of the gastrointestinal tract cannot be excluded. On exposure with nanomaterials from textiles, vital biological endpoints can be defined as inflammation, acute toxicity, oxidative stress, DNA damage, crossing and damage to tissue barriers.

For human health the following criteria were defined: (1) acute toxicity, (2) chronic toxicity, (3) impairment of DNA, (4) crossing and damaging of tissue barriers, (5) brain damage and translocation and effects of ENM in the (6) skin, (7) gastrointestinal or (8) respiratory tract. These criteria comprise toxic effects of ENM, essential biological endpoints and potential for uptake into the blood. The exposure of electromagnetic radiation on the human body may lead to changes in the nerve cells, stimulating muscles and physical damage [21]. Because of vibration and heat generation by electromagnetic radiation in the human's body, RNA and DNA in cells may stop production, which will lead to abnormal chemical activity of cells and cell cancer [152]. One of the key approaches used in protection from electromagnetic radiation is shielding. Shielding can be defined as dipping the electromagnetic field in a space by obstructing the field with barriers comprising of magnetic or conductive materials. In recent years, conductive polymers (conventional textile fibers) are used to defend the body against electromagnet waves. The elasticity and lightweight of conductive fabrics have engrossed the attention of many scientists to fabricate protective textiles against electromagnetic waves [153]. Conductive fabrics are synthesized by special finishing methods using a small amount of conductive materials like carbon. The following approaches can be used to achieve the conductive textiles:

1. By adding conductive fillers, like conductive carbon black, carbon fiber, CNTs, metal powder, and flake to the spinning dope of synthetic fibers.
2. Incorporation of yarns or conductive fibers into the fabric [154].

Lamination of conducting layers onto the fabric surface, conductive coating, zinc arc spraying, ionic plating, vacuum metallization, and sputtering.

6.3 Development of Resistance

High washing resistance of textiles was achieved when new chemicals were applied on polymeric dendrimers and their performance was dependent on the concentration of chemicals. To enhance self-extinguishing of fibers, scientists developed flame-retardant agents for finishing of fabric and this flame-retardant finishing enhanced the resistance of the fabric against igniting. Also, the flame spread rate reduced because of using flame retardant finishing. Aromatic structured polymeric chains such as aramids and the poly(meta-aramid) fiber can be used to produce fire resistance types of regular fibers. Kevlar (Du Pont), has high modulus and tenacity, besides its great heat and fire resistance [155]. Modacrylic and flame-retardant viscose fibers were prepared for inherently fire-retardant fabrics. These fibers were typically used for the manufacture of firefighting clothes and anti-flash arc work wears. Flame-retardant agents, containing various metals phosphorus, boron, nitrogen, silicon, and other nanostructures, have been explored to produce the flame resistance of fabrics and polymers [156]. Phosphorus-based agents among various other halogen-free flame retardants display high flame-retardant properties but these compounds typically have low water resistance to cause significant reductions in insulation properties and high humidity and appearance in high temperatures [157].

So, using non-halogen and non-phosphorous based flame-retardant agents in various applications are favored. Gashti et al. prepared coating with water-repellent and flame-retardant properties from cellulose fibers using a polycarboxylic acid/hydrophobic silica nanocomposite. The thermal properties of the substrate were increased by the incorporation of nanoparticles because the nanoparticles modify the surface of fibers, which have high heat resistance, heat insulation effect, and mass transport barrier toward cotton molecular chains [158]. Carosio et al. produced a new technique to improve flame-retardant properties of textile fabric by using multilayered thin films. Polyethylene terephthalate fabrics were covered with silica nanoparticles using the layer-by-layer assembly method. Bilayers of positively and negatively charged colloidal silica nanoparticles increased time to ignition and decreased heat release rate peak of polyethylene terephthalate fabric. This study establishes the capability to impart flame-retardant behavior using a water-based, environment-friendly protective coating [159]. Resins based on formaldehyde such as phenol formaldehyde, urea formaldehyde, melamine formaldehyde, and cross-linking agents such as glutaraldehyde and poly-carboxylic acid cause an increase in the crease-resistance properties of the textile. Various nano materials are used to enhance the wrinkle resistance of textiles [160]. Lam et al. explored the wrinkle-resistant property of cotton specimens treated by butane tetra-carboxylic acid and catalyzed by SHP in the presence of nano-TiO_2. They revealed that the addition of nano-TiO_2 further enhance the wrinkle resistance of BTCA/SHP-treated cotton fabrics [161]. TiO_2 nanoparticles restrict the molecular movement of cellulose chains and leads to the improvement of crease resistance. In situ preparation of silver nanoparticles on the cotton fabric using silver nitrate with a reducing agent and stabilizer citric acid/sodium hypophosphite has been reported by various scientists to enhance wrinkle resistance properties [162, 163].

6.4 Biological Degradation

Biological degradation *i.e.* microbial metabolism offers an exceptional alternative to different conventional physical/chemical treatments approaches used for the decontamination of toxic from the wastes of textiles. Usually the off-site biological treatment system for hazardous waste management provides an exceptional alternative to this problem. This is the only means to totally mineralize numerous toxic compounds from textile industry [164] and have the following advantages:

- It is an ecologically sound, and a natural process.
- The toxic chemicals are destroyed or detoxified into harmless intermediates and finally assimilating them forming carbon dioxide and water.
- This method is reported to be less costly as it employs growing the microorganisms at the outlay of the toxic chemicals.
- Functioning conditions are less extreme and controls are not required.
- This can often be accomplished where the problem is located, eliminating the necessity to transport large quantities of contaminated wastes off site.

In biological degradation of textile effluent, various microorganisms have been found that are capable of degrading different toxic effluents include bacteria [165–167], fungi [168–170] and algae [171]. To impact the efficiency of biodegradation treatment process, various essential factors are required. To speed up the biological degradation process, seeding of contaminated wastewater of textile effluents with competent microflora that are capable to destroy hazardous waste is used in the treatment. The injected microorganisms either may be prepared in the laboratory or naturally occurring types to attack the target waste [172]. Temperature has a key and important influence on the microbial growth rate. Cellular activity, particularly enzyme systems, retorts to heat so that the rate of cell growth increases abruptly with increasing temperature until the optimum growth temperature is reached. Cell growth can slow down dramatically by an increase in temperature only a few degrees above an organism's optimum temperature and cell death can occur by continued exposure to high temperature. The biological degradation process also depends on the pH of the surrounding environment which is important for microbial growth and metabolism with maximum growth rate in a relatively narrow pH range around neutrality (pH 6–8) [173].

7 Conclusion

The present status of the global textile market is based on innovations in high-tech textile products, and nanotechnology is one of the prominent driving forces, which helps the technical textiles stay ahead of the competition. The textile industry is affected by two main factors, that is, market volatility and world-wide competition. Therefore, the industry needs to enhance its capability to produce and market high quality and added value products. Recently, with dramatic growth in nanotechnology, increased volumes of engineered nanoparticles have been produced which represent the potential functional benefits in a wide range of applications by integrating into textiles. Inorganic and metallic-based nano-structured materials have created a new interesting field in all sciences for the continuous investigations due to their undeniably unique properties. Their applications have already led to the development of new practical productions. However, high release of AgNPs from textiles after washing is still a major challenge in this particular area and becomes a potential new pollutant source.

Acknowledgements Financial support provided by Southwest University Chongqing, China as Postdoctoral Fellow for Luqman Jameel Rather and Showkat Ali Ganie is also thankfully acknowledged.

References

1. Shahid M, Islam S, Mohammad F (2013) Recent advancements in natural dye applications: a review. J Clean Prod 53:310–331
2. Islam S, Shahid M, Mohammad F (2013) Green chemistry approaches to develop antimicrobial textiles based on sustainable biopolymers—a review. Ind Eng Chem Res 52:5245–5260
3. Dastjerdi R, Montazer M (2010) A review on the application of inorganic nano-structured materials in the modification of textiles: focus on anti-microbial properties. Colloids Surf B 79:5–18
4. Smijs TJ, Pavel S (2011) Titanium dioxide and zinc oxide nanoparticles in sunscreens: focus on their safety and effectiveness. Nanotechnol Sci Appl 4:95–112
5. Chen Q, Shen X, Gao H (2006) One-step synthesis of silver-poly(4-vinylpyridine) hybrid microgels by γ-irradiation and surfactant-free emulsion polymerization, the photoluminescence characteristics. Colloids Surf A Physicochem Eng Asp 275:45–49
6. Dimitrov DS (2006) Interactions of antibody-conjugated nanoparticles with biological surfaces. Colloids Surf A Physicochem Eng Asp 282–283:8–10
7. Bashari A, Shakeri M, Shirvan AR, Najafabadi SAN (2018) Functional finishing of textiles via nanomaterials. In: Islam S, Butola BS (eds) Nanomaterials in the wet processing of textiles. Scrivener Publishing LLC, pp 1–70

8. Mucha H, Hofer D, Aßflag S, Swere M (2002) Antimicrobial finishes and modification. Melliand Text Berichte 83(4):53–56
9. Perelshtein I, Perkas N, Gedanken A (2016) Ultrasonic coating of textiles by antibacterial and antibiofilm nanoparticles. In: Handbook of ultrasonics and sonochemistry, vol 967
10. Syafiuddin A (2019) Toward a comprehensive understanding of textiles functionalized with silver nanoparticles. J Chin Chem Soc 1–22
11. Quincy III RB, Karandikar BM, MacDonald JG, Elshani T (2011) Dual element odor control in personal care products. Google patents
12. Cai K, Bossert J, Jandt KD (2006) Does the nanometre scale topography of titanium influence protein adsorption and cell proliferation? Colloids Surf B 49:136–144
13. Han K, Yu M (2006) Study of the preparation and properties of UV-blocking fabrics of a PET/TiO_2 nanocomposite prepared by in situ polycondensation. J Appl Polym Sci 100:1588–1593
14. Ikezawa S, Homyara H, Kubota T, Suzuki R, Koh S, Mutuga F, Yoshioka T, Nishiwaki A, Ninomiya Y, Takahashi M, Baba K, Kida K, Hara T, Famakinwa T (2001) Applications of TiO film for environmental purification deposited by controlled electron beam-excited plasma. Thin Solid Films 386:173–176
15. Li D, Haneda H, Hishita S, Ohashi N (2005) Visible-light-driven N–F–Co doped TiO_2 photocatalysts. 2. Optical characterization, photocatalysis, and potential application to air purification. Chem Mater 17:2596–2602
16. Weibel A, Bouchet R, Knauth P (2006) Electrical properties and defect chemistry of anatase (TiO_2). Solid State Ionics 177:229–236
17. Verran J, Sandoval G, Allen NS, Edge M, Stratton J (2007) Variables affecting the antibacterial properties of nano and pigmentary titania particles in suspension. Dyes Pigm 73:298–304
18. Reidy DJ, Holmes JD, Morris MA (2006) Preparation of a highly thermally stable titania anatase phase by addition of mixed zirconia and silica dopants. Ceram Int 32:235–239
19. Wong YWH, Yuen CWM, Leung MYS, Ku SKA, Lam HLI (2006) Selected applications of nanotechnology in textiles. Autex Res J 6:1–8
20. Michna A, Morga M, Adamczyk Z, Kubiak K (2019) Monolayers of silver nanoparticles obtained by green synthesis on macrocation modified substrates. Mater Chem Phys 227:224–235
21. Som C, Wick P, Krug H, Nowack B (2011) Environmental and health effects of nanomaterials in nanotextiles and facade coatings. Environ Int 37:1131–1142
22. Syafiuddin A, Salmiati S, Hadibarata T, Salim MR, Kueh ABH, Sari AA (2017) A purely green synthesis of silver nanoparticles using *Carica papaya*, *Manihot esculenta*, and *Morinda citrifolia*: synthesis and antibacterial evaluations. Bioprocess Biosyst Eng 40:1349–1361
23. Syafiuddin A, Salmiati S, Jonbi J, Fulazzaky MA (2018) Application of the kinetic and isotherm models for better understanding of the behaviors of silver nanoparticles adsorption onto different adsorbents. J Environ Manage 218:59–70
24. Ullah H, Wilfred CD, Shaharun MS (2017) Synthesis of silver nanoparticles using ionic-liquid-based microwave-assisted extraction from polygonum minus and photodegradation of methylene blue. J Chin Chem Soc 64:1164–1171
25. El-Rafie M, Mohamed A, Shaheen TI, Hebeish A (2010) Antimicrobial effect of silver nanoparticles produced by fungal process on cotton fabrics. Carbohyd Polym 80:779–782
26. Montazer M, Alimohammadi F, Shamei A, Rahimi MK (2012) Durable antibacterial and cross-linking cotton with colloidal silver nanoparticles and butane tetracarboxylic acid without yellowing. Colloids Surf B Biointerfaces 89:196–202
27. Arnold MS, Avouris P, Pan ZW, Wang ZL (2003) Field-effect transistors based on single semiconducting oxide nanobelts. J Phys Chem B 107:659–663
28. Becheri A, Durr M, Nostro PL, Baglioni P (2007) Synthesis and characterization of zinc oxide nanoparticles: application to textiles as UV-absorbers. J Nanopart Res 10:679–689
29. Behnajady MA, Modirshahla N, Hamzavi R (2006) Kinetic study on photocatalytic degradation of C.I. acid yellow 23 by ZnO photocatalyst. J Hazard Mater 133:226–232

30. Pan ZW, Dai ZR, Wang ZL (2001) Nanobelts of semiconducting oxides. Science 291:1947–1949
31. Tang E, Cheng G, Ma X, Pang X, Zhao Q (2006) Surface modification of zinc oxide nanoparticle by PMAA and its dispersion in aqueous system. Appl Surf Sci 252:5227–5232
32. Xu T, Xie CS (2003) Tetrapod-like nano-particle ZnO/acrylic resin composite and its multi-function property. Prog Org Coat 46:297–301
33. Vigneshwaran N, Kumar S, Kathe AA, Varadarajan PV, Prasad V (2006) Functional finishing of cotton fabrics using zinc oxide-soluble starch nanocomposites. Nanotechnology 17:5087–5095
34. Li Q, Chen SL, Jiang WC (2007) Durability of nano ZnO antibacterial cotton fabric to sweat. J Appl Polym Sci 103:412–416
35. Wang ZL (2004) Zinc oxide nanostructures: growth, properties and applications. J Phys Condens Matter 16:829–858
36. Wang ZL (2004) Nanostructures of zinc oxide. J Phys Condens Matter 7(6):26–33
37. Wang ZL (2007) Novel nanostructures of ZnO for nanoscale photonics, optoelectronics, piezoelectricity, and sensing. Appl Phys A Mater Sci Process 88:7–15
38. Cubillo E, Pecharroman C, Aguilar E, Santaren J, Moya JS (2006) Antibacterial activity of copper monodispersed nanoparticles into sepiolite. J Mater Sci 41:5208–5212
39. Pape HL, Serena FS, Contini P, Devillers C, Maftah A, Leprat P (2002) Evaluation of the anti-microbial properties of an activated carbon fibre supporting silver using a dynamic method. Carbon 40:2947–2954
40. Wel Q, Yu L, Wu N, Hong S (2008) Preparation and characterization of copper nanocomposite textiles. J Ind Text 37(3):275–283
41. Grace AN, Pandian K (2007) Antibacterial efficacy of aminoglycosidic antibiotics protected gold nanoparticles: a brief study. Colloids Surf A 297:63–70
42. Park SH, Oh SG, Munb JY, Han SS (2006) Loading of gold nanoparticles inside the DPPC bilayers of liposome and their effects on membrane fluidities. Colloids Surf Biointerfaces 48:112–118
43. Yonezawa T, Kunitake T (1999) Practical preparation of anionic mercapto ligand stabilized gold nanoparticles and their immobilization. Colloids Surf A 149:193–199
44. Zhang Y, Peng H, Huanga W, Zhou Y, Yan D (2008) Facile preparation and characterization of highly antimicrobial colloid Ag or Au nanoparticles. J Colloid Interface Sci 325:371–376
45. Jimenez G, Ogata N, Kawai H, Ogihara T (1997) Structure and thermal/mechanical properties of poly(γ-caprolactam)–clay blend. J Appl Polym Sci 64:2211–2220
46. Kodgire P, Kalgaonkar R, Hambir S, Bulakh N, Jog JP (2001) PP/clay nanocomposites: effect of clay treatment on morphology and dynamic mechanical properties. J Appl Polym Sci 81:1786–1792
47. Mlynarcikova Z, Brsig E, Legen J, Marcincin A, Alexy P (2005) Influence of the composition of polypropylene/organoclay nanocomposite fiber on their tensile strength. J Macromol Sci Part A Pure Appl Chem 42(5):543–554
48. Pavlikova S, Thomann R, Reichert P, Mulhaupt R, Marcincin A, Borsig E (2003) Fiber spinning from poly(propylene)-organoclay nanocomposite. J Appl Polym Sci 89(3):604–611
49. Razafimahefa L, Chlebicki S, Vroman I, Devaux E (2005) Effect of nanoclay on the dyeing ability of PA6 nanocomposite fibers. Dyes Pigm 66(1):55–60
50. Williams LB, Holland M, Eberl DD, Brunet T, De Courrsou LB (2004) Killer clays! Natural antibacterial clay minerals. Miner Soc Bull 139:3–8
51. Haydel SE, Remenih CM, Williams LB (2008) Broad spectrum in vitro antibacterial activities of clay minerals against antibiotic susceptible and antibiotic-resistant bacterial pathogens. J Antimicrob Chemother 61(2):353–361
52. Seckin T, Onal Y, Yesilada O, Gultek A (1997) Preparation and characterization of a clay–polyvinylpyridinium matrix for the removal of bacterial cells from water. J Mater Sci 32:5993–5999
53. Hu CH, Xu ZR, Xia MS (2005) Antibacterial effect of Cu^{2+}-exchanged montmorillonite on aeromonashydrophlia and discussion on its mechanism. Vet Microbiol 109:83–88

54. Zhou NL, Liu Y, Li L, Meng N, Huang YX, Zhang J, Wei SH, Shen J (2008) A new nanocomposite biomedical material of polymer/clay–Cts–Ag nanocomposites. J Mol Catal A Chem 281:192–199
55. Singh M, Sharma R, Banerjee UC (2002) Biotechnological applications of cyclodextrins. Biotechnol Adv 20:341–359
56. Voncina B, Vivod V, Jausovec D (2007) B-cyclodextrin as retarding reagent in polyacrylonitrile dyeing. Dyes Pigm 74(3):642–646
57. Martin Del Valle EM (2004) Cyclodextrins and their uses: a review. Process Biochem 39:1033–1046
58. Martel B, Morcellet M, Ruffin D, Ducoroy L, Weltrowski M (2002) Finishing of polyester fabrics with cyclodextrins and polycarboxylic acids as crosslinking agents. J Incl Phenom Macrocycl Chem 44:443–446
59. Savarino P, Viscardi G, Quagliotto P, Montoneri E, Barni E (1999) Reactivity and effects of cyclodextrins in textile dyeing. Dyes Pigm 42(2):143–147
60. Martel B, Ruffin D, Weltrowski M, Lekchiri Y, Morcellet M (2002) Water soluble polymers and gels from the polycondensation between cyclodextrins and poly(carboxylic acid)s: a study of the preparation parameters. J Appl Polym Sci 97:433–442
61. Ducoroy L, Martel B, Bacquet B, Morcellet M (2007) Ion exchange textile from the finishing of PET fabrics with cyclodextrins and citric acid for the sorption of metallic cations in water. J Incl Phenom Macrocycl Chem 57:271–277
62. Volodkin DV, Ball V, Voegel JC, Mohwald H, Dimova R, Marchi-Artzner V (2007) Control of the interaction between membranes or vesicles: adhesion, fusion and release of dyes. Colloids Surf A Physicochem Eng Asp 303:89–96
63. Mao J (2002) Durable antimicrobial finish for cotton with new technology. AATCC Rev 2(12):15–18
64. Benita S (1996) Microencapsulation: methods and industrial application. Marcel Dekker, New York
65. Karsa DR, Stephensone RA (1993) Encapsulation and controlled release. The Royal Society of Chemistry, Cambridge
66. Watnasirichaikul S, Davies NM, Rades T, Tucker IG (2000) Preparation of biodegradable insulin nanocapsules from biocompatible microemulsions. Pharm Res 17:6
67. Zhang XX, Fan YF, Tao XM, Yick KL (2004) Fabrication and properties of microcapsules and nanocapsules containing n-octadecane. Mater Chem Phys 88:300–307
68. Shim JW, Kim JW, Han SH, Chang IS, Kim HK, Kang HH, Lee OS, Suh KD (2002) Zinc oxide/polymethylmethacrylate composite microspheres by in situ suspension polymerization and their morphological study. Colloids Surf A 207(1–3):105–111
69. Oku T, Kusunose T, Niihara K, Suganuma K (2000) Chemical synthesis of silver nanoparticles encapsulated in boron nitride nanocages. J Mater Chem 10:255–257
70. Huwyler J, Yang JJ, Pardridge WM (1997) Receptor mediated delivery of daunomycin using immunoliposomes: pharmacokinetics and tissue distribution in the rat. J Pharmacol Exp Ther 282(3):1541–1546
71. Barani H, Montazer M (2008) A review on applications of liposomes in textile processing. J Liposome Res 18:249–262
72. Montazer M, Taghavi FA, Toliyat T, Moghadam MB (2007) Optimization of dyeing of wool with madder and liposomes by central composite design. J Appl Polym Sci 106:1614–1621
73. Montazer M, Zolfaghari AR, Toliat T, Moghadam MB (2009) Modification of wool surface by liposomes for dyeing with weld. J Liposome Res 19(3):173–179
74. Park SH, Oh SG, Mun JY, Han SS (2005) Effects of silver nanoparticles on the fluidity of bilayer in phospholipid liposome. Colloids Surf B 44:117–122
75. Balogh L, Tomalia DA, Hagnarue GL (2000) A revolution of nanoscale proportions, chemical innovation. Am Chem Soc 30(3):19–26
76. He JA, Valluzzi R, Yang K, Dolukhanyan T, Sung C, Kumar J, Tripathy SK, Samuelson L, Balogh L, Tomalia DA (1999) Electrostatic multilayer deposition of a gold-dendrimer nanocomposite. Chem Mater 11:3268–3274

77. Tan NB, Tomalia DA, Linm JS (1999) A small angle scattering of dendrimer-copper sulfide nanocomposite. Polymer 40:2537–2545
78. Balogh L, Swanson DR, Tomalia DA, Hagnauer GL, Manus ATM (2001) Dendrimer-silver complexes and nanocomposites as antimicrobial agents. Nano Lett 1:8–21
79. Raveendran P, Goyal A, Blatchford MA, Wallen SL (2006) Stabilization and growth of silver nanocrystals in dendritic polyol dispersions. Mater Lett 60:897–900
80. Lacasse K, Baumann W (2004) Textile chemicals: environmental data and facts. Springer, Berlin
81. Heine E, Knops HG, Schaefer K, Vangeyte P, Moeller M (2007) Antimicrobial functionalization of textile materials. In: Duquesne S, Magniez C, Camino G (eds) Multifunctional barriers for flexible structure: textile, leather and paper. Springer, Berlin Heidelberg, pp 125–138
82. Maillard JY (2005) Antimicrobial biocides in the healthcare environment: efficacy, usage, policies, and perceived problems. Ther Clin Risk Manag 1:307–320
83. NPIRS (2012) National Pesticide Information Retrieval System
84. PAN (2012) PAN pesticide database
85. Sun G, Worley SD (2005) Chemistry of durable and regenerable biocidal textiles. J Chem Educ 82:60–64
86. Benn TM, Westerhoff P (2008) Nanoparticle silver released into water from commercially available sock fabrics. Environ Sci Technol 42:4133–4139
87. Benn T, Cavanagh B, Hristovski K, Posner JD, Westerhoff P (2010) The release of nanosilver from consumer products used in the home. J Environ Qual 39:1875–1882
88. Geranio L, Heuberger M, Nowack B (2009) The behavior of silver nanotextiles during washing. Environ Sci Technol 43:8113–8118
89. Lorenz C, Windler L, Lehmann RP, Schuppler M, Von Goetz N, Hungerbühler K, Nowack B (2012) Characterization of silver release from commercially available functional (nano) textiles. Chemosphere 89:817–824
90. KEMI (2012) Antibacterial substances leaking out with the washing water—analyses of silver, triclosan and triclocarban in textiles before and after washing. Swedish Chemicals Agency, Sundbyberg
91. Orhan M, Kut D, Gunesoglu C (2007) Use of triclosan as antibacterial agent in textiles. Indian J Fibre Text Res 32:114–118
92. Erdem AK, Yurudu NOS (2008) The evaluation of antibacterial activity of fabrics impregnated with dimethyltetradecyl (3-(trimethoxysilyl) propyl) ammonium chloride. J Biol Chem 2:115–122
93. USEPA (2008) Memorandum: zinc 2-pyridinethiol-1-oxide (Zinc Omadine®): occupational and residential exposure risk assessment for new uses (textiles) on EPA Reg Numbers 1258–840 and 1258–841. United States Environmental Protection Agency, Washington, DC
94. Papaspyrides CD, Pavlidou S, Vouyiouka SN (2009) Development of advanced textile materials: natural fibre composites, anti-microbial, and flame-retardant fabrics. Proc Inst Mech Eng Part L J Mater Des Appl 223:91–102
95. Biba E (2014) The sunscreen pill. Nature 515:124–125
96. Damiani E, Baschong W, Greci L (2007) UV-filter combinations under UV-A exposure: concomitant quantification of overall spectral stability and molecular integrity. J Photochem Photobiol B 87:95–104
97. Kockler J, Oelgemoller M, Robertson S, Glass BD (2012) Photostability of sunscreens. J Photochem Photobiol C 13:91–110
98. Oresajo C, Yatskayer M, Galdi A, Foltis P, Pillai S (2010) Complementary effects of antioxidants and sunscreens in reducing UV-induced skin damage as demonstrated by skin biomarker expression. J Cosmet Laser Ther 12:157–162
99. Deng Y, Ediriwickrema A, Yang F, Lewis J, Girardi M, Saltzman WM (2015) A sunblock based on bioadhesive nanoparticles. Nat Mater 14:1278–1285
100. Tolbert SH, McFadden PD, Loy DA (2016) New hybrid organic/inorganic polysilsesquioxane-silica particles as sunscreens. ACS Appl Mater Interfaces 8:3160–3174

101. Selishchev DS, Karaseva IP, Uvaev VV, Kozlov DV, Parmon VN (2013) Effect of preparation method of functionalized textile materials on their photocatalytic activity and stability under UV irradiation. Chem Eng J 224:114–120
102. Saravanan R, Karthikeyan S, Gupta VK, Sekaran G, Narayanan V, Stephen A (2013) Enhanced photocatalytic activity of ZnO/CuO nanocomposites for the degradation of textile dye on visible light illumination. Mater Sci Eng C 33:91–98
103. Bazant P, Kuritka I, Munster L, Kalina L (2015) Microwave solvothermal decoration of the cellulose surface by nanostructured hybrid Ag/ZnO particles: a joint XPS, XRD and SEM study. Cellulose 22:1275–1293
104. El Shafei A, Abou-Okeil A (2011) ZnO/carboxymethyl chitosan bionano-composite to impart antibacterial and UV protection for cotton fabric. Carbohyd Polym 83:920–925
105. Montazer M, Amiri MM, Malek RMA (2013) In situ synthesis and characterization of nano ZnO on wool: influence of nano photo reactor on wool properties. J Photochem Photobiol A 89:1057–1063
106. Ibrahim NA, El-Zairy EM, Eid BM, Emam E, Barkat SR (2017) A new approach for imparting durable multifunctional properties to linen-containing fabrics. Carbohyd Polym 157:1085–1093
107. Hsu CH, Chen LC, Lin YF (2013) Preparation and optoelectronic characteristics of ZnO/CuO-Cu_2O complex inverse heterostructure with GaP buffer for solar cell applications. Materials 6:4479–4488
108. Jiang T, Xie T, Chen L, Fu Z, Wang D (2013) Carrier concentration-dependent electron transfer in Cu_2O/ZnO nanorod arrays and their photocatalytic performance. Nanoscale 5:2938–2944
109. Prasad V, Arputharaj A, Bharimalla AK, Patil PG, Vigneshwaran N (2016) Durable multifunctional finishing of cotton fabrics by in situ synthesis of nano-ZnO. Appl Surf Sci 390:936–940
110. Shaheen TI, El-Naggar ME, Abdelgawad AM, Hebeish A (2016) Durable antibacterial and UV protections of in situ synthesized zinc oxide nanoparticles onto cotton fabrics. Int J Biol Macromol 83:426–432
111. Montazer M, Seifollahzadeh S (2011) Enhanced self-cleaning, antibacterial and UV protection properties of nano TiO_2 treated textile through enzymatic pretreatment. J Photochem Photobiol 87:877–883
112. Zhang M, Tang B, Sun L, Wang X (2014) Reducing photoyellowing of wool fabrics with silica coated ZnO nanoparticles. Text Res J 84:1840–1848
113. Barik S, Khandual A, Behera L, Badamali SK, Luximon A (2017) Nano-Mg–Al-layered double hydroxide application to cotton for enhancing mechanical, UV protection and flame retardancy at low cytotoxicity level. Cellulose 24:1107–1120
114. Lee S (2009) Developing UV-protective textiles based on electrospun zinc oxide nanocomposite fibers. Fibers Polym 10:295–301
115. Dadvar S, Tavanai H, Morshed M (2011) UV-protection properties of electrospun polyacrylonitrile nanofibrous mats embedded with MgO and Al_2O_3 nanoparticles. J Nanopart Res 13:5163–5169
116. Pu X, Li L, Liu M, Jiang C, Du C, Zhao Z, Hu W, Wang ZL (2016) Wearable self-charging power textile based on flexible yarn supercapacitors and fabric nanogenerators. J Adv Mater 28:98–105
117. Higginbotham AL, Lomeda JR, Morgan AB, Tour JM (2009) Graphite oxide flame-retardant polymer nanocomposites. ACS Appl Mater Interfaces 1:2256–2261
118. Cai G, Xu Z, Yang M, Tang B, Wang X (2017) Functionalization of cotton fabrics through thermal reduction of graphene oxide. Appl Surf Sci 393:441–448
119. Luoma SN (2008) Silver nanotechnologies and the environment: old problems or new challenges. Project on Emerging Nanotechnologies of the Woodrow Wilson International Center for Scholars, Washington, DC
120. Impellitteri CA, Tolaymat TM, Scheckel KG (2009) The speciation of silver nanoparticles in antimicrobial fabric before and after exposure to a hypochlorite/detergent solution. J Environ Qual 38:1528–1530

121. Choi O, Cleuenger TE, Deng BL, Surampalli RY, Ross L, Hu ZQ (2009) Role of sulfide and ligand strength in controlling nanosilver toxicity. Water Res 43:1879–1886
122. Nowack B, Ranville JF, Diamond S, Gallego-Urrea JA, Metcalfe C, Rose J, Horne N, Koelmans AA, Klaine SJ (2012) Potential scenarios for nanomaterial release and subsequent alteration in the environment. Environ Toxicol Chem 31:50–59
123. Hagfeldt A, Gratzel M (1995) Light-induced redox reactions in nanocrystalline systems. Chem Rev 95:49–68
124. Mills A, Hunte SL (1997) An overview of semiconductor photocatalysis. J Photochem Photobiol A Chem 108:1–35
125. Brunet L, Lyon DY, Hotze EM, Alvarez PJJ, Wiesner MR (2009) Comparative photoactivity and antibacterial properties of C_{60} fullerenes and titanium dioxide nanoparticles. Environ Sci Technol 43:4355–4360
126. Auffan M, Pedeutour M, Rose J, Masion A, Ziarelli F, Borschneck D, Chaneac C, Botta C, Chaurand P, Labille J, Bottero JY (2010) Structural degradation at the surface of a TiO_2-based nanomaterial used in cosmetics. Environ Sci Technol 44:2689–2694
127. Carp O (2004) Photoinduced reactivity of titanium dioxide. Prog Solid State Chem 32:33–177
128. Stumm W, Morgan JJ (1995) Aquatic chemistry: chemical equilibria and rates in natural waters, 3rd edn. Wiley, NewYork, NY, USA
129. Liu JY, Hurt RH (2010) Ion release kinetics and particle persistence in aqueous nano-silver colloids. Environ Sci Technol 44:2169–2175
130. Borm P (2005) Research strategies for safety evaluation of nanomaterials, part V: role of dissolution in biological fate and effects of nanoscale particles. Toxicol Sci 90:23–32
131. Costa P, Sousa Lobo JM (2001) Modeling and comparison of dissolution profiles. Eur J Pharm Sci 13:123–133
132. Miller-Chou BA, Koenig JL (2003) A review of polymer dissolution. Prog Polym Sci 28:1223–1270
133. Metz KM, Mangham AN, Bierman MJ, Jin S, Hamers RJ, Pedersen JA (2009) Engineered nanomaterial transformation under oxidative environmental conditions: development of an in vitro biomimetic assay. Environ Sci Technol 43:1598–1604
134. Fabrega J, Luoma SN, Tyler CR, Galloway TS, Lead JR (2011) Silver nanoparticles: behaviour and effects in the aquatic environment. Environ Int 37:517–531
135. Buffle J, Wilkinson KJ, Van Leeuwen HP (2009) Chemodynamics and bioavailability in natural waters. Environ Sci Technol 43:7170–7174
136. Diebold U (2003) The surface science of titanium dioxide. Surf Sci Rep 48:53–229
137. Talapin DV, Yin Y (2001) Themed issue: chemical transformations of nanoparticles. J Mater Chem 21:11454
138. Dabrowski A (2001) Adsorption: from theory to practice. Adv Coll Interface Sci 93:135–224
139. Pan B, Xing B (2008) Adsorption mechanisms of organic chemicals on carbon nanotubes. Environ Sci Technol 42:9005–9013
140. Rabe M, Verdes D, Seeger S (2011) Understanding protein adsorption phenomena at solid surfaces. Adv Coll Interface Sci 162:87–106
141. Saleh NB, Pfefferle LD, Elimelech M (2010) Influence of biomacromolecules and humic acid on the aggregation kinetics of single-walled carbon nanotubes. Environ Sci Technol 44:2412–2418
142. Hassellov M, Readman JW, Ranville JF, Tiede K (2008) Nanoparticle analysis and characterization methodologies in environmental risk assessment of engineered nanoparticles. Ecotoxicology 17:344–361
143. Mueller NC, Nowack B (2010) Nanoparticles for remediation—solving big problems with little particles. Elements 6:395–400
144. Parida SK, Dash S, Patel S, Mishra BK (2006) Adsorption of organic molecules on silica surface. Adv Coll Interface Sci 121:77–110
145. Walser T, Demou E, Lang DJ, Hellweg S (2011) Prospective environmental life cycle assessment of nanosilver T-shirts. Environ Sci Technol 45:4570–4578

146. BMU (Hrsg.) (2011) Verantwortlicher Umgang mit Nanotechnologien. Bericht der Themengruppen der NanoKommission der deutschen Bundesregierung. http://www.bmu.de/service/publikationen/downloads/details/artikel/ergebnisse-aus-der-zweitennanodialog-phase-2009
147. Glover RD, Miller JM, Hutchison JE (2011) Generation of metal nanoparticles from silver and copper objects: nanoparticle dynamics on surface and potential sources of nanoparticles in the environment. ACS Nano 5:8950–8957
148. Limbach LK, Bereiter R, Müller E, Krebs R, Galli R, Stark WJ (2008) Removal of oxide nanoparticles in a model wastewater treatment plant: influence of agglomeration and surfactants on clearing efficiency. Environ Sci Technol 42:5828–5833
149. Burkhardt M, Zuleeg S, Kagi R, Eugster J, Boller M, Siegrist H (2010) Verhalten von Nanosilber in Klaranlagen und dessen Einfluss auf die Nitrifikationsleistung von Belebtschlamm. Umweltwiss Schadst Forsch 22:529–540
150. Nickel C, Hellack B, Gartiser S, Flach F, Schiwy A, Maes H, Schaffer A, Gabsch S, Stintz M, Erdinger L, Kuhlbusch TAJ (2012) Fate and behaviour of TiO_2 nanomaterials in the environment, influenced by their shape, size and surface area. UBA-Text 25/2012. Umweltbundesamt
151. Walser T, Limbach LK, Brogioli R, Erismann E, Flamigni L, Hattendorf B, Juchli M, Krumeich F, Ludwig C, Prikopsky K, Rossier M, Saner D, Sigg A, Hellweg S, Günther D, Stark WJ (2012) Persistence of engineered nanoparticles in a municipal solid-waste incineration plant. Nat Nanotechnol 78:520–524
152. Motojima S, Noda Y, Hoshiya S, Hishikawa Y (2003) Electromagnetic wave absorption property of carbon microcoils in 12–110 GHz region. J Appl Phys 944:2325–2330
153. Rosace G (2015) Radiation protection finishes for textiles. In: Paul R (ed) Functional finishes for textiles. Woodhead Publishing, pp 487–512
154. Chen HC, Lee KC, Lin JH, Koch M (2007) Fabrication of conductive woven fabric and analysis of electromagnetic shielding via measurement and empirical equation. J Mater Process Technol 184:124–130
155. Horrocks AR (2011) Flame retardant challenges for textiles and fibres: new chemistry versus innovatory solutions. Polym Degrad Stab 96:377–392
156. Yang JC, Liao W, Deng SB, Cao ZJ, Wang YZ (2016) Flame retardation of cellulose-rich fabrics via a simplified layer-by-layer assembly. Carbohyd Polym 151:434–440
157. Soutter W (2012) Nanomaterials for environmentally friendly flame retardants. https://www.azonano.com/article.aspx?ArticleID=3087
158. Gashti MP, Alimohammadi F, Shamei A (2012) Preparation of water-repellent cellulose fibers using a polycarboxylic acid/hydrophobic silica nanocomposite coating. Surf Coat Technol 206:3208–3215
159. Carosio F, Laufer G, Alongi J, Camino G, Grunlan JC (2011) Layer-by-layer assembly of silica-based flame-retardant thin film on PET fabric. Polym Degrad Stabil 96:745–750
160. Yang CQ, Xu L, Li S, Jiang Y (1998) Non-formaldehyde durable press finishing of cotton fabrics by combining citric acid with polymers of maleic acid. Text Res J 68:457–464
161. Lam Y, Kan C, Yuen C (2010) Effect of concentration of titanium dioxide acting as catalyst or co-catalyst on the wrinkle-resistant finishing of cotton fabric. Fibers Polym 11:551–558
162. Hebeish A, El-Bisi M, El-Shafei A (2015) Green synthesis of silver nanoparticles and their application to cotton fabrics. Int J Biol Macromol 72:1384–1390
163. Montazer M, Alimohammadi F, Shamei A, Rahimi MK (2012) Durable antibacterial and cross-linking cotton with colloidal silver nanoparticles and butane tetracarboxylic acid without yellowing. Colloids Surf B 89:196–202
164. Atlas RM, Pramer D (1990) Focus on bioremediation. ASM News 56:7–15
165. Chung KT, Stevens SEJ (1993) Degradation of azo dyes by environmental microorganisms and helminthes. Environ Toxicol Chem 12:2121–2132
166. Wong PK, Yuen PY (1996) Decolorization and degradation of methyl red by *Klebsiella pneumoniae* RS-13. Water Res 30:1736–1744
167. Sharma MK, Sobti RC (2000) Rec effect of certain textile dyes in *Bacillus subtilis*. Mutat Res 465:27–38

168. Banat IM, Nigam P, Singh D, Marchant R (1996) Microbial decolorization of textile-dye-containing effluents: a review. Bioresour Technol 58:217–227
169. Shin KS, Oh IK, Kim CJ (1997) Production and purification of remazol brilliant blue R decolorizing peroxidase from the culture filtrate of *Pleurotus ostreatus*. Appl Environ Microbiol 63:1744–1748
170. Swamy J, Ramsay JA (1999) The evaluation of white rot fungi in the decoloration of textile dyes. Enzyme Microb Technol 24:130–137
171. Dilek FB, Taplamacioglu HM, Tarlan E (1999) Color and AOX removal from pulping effluents by algae. Appl Microbiol Biotechnol 52:585–591
172. Murthy DSV, Levine RL, Hallas LE (1988) Principles of organism selection for the degradation of glyphosate in a sequencing batch reactor. In: Proceedings of the 43rd industrial waste conference, 10–12 May. Lewis Publisher, pp 267–274
173. Jilani S (2015) Bioremediation application for textile effluent treatment. Middle-East J Sci Res 231:26–34

Biotechnology: An Eco-friendly Tool of Nature for Textile Industries

Shahid Adeel, Shagufta Kamal, Tanvir Ahmad, Ismat Bibi, Saima Rehman, Amna Kamal and Ayesha Saleem

Abstract Biotechnology has impacted the textile industry through the improvement of more proficient and eco-friendly manufacturing processes, as well as by facilitating the amended designs of textile materials. Traditionally, the growing textile industry requires harsh chemicals, a lot of costs, labor, and energy for processing. Efforts due to heavy energy costs and water deficit are being made to substitute the conventional chemical processes with eco-friendly and economically alluring bioprocesses. Applications of laccases for nape removing, fabric processing, bio bleaching, dyeing, printing, and cellulases for denim finishing are the recent commercial advancements. Latterly, tools of biotechnology also include the modification of synthetic and natural fibers. This chapter represents the application of biotechnological tools involved in the textile industry to make it more clean and global friendly. This manuscript is about to explore the design and engineering of novel enzymes for textile applications. Hopefully, this chapter will give a new guideline to the textile community, academic researchers and traders to move towards the applications of biotechnology for improvement of textile processing.

Keywords Engineering of enzymes · Laccases · Cellulases · Green technologies

S. Adeel · S. Rehman · A. Saleem
Department of Chemistry, Government College University, Faisalabad 38000, Pakistan

S. Kamal (✉)
Department of Biochemistry, Government College University, Faisalabad 38000, Pakistan
e-mail: shaguftakamal@gcuf.edu.pk

T. Ahmad
Department of Statistics, Government College University, Faisalabad 38000, Pakistan

I. Bibi
Department of Chemistry, The Islamia University, Bahawalpur 63000, Pakistan

A. Kamal
Department of Chemistry, University of Agriculture, Faisalabad 38000, Pakistan

M. Shahid and R. Adivarekar (eds.), *Advances in Functional Finishing of Textiles*, Textile Science and Clothing Technology, https://doi.org/10.1007/978-981-15-3669-4_4

1 Introduction

Textile industries have been considered among major sectors with a striking in the expansion of human civilization since the discovery of synthetic dyes. Technological developments have led to a rapid extension of cotton output, throughout the globe [1]. White or industrial biotechnology involving enzymes or biomolecules applications in textile industries acquiesces the development of eco-friendly technologies in fiber processing whereas elaborates strategies for the final product quality improvement [2]. Rising awareness about environmental concerns due to the disposal of chemicals, effluents or smoke into the landfills produced by the utilization of expensive raw material, huge amounts of energy has compelled the scientists to apply white biotechnology in the textile industry [3]. Besides the environmental burdens, chemical processing of textile items are concerned with economic as well as fabric quality loses [4]. In addition to chemical processing, expensive and massive chemicals are being practiced for the neutralization of textile effluents [5]. The obstinate temperament of dye-based diverse preparations or different synthetic dyes has influenced the liabilities to withstand environmental regulations and legislation [6]. Biomolecules or green catalysts play an alternative role in chemical processing and accouters the perfect elucidation for pollution control. Undoubtedly, lowering the overall cost, these biomolecules or green catalysts are processing convivial and environment-friendly [7]. Application of green catalysts or biomolecules in textiles is not considered a new; anachronized in the mid of the nineteenth century [8]. In 1857, biomolecules or green catalysts were the first time used for de-sizing in the textile industry. However, due to its poor efficiency, acid-based desizing was preferred over it for a long time. In 1900s, malt extract was introduced in wet processing, while later in 1912, many textile industries successfully adopted bio-engineered green catalysts [9]. Biotechnology revolutionized the finishing sector by reducing water consumption by 17–18%. Overall 30–50% of water consumption has been reduced in the textile sector by using biotechnology [10]. Hence these days' green catalysts are considered as an integral component of the textile sector [11].

There are two well-established sectors for the applications of green catalysts:

(i) Preparatory sector; amylases are commonly applied in this sector
(ii) Finishing sector; cellulases are commonly applied for bio-stoning, softening and to reduce pilling propensity [12].

Recently lipases, catalases, xylanases, pectinase, etc. (Table 1) are commonly applied in textile industries for finishing, bio-scouring, fading, decolorization, and bio-polishing, etc. [13]. Ambitious research to explore new green catalysts that cover almost all cotton processing steps in the textile industry is the need of the hour [14]. The schematic of processing involved in the woven sector of cotton textile at the commercial level has been designed in Fig. 1.

This chapter explains the sources of green catalysts, their mode of action, thermodynamics, specificity, retention, engineered green using biotechnological tools and promising areas concerning textile industries. Table 1 gives an overview of Overall processing steps and t related enzymes in enzymatic processing of fibers.

Table 1 Overall processing steps and t related enzymes in enzymatic processing of fibers

Sr. No.	Enzymes	Substrate	Area of application	Working conditions
1	Amylases	Starch	Desizing of handmade fabrics as well as woven cotton	pH 4.00–11.00; 25–130 °C
2	Cellulases	Cellulose	Bio scouring, bio polishing, bio stone washing, wool carbonization	pH 4.5–7.00; 9–10; 30–60 °C
3	Catalases	Peroxides	Disruption of residual H_2O_2	50–60 °C, 7.4 pH
4	Glucoxidase	β-D-glucose	Bio-bleaching	pH 6, 25–45 °C
5	Lipases	Oils and fats	Remove waxes and Greece; desizing	pH 8; 40 °C
6	Laccases	Indigo dye	Bio-bleaching; mercerization; denim finishing	pH 4.00–6.00; 50–75 °C
7	LiP	Peroxide	Color striping	pH 4.00–6.5; 40 °C
8	MVP	Peroxide	Color striping	pH 5, 40 °C
9	Pectinases	Pectin	Bio scouring; flax retting	pH 2.5–11.00; 30–75 °C
10	Peroxidase	Peroxide	Bio-bleaching	pH 5, 50 °C
11	Proteases	Proteins	Bio-degumming of silk; dyeing; wool finishing	7 pH; 35 °C
12	Transglutaminase	γ-carboxamide; ε-NH_2 of lysine	Recovers proteolytic treatment damage	pH 5–9; 10–55 °C
13	Xylanase	Hemicellulose, xylan	Bleaching and scouring	pH 5–10; 70 °C

2 Green Catalysts

Green or biocatalysts usually belong to the family of globular proteins with linear chains of amino acids that fold into a specific manner to give its unique 3D structures with distinct properties. Green or biocatalysts are enzymes due to their mere presence; speed up the chemical reaction without being used. The enzyme market (Fig. 2) is divided according to their applications into three segments: (i) food enzymes (ii) animal feed enzymes (iii) technical enzymes. The major enzymes involved in the technical category are cotton or cellulosic processing enzymes following fur and leather processing.

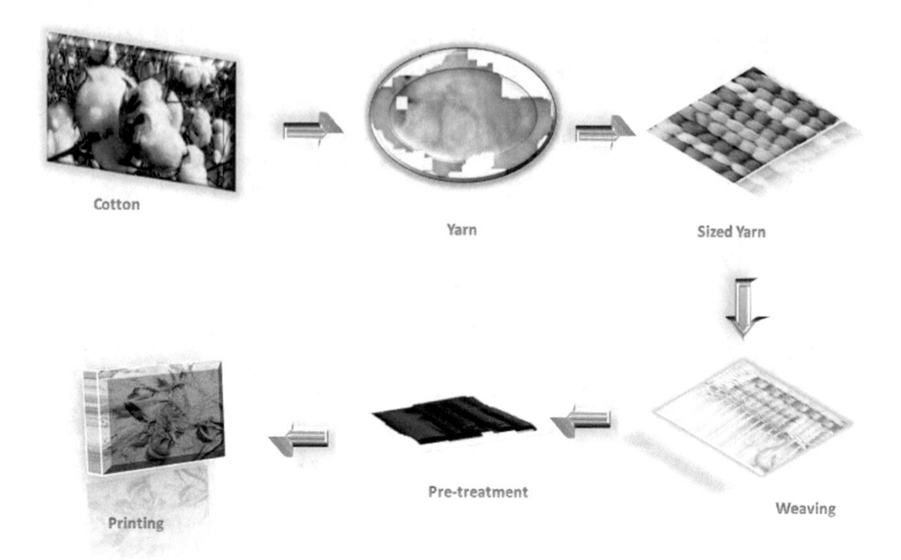

Fig. 1 Typical representation of the processing steps of cotton in textiles

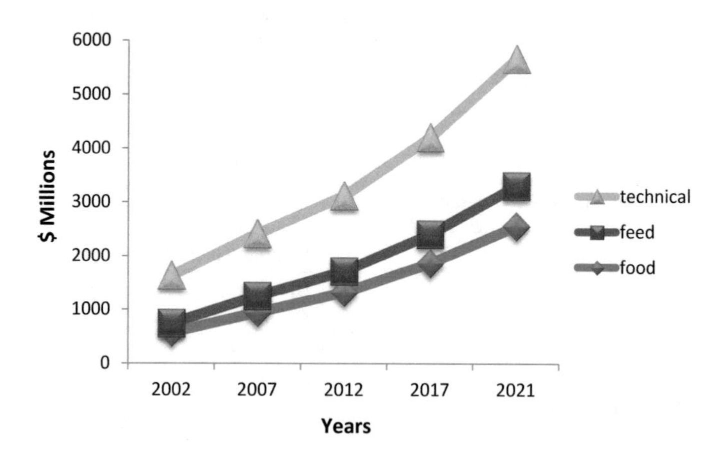

Fig. 2 Global market of enzymes through 2020

The applications of green catalysts in textile industries have many advantages as compared to non-enzymatic or conventional methods as they are active even at neutral pH or low temperatures, rarely form by-products and 10^6–10^{13} times speed up the reaction [15].

3 Textile Processing: Applications of Green Catalysts

The complete process of textile manufacturing and application of enzymes is presented in Fig. 3.

3.1 Wet Processing of Textiles

The word textile is coined from the Latin word "Texere" meaning to weave [16]. Currently, textile fulfills "livery" the basic necessity of humankind's yet permits people to accomplish a fashion statement. Wet processing involves the chemical treatments of fabric after manufacturing [17]. Without chemical processing, the fabric cannot be used for clothing as it doesn't have properties like softness, absorbency, etc., and have dirty pale yellow appearance [18]. After the whole process grey fabric is analyzed to find out any weaving faults. The true wet process is a defined process that uses approximately more than 8000 chemicals for textile processing. Most of these chemicals are toxic to the global environment and community. It is estimated 50% of the global pollution is due to effluents load sheds by the textile-related activities. The environment of ancillary industries that is responsible for supplying detergents, chemicals, dyes, machinery, etc. to textile industries which also play their role in

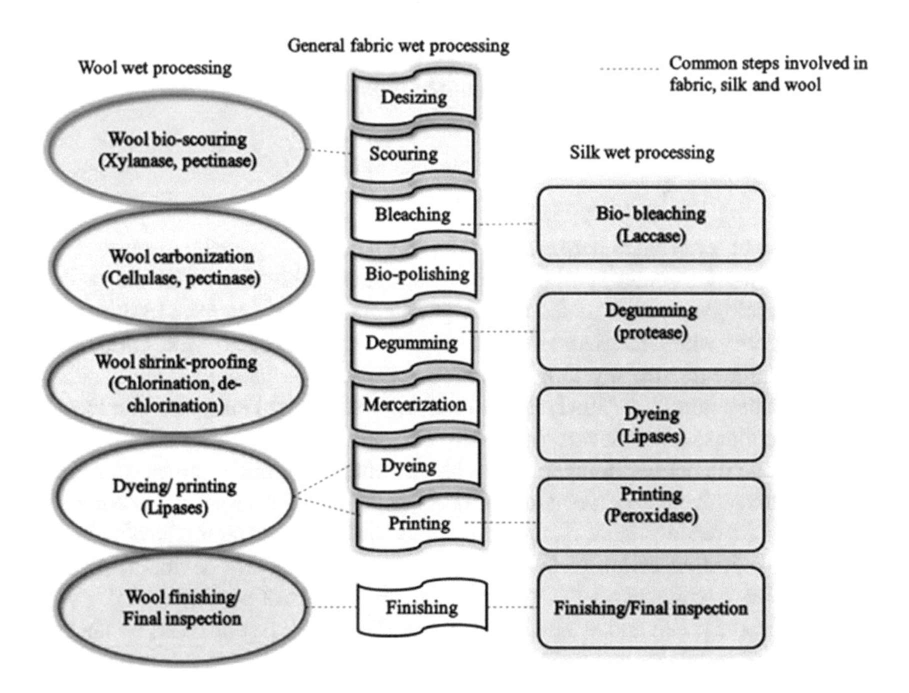

Fig. 3 Bioprocessing of fabrics for textiles

distributing the eco-balance. Biotechnology; a cleaner alternative for most the wet process is achieving contemplation [19]. Thus, collaborative and vigorous research between biotechnological and textile communities to replace conventional processes is the need of time.

3.1.1 Bio-desizing

Removal of sizing agents from cotton fabric to make it suitable for further processing is known as "desizing" [16]. Starch and its derivatives due to their low cost, easy availability with excellent capacity of film formation are considered the most common substrates [17]. Processes involved in conventional desizing cause's excessive damage by intensive rinsing to neutralize the substrates and also increase the pollution load [20]. Bio-desizing by α-amylase has been considered as state of the art for several years [21]. Yet commercially available α-amylase doesn't meet with extreme environmental conditions [17]. Because high desizing efficiency with larger penetration capacity is usually achieved at a maximum temperature [22]. Therefore different biotechnological and statistical techniques have has been used to adopt to make suitable utilization of α-amylase at the industrial level [20, 23]. Treatment of α-amylase with different ultrasound techniques may enhance the desizing efficiency in extreme environmental conditions [24, 25]. Literature reported that α-amylase and glucoamylase could be collectively used for acid-demineralization and desizing simultaneously [26]. Bio-desizing due to its overall quick efficiency to remove starch makes it the most suitable choice for wet processing [2]. The entire process of bio-desizing is eco-friendly with minimum production of wastewater [27].

3.1.2 Bio-scouring

The process of scouring is performed to remove natural or synthetic impurities like esters from the thin outer coating of cuticles, hydrophobic non-cellulosic components especially wax, high molecular weight alcohols [28]. Scouring improves the fiber wettability which facilitates uniform finishing and dying [29]. Conventional hot aqueous alkaline treatment yet high resource-consuming techniques are used at high pH and temperature [30]. Whereas bio-scouring with several enzymes like lipases, cellulase, cutinases, pectinases, proteases, etc. offers superior flaccidity having a broader range of pH and temperature without influencing environment and fabric [31]. Moreover, the expensive nature of chemicals, low effectiveness and comparatively lethargic reaction rates hamper the application of scouring methods in textile sectors [20]. Pectinases are most effectively applied to remove pectins from both acidic and basic media at 50–60 °C depending on the type of pectinase [32].

However; the speed and evenness of bio-scouring could be enhanced by the synergistic use of cellulases and pectinases [33]. Bio-scouring of cotton using pectinase in $C_{20}H_{36}Na_2O_7S$ reversing micellar system is found to be better than conventional scouring of cotton indicating that bio-scouring is a promising and efficient alternative in cotton textiles [34]. Japanese scientists submitted a patent on scoured yarns with out-standing tensile strength retention after treating for 18 h at 40 °C cotton fibers or their blends with proto-pectinases [2].

3.1.3 Bio-bleaching

The pure white appearance of the fabric is a fundamental requirement for the dyeing process [35]. H_2O_2 in alkaline pH and near to 100 °C is the most commonly applied agent in industries for the bleaching process. The whole process of chemical bleaching demands careful control to protect any damage of fiber from radical reaction [36]. Moreover, a huge amount of water is needed for the removal of excess peroxide for the smooth dyeing process [2]. Although, in biological processing, H_2O_2 is generated from $C_6H_{12}O_6$ in aqueous media by glucose oxidize or at controlled rates by the combination of any detergent [37]. So this process is preferred because of its feasibility to reuse de-sizing waste baths with minimum utilization of sources [38]. Although many processes of bleaching are based on the enzymatic generation of H_2O_2 is still in practice [39]. However, laccase enzymes are used as bio-bleaching agents, attached to hydroxyl groups of phenolics to fade colored flavonoids [13]. To overcome the limitations of the bleaching process or its related concerns like processing costs, bio-bleaching or integration of the enzymatic process is a lucrative substitute [35]. However, more benefits like temperature reduction, duration shortening and lower doses of hydrogen peroxide can be achieved by the integration of chemical and bio-bleaching [14].

3.1.4 Enzymatic Cleansing

Shade variation, especially to those fabrics that are sensitive to oxidation, has been observed due to the presence of residual peroxide after bleaching [10]. Extensive washing generating the bulk volume of alkaline wastewater or producing high salts in-stream processes due to treatment with $Na_2S_2O_4$ or $NaHSO_3$ are commonly applied through the classical method [17]. Catalases are well known alternate bio-catalyst having ability to remove residual peroxides by converting it into gaseous oxygen and H_2O with minimum pollution load [40]. The flaws of catalase in commercial applications like high cost, low ability to withstand with high temperatures, alkaline medium or its reactive nature towards dyes could be overcome by immobilizing catalases or microbial cells on solid supports [41]. In this way, it is suggested that the immobilizing of alumina or clay will not only help to control its action with increased recycling ability [42].

3.1.5 Bio-stoning

In the early 1980s or late 1970s, pumice stone washing was developed by denim garments to give fashionable contrast and worn look [43, 44] while the worn look of blue jeans was obtained by non-homogenous removal of indigo dye [45]. The consequences of pumice stone washing were very harsh for garments, equipment as well as for machine [46, 47]. Introduction of microbial cellulases for denim washing in the mid of 1980s presented an alternate of pumice stone washing [48]. It has been proved that cellulase has produced high productivity, reduce time and minimized the damage to a machine with less pumice dust through less use of stones [49–52] To achieve better or lighter shades of indigo-dyed denim, stone washing can be combined with laccase mediated bleaching [53]. Nowadays bio-washing and bio-desizing can be successfully performed in one step by using laccases, cellulase, and amylases without influencing the color of garment [35, 50].

The aged appearance of denim garment could be produced by combining stone and cellulase washing [54]. The surface of cellulose plays a major role in enzymatic decolorization treatment because enzymes, give softness and brightness to the fabric by hydrolyzing cellulose and removing surface naps [55, 56]. Traditional stone washing has a problem in disposing of sands produced by stones eroding therefore; it is usually performed near wastewater streams (Fig. 4) [57, 58]. Moreover, traditional

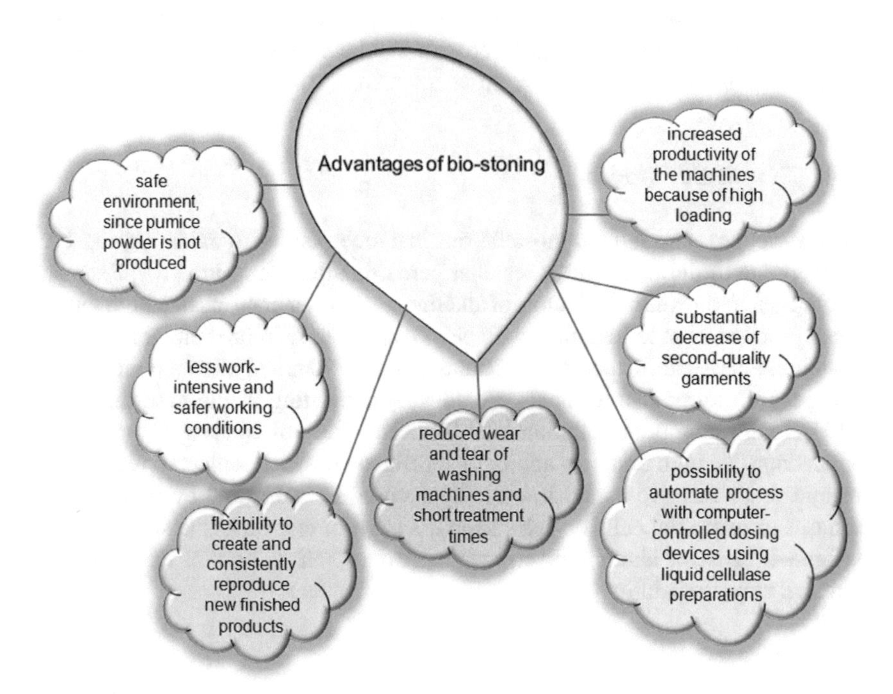

Fig. 4 Advantages of bio-stoning

stone washing produces large amounts of sludge approximately 18 metric tons per weak which clog up the wastewater streams by causing water bodies imbalance [59]. Substitution of stone washing by enzymatic washing raising the environmental issues has taken worldwide fame where since last decade do most of the finishing laundries have been switched to bio-washing [60]. Hence for bio-stoning, an ideal cellulase should have some exposed site capable of binding with indigo dye with controlled or minimum back staining especially during white/blue contrasts are supposed to appear with no post-wash bleaching steps [61].

3.1.6 Bio-polishing

It is an optional process necessary for superior quality brightness, lessen pilling, soothing of fabric and softener feel of fabric surface is a finishing or bio-polishing process because most of the cotton fabrics or its blends become dull and fluffy after continuous wet processing [20, 56]. However, the cotton fabric itself tends to form a pilling as well as fuzz formation [25]. Therefore permanent removal of pilling or fuzz is required to enhance the market values of cotton fabric [62]. The family of cellulases used for the bio-polishing process is (i) endo-cellulases (ii) exo-cellulases (iii) β-glucosidase (iv) cellulose phosphorylases (v) oxidative cellulases, which are commonly utilized for hydrolyzing micro-fibrils of cellulose that gives a glossier and smoother appearance to fabric [55, 63, 64]. The tendency of cotton clothes to form fuzz has been much reduced and has become more fragile after bio-polishing [51, 65]. Other benefits of bio-polishing include:

(i) improvement in breaking strength
(ii) improves elongation [43]
(iii) reduce rugosity [50]
(iv) reduce energy and water consumptions [46, 47]
(v) don't need any chemical coating [66]
(vi) eco-friendly process [67]
(vii) Generally, bio-polishing is performed after bleaching and pre-treatment are necessary for inductions of enzyme attack that becomes minimum due to packing [44].

Cellulases in bio-polishing produce substantial changes in structure that improve consequent chemical processes [43, 44, 68]. Kan and Au [69] reported that bio-polishing and UV-absorber treatment act synergistically. The degree of surface disruption is directly proportional to the binding capacity of cellulase with crystalline and insoluble cellulose [70]. Cellulase extracted from thermophile *T. maritime* didn't influence the surface of cotton fabric, indicating maximum hydrolytic activity with soluble cellulose [71]. In contrary to this, endoglucanase (EG-II) and cellobiohydrolase (CBH-I) extracted from *T. reesei* disrupt the surface of cotton fabric when worked simultaneously [70].

Bio-polishing is always performed during treatments like jets and rotating drum washers in which fabric is supplied with strong mechanical agitation. The balance between mechanical agitation and cellulase activity is vital for pill removing without any strength loss to get valuable fuzz fiber [72]. Enzymatic softening or bio-polishing of fabric lasts over several washes [73].

3.1.7 Bio-mercerizing

The next step to improve dye affinity, luster and strength of the fabric is mercerizing which involves conventional process i.e., treatment with NaOH [74]. It is not a fiber purification process, but it does induce desirable changes in cotton yarn and fabric properties [75]. Mercerization is a well-established finishing process for cotton which affects its fine structure, morphology, mechanical properties and reactivity [76]. This step is usually performed after bleaching on gray cloth [77]. For bio-mercerizing few enzymes like proteases, lipases, pectinases, and cellulases have been reported for mercerizing process yet detailed studies are needed [78].

3.1.8 Bio-dyeing

Enzymes are continuously used since the last decade for bio-preparation of silk, leather, wool and cotton fabric [79]. Diastasis, lipases proteases, and amylase, all four enzymes in the conjugation of tannic acid enhance the dye-ability [80]. Diastasis with natural colorant like *Punicagranatum* (pomegranate), lipase with *Rheum emodi* (red wined pie plant), amylase and proteases with *Terminalia arjuna* (Arjun tree) produced best colorimetric picture regarding dyeing [81]. Markedly, two well-known textiles Sir Naturals and Jet Hosiery have established a sonicator for dyeing cotton and silk naturally on the industrial level [2].

3.1.9 Bio-color Stripping

Destructive stripping, back stripping or color stripping is a dye removal process. Enzymes belong to oxidoreductases like lignin peroxidase, laccases, manganese peroxidase perform color stripping using mechanisms of decolorization, bioremediation, mineralization or degradation [82–85] Color stripping potential of different microbial strains with non-specific nature of enzymes has also been investigated [86]. Indigenously isolated species of white-rot fungi i.e., *Ganoderma lucidum* exhibited an excellent color stripping potential reactive black dyed cotton fabric [87]. Although bio-stripping is eco-friendly with better results than chemical color stripping yet bio-stripping required larger time [88]. Limitation of time can be overcome by the use of different biotechnological tools.

4 Bast Fibers Bioprocessing

4.1 Retting

Bast fibers extraction process from plant stem is called "retting" [89]. Jute, linen, ramie, and hemp are the types of natural bast fibers with more than 50% cellulose and 15–30% non-cellulosic impurities [90, 91]. Natural bast fibers are extracted from plants using specified methods. Retting is the process of bast fiber extraction from plants [89]. Conventional alkali treatments are used as pre-treatment to remove impurities. A single component of impurities can be removed by an individual enzyme such as proteases, hemicellulases, and pectinases [79]. However, a combination of these enzymes results in much better, efficient and faster pre-treatment. Cellulases in multi-enzyme complex remove cellulosic components and facilitate other enzymes to attack the inner layer of the fiber [92]. Bio-finishing treatment conditions influence differently on different fabrics e.g., linen may be destroyed in extreme conditions [93]. The first biotechnological process used since BC in the textile sector is microbial retting [94]. The most common methods of retting (Fig. 5) applied commercially are dew retting and anaerobic retting [95]. Bio-retting principally using pectinases is a comparatively novel and new method to improve the classical retting with controlled process conditions like pH, temperature, enzyme to fiber ratio, etc. [20]. The efficiency of bio-retting could be increased by adding chelators like EDTA with enzyme mixture. Spray enzyme retting is another promising strategy where using negligible amounts of the enzyme has been tested at the pilot-scale [96]. The advantages of bio-retting are explained in Table 2.

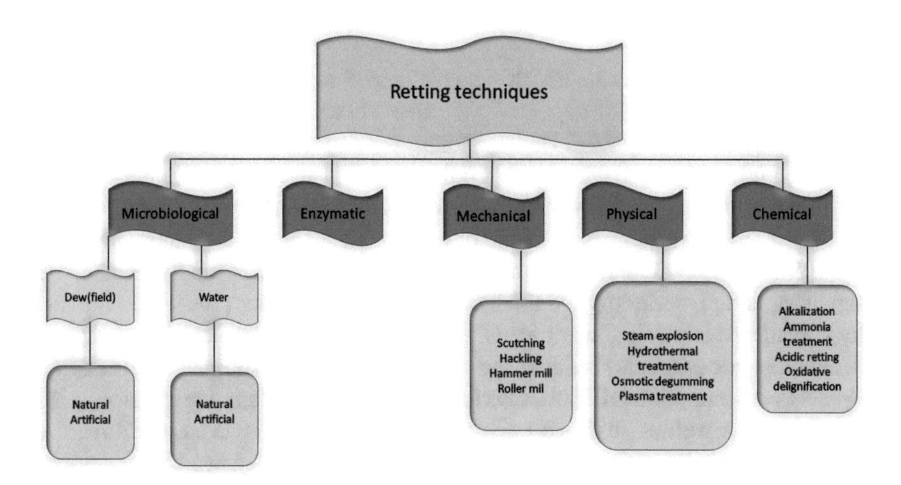

Fig. 5 Different types of retting techniques

Table 2 Comparison of conventional and bio-retting

Sr. No.	Particulars	Conventional retting	Improved retting
1	Source of power	Manual	Partly mechanical
2	The process involved after harvest	Canes-defoliation, bundling, transport steeping in water, steeping-fiber	Machine extraction, steeping of ribbons in water
3	Retting duration	2–3 Weeks	5–7 days
4	Water requirement	Large-volume	20% of conventional
5	Drudgery	Drudgery in steeping and stripping	No drudgery
6	Thin plan	Lost during processing	Salvage's fiber (15% more)
7	Extraction capacity	4–5 kg dry fiber/man-hour	25 kg of dry fiber/machine-hour
8	Crop harvest	Availability of water regulated the time	After 100 days any time
9	Time requirement	4 weeks	1 week
10	Quality	Roots, knots, and specks are present	Strong free from defects
11	Grade	More variation	Less variation
12	Wood stick	Entire	Small species
13	Extraction	Uncontrolled condition	Controlled condition

4.2 Bio-degumming

Degumming is a process in which heavily coated non-cellulosic nature gummy material is removed cellulosic part of plant fibers because decorticated ramie fibers contained approximately 20–35% hemicelluloses and pectin constituting gum [33, 97]. Conventionally these gummy materials were removed by hot alkaline solutions with or without applying pressure [98]. In conventional chemical treatment, 12–20% NaOH is used to treat fibers of ramie and 2% NaOH for fibers of sunn hemp for 24 h [99]. Then upon neutralization, dried over charcoal, it is finally treated with soap, wax, and glycerin to avoid the brittle appearance of fiber [100]. This methodology results in lots of pollutant loads, creating severe environmental threats, non-degradable toxic effluents shedding and causing biological disturbances and exploiting energy consumption [101]. An alternate eco-friendly process due to power crisis, lots of labor as well as stringent environmental conditions is an essential requirement [102, 103]. Bio-degumming relying on proteolytic enzymes for bio-process and whole-cell microbes is a potential substitute of conventional de-gunning. The process of Bio-degumming offers many advantages like energy-saving, eco-friendly;

don't influence the chemical nature, tenacity as well as strength of the fiber. Furthermore, bio-degumming improves the quality of fiber, also elevates its flexibility with less odor [104].

The potent harnessing of microorganisms possessing polysaccharides degrading enzyme systems, O_2 plasma aided bio-degumming and pectinases based bio-degumming are becoming eternally authoritative these days due to limitations of conventional degumming [105]. Pectinases or combinations of pectinases with xylanases due to their specific nature are playing a leading role to remove gummy material as pectin constitutes 40% dry weight of plants [103, 106]. This enzyme is responsible for the degradation of pectin present in the primary cell wall and middle lamella [107]. Degumming, maceration, retting of hemp, ramie and jut bast fibers is effectively completed with the assistance of pectinases. Thermo stable bacterial polygalacturonase exhibited excellent degumming of sunn hemp (*C. juncea*) and ramie (*B. nivea*) bast fibers [106, 108]. To improve the results of bio-degumming, it is performed in combination with chemicals such as urea, NaCl, Na_2CO_3 and NaOH [109–111]. Analytical analysis of sunn and ramie hemp fiber by scanning electron microscopy (SEM) indicates the complete loss of non-cellulosic gummy materials [112–114] supported that soft alkali treatment enhances the bio-degumming properties. Optimum pH is a fundamental requirement of pectinases to avoid any contamination in adopted open fermentation system [112]. Furthermore, to improve efficiency and to lower enzyme doses, enzyme balances should be developed [35]. Only a small amount of pectin for hydrolysis is required to attain maximum strength of retted flax [115].

5 Silk Wet Processing

Silk is one of the strongest natural fibers with approximately 11% moisture regains capacity [116, 117]. Silk loses its strength up to 20% during wetting, possesses moderate to poor elasticity and faints in excessive sunlight [118]. Silk due to fiber macrostructure upon relaxation shrinks up to 4% during wet or dry washing that reversed with pressing [119]. Following these chemical processes are involved in silk processing:

(i) degumming
(ii) bleaching
(iii) dyeing
(iv) finishing.

are involved in silk processing [12]. Sericin (approximately 25%) and fibroin (approximately 75%) are the functional moieties of silk fiber [16]. Sericin is responsible for gum formation, the main impurity which is removed by degumming that is synonymous with cotton or wool scoring [120]. Besides sericin, other impurities like gum tallow, starch or carboxymethyl cellulose (CMC), etc. are also present in silk fiber which are removed by desizing followed by a degumming process. Degumming can be performed by

(i) soap
(ii) enzymes
(iii) water
(iv) alkali [121].

Clinical uses of silk-based biomaterials (SBBs) are being progressively used as potential goods for biomedical textiles. Bio-compatibility, extra-ordinary mechanical properties, controllable degradability and comfort processing nature, the silk-based biomaterials (SBBs) have been explored for different medical materials such as heart valves, arterial grafts, tendons, etc. [122]. To further expand the applications of SBBs or to fulfill the demands of an increasing population, there is growing attentiveness for bio-engineered silk fabrics. Furthermore, the biological properties of bio-engineered silk-like thrombogenicity, biocompatibility, and antimicrobial behavior have been improved [123]. The complete bio-processing of silk has been shown in Fig. 6.

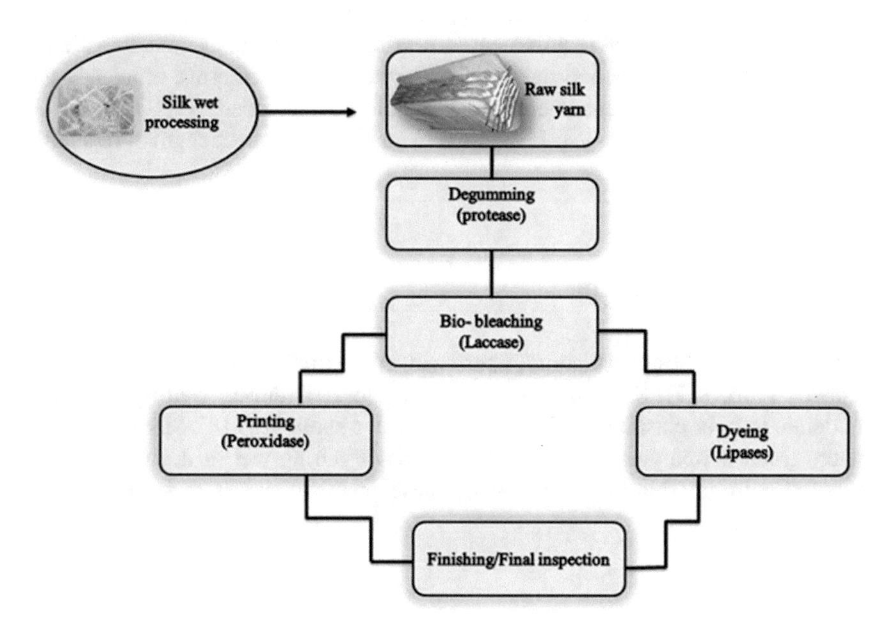

Fig. 6 Different processes for silk wet processing

6 Wool Bio-processing

6.1 *Bio-carbonization of Wool*

The process of removing the vegetable or cellulosic material from the surface of the wool is called carbonizing [124]. The carbonization process involved the use of acidic conditions and high temperature there is a need to replace it with eco-friendly parameters [125]. Bio-carbonization solves the problem by controlling the harmful and destructive effect of process i.e., treatment with pectinases and cellulases overcome all limitations of conventional wool processing [126].

6.2 *Wool Shrink-Proofing*

Commercial finishing processes of wool are categorized into three groups (i) Oxidation/reduction (subtractive) (ii) Synthetic resin layer (additives) (iii) Hercosett/chlorine method (combined process) [127]. Hercosett/chlorine causes oxidation of cysteine residue to form cysteic acid constituted de-chlorination and chlorination process. Though the commercial process is efficient, its effluents resulted in the production of absorbable organic compounds (AOX) yet cause toxicity to the whole food chain [128]. Commercially use of keratinase and laccase produced anti-shrink wool fabric without loss in weight [129].

6.3 *Wool Finishing*

Finishing industries of wool are under continuous strain to adopt eco-friendly finishing processes and to explore novel techniques of wool garments preparation with extra viability in the global market. Applications of disulfide isomerases, lipases, proteases, lipases can modify wool properties [130, 131]. Currently used enzymatic methods were failed to fulfill the final premise, therefore acidic chlorination, per mono sulphuric acid (H_2SO_5) treatment are applied commercially [132]. These chemical treatments significantly improved shrink resistance property but adversely influence the handling of hazardous effluents [133]. Transglutaminase revolutionized the wool finishing industries by fulfilling the alternate need for chemical methods [134].

7 Advantages of Enzymatic Textile Processing

The value of the global market for industrial enzymes was 7082 million USD in 2017 and estimated to increase by 10,519 million USD in 2024 [135]. The detailed comparison of both chemical versus enzymatic processing is made in Fig. 7. The advantages of enzymatic processes in textile industries are

- Improved process competence i.e., permits 50% more jeans load with desired results
- Increased amounts of energy per ton of fabric i.e., production of T-shirts recovers substantial amounts of energy
- Less severe processing conditions i.e., lower doses of H_2O_2 for a short period at low temp
- Reduced source consumption i.e., 20–50% reduction water usage in bio-scouring
- Less consumption of CO_2 i.e., *GHG saving* 990 kg and 410 CO_2 equivalents/FU are produced in bio-scouring as well as in catalase mediated bleaching
- Producing superior quality products i.e., types of denim with cellulases
- Shorter processing time i.e., no neutralization step in bio-scouring
- Low effluent load production i.e., reduces BOD and COD in wastewater [14].

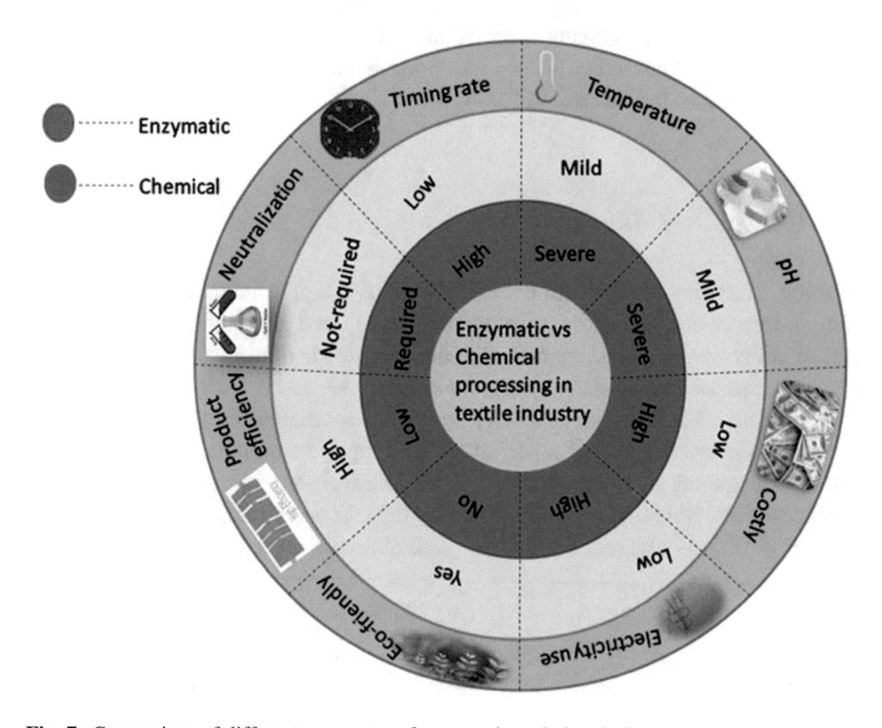

Fig. 7 Comparison of different parameter of enzymatic and chemical processes

8 Enzymes Engineering as a Tailoring Tool

Re-designing of enzymatic traits are called enzyme tailoring [136]. It is necessary to make the most suitable candidates for industrial processes [137]. The engineering of 1st enzyme "Subtilisins" in the early twenty-first century began the revolutionized era of protein engineering [138]. Currently, numerous optimized enzymes such as amylases, cellulases, laccases, proteases, pectinases, and lipases are available to meet industrial applications [139]. Following strategies are commonly applied in the tailoring process:

(i) Directed evolution technique
(ii) High throughput screening
(iii) Computational designs
(iv) Rational designs are commonly applied [13, 140].

Rational designs and directed evolution (Fig. 8) are exclusively unique designs based on the principle of genetic recombination, natural evolutions and computational designs [141]. Whereas data regarding structure-function relationship is required in rational designs to modify one or more amino acid residues for the enhancement of fabric functional properties [142]. The combination of rational and directed evolution is the latest, most powerful and efficient process for positive mutations [143]. The advancement in computer algorithms further simples the decision process and

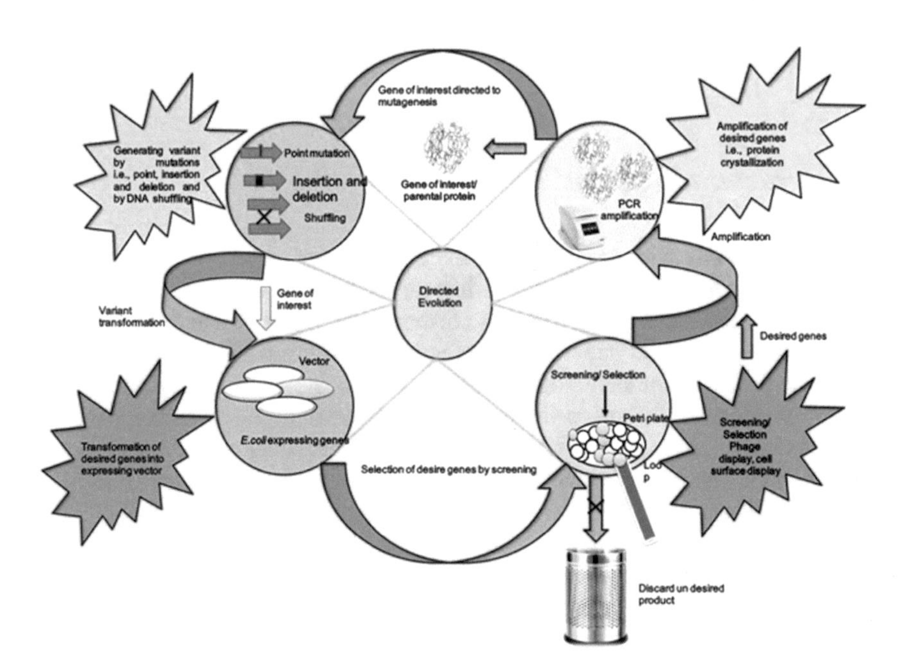

Fig. 8 Methods evolve in directed evolution for enzyme engineering

permits silico identification of useful mutations [140]. Other techniques like microfluidic platforms, ultra-high screening and automated liquid handling combining with fluorescence-activated cell sorting (FACS) have been successfully employed [144, 145]. Recently, the advancement in computational biology has enabled the de novo designing of enzymes for a particular chemical reaction, novel activities for unusual substrates and molecular dynamics [146]. Till now, numerous approaches have been commonly applied to amend properties or even designs of enzymes, where each strategy has some limitations [147]. Therefore, the choice of enzyme engineering techniques should be used based on detailed knowledge of SAR (structure-function relationship), computer algorithms and the opportunity of experimental tools [148, 149]. As the demands of enzymes to catalyze new chemical reactions are drastically increasing, therefore, the achievement in developing beneficial proteins is largely a matter of reliable optimization of strategies and persistent trials [150].

9 The Latest Development in the Textile Sector by Bio-copying of Nature

Techniques for bio-copying of nature like bio-mimetics and bionics can be employed to synthesize (i) structurally colored fibers (ii) lotus effect for self-cleaning textile surfaces [66]. It is a fact that chemist learns significantly from nature while the combination of chemistry and nature results in massive synergistic compensations. Bio-copying is regularly practicing in all steps of textile processing [151, 152].

9.1 Structurally Colored Fibres

Insects particularly *Morphosul-kowskyi* characteristically presents diverse, clear and bright colors whereas some species of insects show different colors by color flopping. It has been investigated that these diverse colors are not due to the presence of any colorants instead due to the interference of reflected light expressed [153]. The same process is observed by thin films of oil on the surface of the water. This natural phenomenon is coined by the cooperation of three Japanese companies to synthetic fibers [154, 155]. Strictly controlled multi-layered structure filaments with calculated thickness have been successfully designed [156]. The reflectance curve of both synthetic and natural fibers is comparable. Morphotic fibers or butterfly (Fig. 9) fibers display flopping of the pure and metallic tone of colors from green to red and violet to blue colors [155]. If this will be successfully launched in the market, then the need for the dying process will be finished. It means lethal ecological and toxicological aspects would no longer perform at an industrial level. Furthermore, bio-colorants possess an unlimited light fastness as compared to conventional dyes a characteristic required for automotive textiles [157].

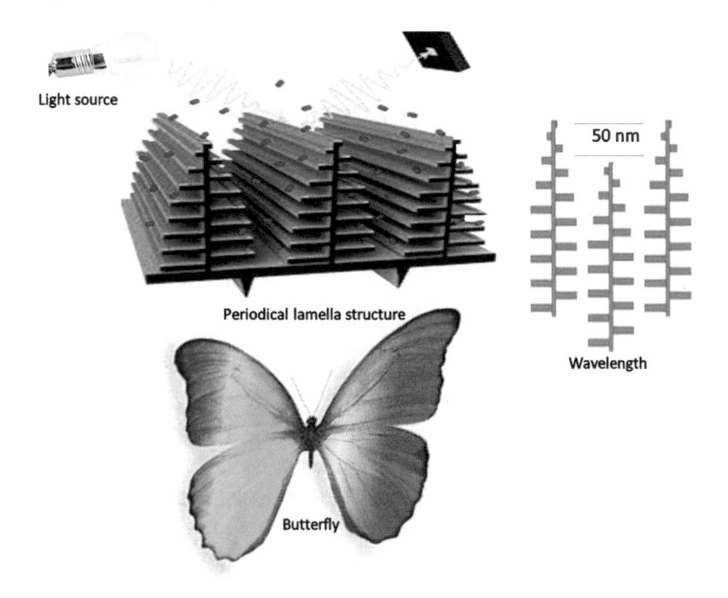

Fig. 9 Color formation following bio-copying of nature

9.2 Lotus Effect for Self Cleaning Textile Surfaces

Lotus effect is another example of bio-copying of nature for the development of a self-cleaning effect on the textile materials. Primarily, leaves of lotus possess two levels:

(i) Nanoscale hair-like structures
(ii) Microscale bumps which are coupled with a waxy chemical composition of leaves.

By utilizing this concept, scientists developed "self-cleaning textiles" that have many advantages:

(i) Don't need any laundry action
(ii) Reduction in time, energy, materials, and cost in production processes
(iii) Eco-friendly and easy to maintain
(iv) Long-lasting by improving the surface purity
(v) Improves the aging of cloth [158].

10 Genetic Manipulation for Bio-based Fibres

Bio-polymers based on natural fibers due to their vast application in the field of food, medicine and agriculture have attracted the attractions of molecular and genetic engineers for (i) induction of entirely new properties by highly specific modifications of natural fibers (ii) marketing of biopolymers from extinct or rarely available natural fibers [159]. The genetic engineer has taken a breakthrough by introducing Bt cotton, herbicide-tolerant cotton, cotton with entirely new characteristics, modified silk fibers large quantities. In spite of these factors, the practical exploitation of rare or extinct species is still pending [160, 161].

10.1 Bt and Herbicide Tolerant Cotton

Bt and herbicide-tolerant cotton are genetically modified or transgenic cotton (Fig. 10) whose cultivation is started in 2000 [162]. The cultivation of transgenic cotton has been considerably increased during the last five years [157]. The global community is now focusing on insect-resistant transgenic cotton varieties [163]. In addition to insect-resistant cotton, herbicide-tolerant cotton species by the transfer of genes have also been developed [164, 165]. These herbicide-tolerant cotton are capable of neutralizing herbicides e.g., bromoxynil or glyphosate. In spite of all these benefits, cultivation of Bt cotton and herbicide-tolerant cotton is not authorized in Europe whereas, in Spain, France and Greece are conducting field trials [166, 167] due to unanswered questions related to health. There is a need to find general-tool for improving the functionality of textile materials. Moreover, genetic manipulation of cell differentiation has been intervened for improvement in the number of fibers per unit area of seed coat [168].

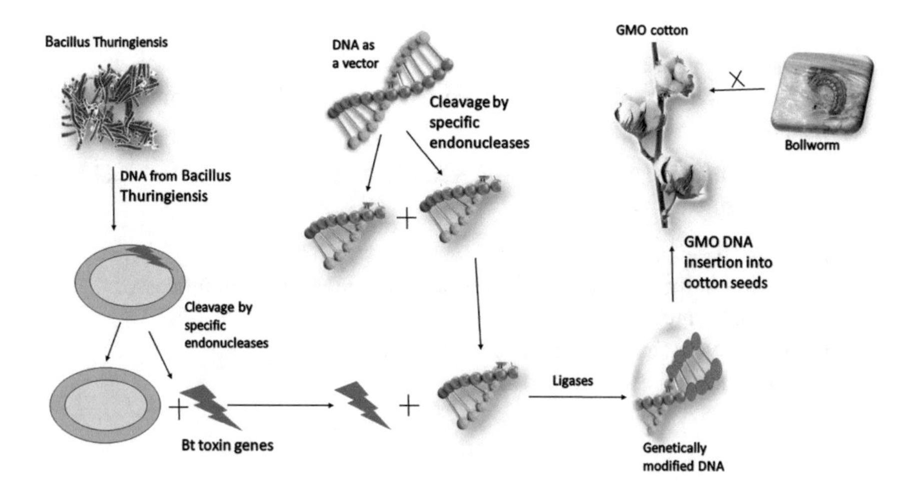

Fig. 10 Method of propagation of Bt cotton and herbicide-tolerant cotton; DNA piece from common soil bacterium B. thuringiensis (Bt) is introduced into a cotton cell that produces (Bt toxin), prevent bollworms attack

10.2 Cotton with Improved Functional Behavior

Genetic engineering techniques for improvement in fundamental properties of cotton e.g., length, strength, absorbance, coloration, thermal properties have been targeted [169]. Genetic modification with the outcome of improved quality and properties of fibers identified by combining indigo and melanin biosynthetic pathways of cotton has been observed [170]. Formation of mohair type fiber by transferring rabbit keratin into the cotton is under process. Insertion of genes containing information for the formation of poly (β-D-hydroxybutyrate) (PHB) enzymes through bombardment into the cotton resulted in better insulating properties than naturally produced cotton [171]. Identification of new genes may lead to developing cotton with improved qualities and desired effects [162, 172].

10.3 Genetically Engineered Silk Fibres

Genetic modification of silkworm (*Bombyxmori*) for the production of new silk proteins or silk fibers is a new promising type of genetic engineering as transgenic silkworm (*Bombyxmori*) was found not to be reproducible and efficient [173]. Different approaches like infection of silkworm larvae with a virus or recombinant DNA, formation by vectors or electroporation methods are commonly applied for the engineering of silk fibers utilizing green fluorescent protein (GFP) as selection marker [174]. Genetic engineering of silkworm has unlocked the gate of engineered silkworm for desired characteristics of silk proteins or virus-resistant silkworms are economical for sericulture [175].

10.4 Spider Fiber

Numerous spiders having the ability to form diverse protein fibers with distinct mechanical properties are known [176]. Most distinguished properties of spider species (Nephilaclavipes or madagascariensis) like high modulus, high strength, a high extension of break, biodegradability, biocompatibility, and high dissipation energy, radial web, spoke thread, dragline thread has been improved by genetic engineering [177, 178]. The viscid spider threads forming fiber cores exhibit excellent elongation break have a high strength level than nylon 6 [179].

Nature continues to be the major source for additional improvements in the fields of textile and fibers. Due to the availability of facilities as well as analytical tools many naturally occurring processes and the phenomenon of a complex system will be revealed and explicated. The results of these novel phenomena can be transferred to the field of textiles by bio-copying nature. Genetic engineering has already revolutionized by commercializing the Bt cotton. However, the success of any genetically modified silk depends on the individual components when these techniques will improve ultimately the field of textile fibers or natural fibers will progress. The key condition for manipulation or copying the existing system is the prime choice of biological kinds. One important factor that will decide which part of the world will lead to copying or manipulating nature in the field of textiles and fiber is biodiversity.

11 Conclusion

The application of biotechnology in textile industry is still limited due to existing technological impediments; the increasing demand of biotechnology-based textile processing applications pose a promising future. Biotechnology proved to be boon to constantly changing circumstances of the economy as well as ecology. Enzymes, in the scenario of the pollution-free process, are the best alternatives for textile processing. Enzymes not only save lots of money by reducing energy, effluent load and labor but they are also effective for ecological orientation. Being non-toxic, ecofriendly and highly efficient moieties, the enhanced use of enzymes in the textile industry can greatly benefit this sector. However, the applications of the enzyme in the textile sector need a better understanding of the effects on the textile substrate and reaction mechanisms. Several advantages such as high specificities, mild operational conditions, reduced process time, reduced waste generation and non-toxicity but have compelled the industrialists to move towards bio-processing. The textile industry has been identified as a key sector with a high potential of biotechnological applications where current awareness related to such applications still needs a lot of improvement. Through the application of these enzymes not only fabrics qualities can be increased but also textile processing improved. In a nutshell, such technologies are the need of the era due to the environmental and energy saving of the globe.

References

1. Rahman H, Biswas PK, Mitra BK, Rakesh MSR (2014) Effect of enzyme wash (cellulase enzyme) on properties of different weft knitted fabrics. Int J Curr Eng Technol 4(6):4242–4246
2. Chatha SAS, Asgher M, Iqbal HM (2017) Enzyme-based solutions for textile processing and dye contaminant biodegradation—a review. Environ Sci Pollut Res 24(16):14005–14018
3. Jaithlia H (2016) Eco-friendly substitute used on polyester-viscose fabric in textile wet processing units of Bhilwara city (Doctoral dissertation, MPUAT, Udaipur)
4. Franco MA (2017) Circular economy at the micro level: a dynamic view of incumbents' struggles and challenges in the textile industry. J Clean Prod 168:833–845
5. Ozturk E, Cinperi NC (2018) Water efficiency and wastewater reduction in an integrated woolen textile mill. J Clean Prod 201:686–696
6. Knowles VL, Hussey CJ (2015) Key choices in developing sustainable apparel for the active ageing population. In: Textile-led design for the active ageing population. Woodhead Publishing, pp 269–281
7. Agarwal S, Gupta KK, Chaturvedi VK, Kushwaha A, Chaurasia PK, Singh MP (2018) The potential application of peroxidase enzyme for the treatment of industry wastes. In *Research advancements*

8. Gamallo M, Moldes-Diz Y, Taboada-Puig R, Lema JM, Feijoo G, Moreira MT (2018) 6 textile wastewater treatment by advanced oxidation processes. Life Cycle Assess Wastewater Treat 5
9. Cavaco-Paulo A, Gübitz G (2003) Catalysis and processing. In: Textile processing with enzymes. Woodhead Publishing Ltd, England, p 86
10. Hasanbeigi A, Price L (2015) A technical review of emerging technologies for energy and water efficiency and pollution reduction in the textile industry. J Clean Prod 95:30–44
11. Asgher M, Parra-Saldivar R, Hu H, Wang W, Zhang X, Iqbal HM (2017) Immobilized ligninolytic enzymes: an innovative and environmental responsive technology to tackle dye-based industrial pollutants–a review. Sci Total Environ 576:646–659
12. Aly AS, Moustafa AB, Hebeish A (2004) Bio-technological treatment of cellulosic textiles. J Clean Prod 12(7):697–705
13. Jemli S, Ayadi-Zouari D, Hlima HB, Bejar S (2016) Biocatalysts: application and engineering for industrial purposes. Crit Rev Biotechnol 36(2):246–258
14. Sheikh J, Bramhecha I (2019) Enzymes for green chemical processing of cotton. In: The impact and prospects of green chemistry for textile technology. Woodhead Publishing, pp 135–160
15. Araujo R, Casal M, Cavaco-Paulo A (2008) Application of enzymes for textile fibres processing. Biocatal Biotransform 26(5):332–349
16. Ando H, Adachi M, Umeda K, Matsuura A, Nonaka M, Uchio R, Tanaka H, Motoki M (1989) Purification and characteristics of a novel transglutaminase derived from microorganisms. Agric Biolog Chem 53:2613–2617
17. Asgher M, Kamal S, Iqbalr HMN (2012) Improvement of catalytic efficiency, thermo-stability and dye decolorization capability of *Pleurotus ostreatus* IBL-02 laccase by hydrophobic sol gel entrapment. Chem Cent J 6(1):110
18. Wamg W, Yu B, Zhong CJ (2012) Use of ultrasonic energy in the enzymatic desizing of cotton fabric. J Clean Prod 33:179–182
19. Sarayu K, Sandhya S (2012) Current technologies for biological treatment of textile wastewater–a review. Appl Biochem Biotechnol 167(3):645–661
20. Baffes J (2005) Cotton: market setting, trade policies, and issues. In: Aksoy MA, Beghin JC (eds), p 259
21. Mojsov K, Andronikov D, Janevski A, Jordeva S, Kertakova M, Golomeova S et al (2018) Production and application of α-amylase enzyme in textile industry. Tekstilna industrija 66(1):23–28
22. Peng S, Gao Z, Sun J, Yao L, Qiu Y (2009) Influence of argon/oxygen atmospheric dielectric barrier discharge treatment on desizing and scouring of poly (vinyl alcohol) on cotton fabrics. Appl Surf Sci 255(23):9458–9462
23. Chand N, Nateri AS, Sajedi RH, Mahdavi A, Rassa M (2012) Enzymatic desizing of cotton fabric using a Ca^{2+}-independent α-amylase with acidic pH profile. J Mol Catal B Enzym 83:46–50
24. Hao L, Wang R, Fang K, Liu J (2013) Ultrasonic effect on the desizing efficiency of α-amylase on starch-sized cotton fabrics. Carbohyd Polym 96(2):474–480
25. Hao L, Wang R, Zhang L, Fang K, Men Y, Qi Z, Liu J (2014) Utilizing cellulase as a hydrogen peroxide stabilizer to combine the biopolishing and bleaching procedures of cotton cellulose in one bath. Cellulose 21(1):777–789
26. Dehabadi VA, Opwis K, Gutmann J (2011) Combination of acid-demineralization and enzymatic desizing of cotton fabrics by using industrial acid stable glucoamylases and α-amylases. Starch-Stärke 63(12):760–764
27. Wang Q, Yuan J, Fan X, Wang P, Cui L (2016) Hydrophobic modification of cotton fabric with octadecylamine via laccase/TEMPO mediated grafting. Carbohyd Polym 137:549–555
28. Ahlawat S, Dhiman SS, Battan B, Mandhan RP, Sharma J (2009) Pectinase production by *Bacillus subtilis* and its potential application in biopreparation of cotton and micropoly fabric. Process Biochem 44(5):521–526

29. Carmen Z, Daniela S (2012) Textile organic dyes–characteristics, polluting effects and separation/elimination procedures from industrial effluents–a critical overview. In: Organic pollutants ten years after the Stockholm convention-environmental and analytical update. Intech Open
30. Agrawal PB, Nierstrasz VA, Klug-Santner BG, Gübitz GM, Lenting HB, Warmoeskerken MM (2007) Wax removal for accelerated cotton scouring with alkaline pectinase. Biotechnol J Healthc Nutr Technol 2(3):306–315
31. Kalantzi S, Mamma D, Kalogeris E, Kekos D (2010) Improved properties of cotton fabrics treated with lipase and its combination with pectinase. Fibres Text Eastern Eur 18(5):86–92
32. Kaur SJ, Gupta VK (2017) Production of pectinolytic enzymes pectinase and pectin lyase by *Bacillus subtilis* SAV-21 in solid state fermentation. Ann Microbiol 67(4):333–342
33. Mukhopadhyay A, Dutta N, Chattopadhyay D, Chakrabarti K (2013) Degumming of ramie fiber and the production of reducing sugars from waste peels using nanoparticle supplemented pectatelyase. Biores Technol 137:202–208
34. Hasanbeigi A, Price L (2012) A review of energy use and energy efficiency technologies for the textile industry. Renew Sustain Energy Rev 16(6):3648–3665
35. Bhatti IA, Adeel S, Parveen S, Zuber M (2016) Dyeing of UV irradiated cotton and polyester fabrics with multifunctional reactive and disperse dyes. J Saudi Chem Soc 20(2):178–184
36. Shatalov MS, Deng J, Dobrinsky A, Hu X, Gaska R, Shur M (2017) U.S. Patent No. 9,595,636. U.S. Patent and Trademark Office, Washington, DC
37. Elhami V, Karimi A (2016) Preparation of Kissiris/TiO2/Fe3O4/GOx biocatalyst: feasibility study of MG decolorization. Adv Environ Technol 3:111–117
38. Buschle-Diller G, Yang XD, Yamamoto R (2001) Enzymatic bleaching of cotton fabric with glucose oxidase. Text Res J 71(5):388–394
39. Zhu D, Mobasher B, Rajan SD (2010) Dynamic tensile testing of Kevlar 49 fabrics. J Mater Civ Eng 23(3):230–239
40. Soares JC, Moreira PR, Queiroga AC, Morgado J, Malcata FX, Pintado ME (2011) Application of immobilized enzyme technologies for the textile industry: a review. Biocatal Biotransform 29(6):223–237
41. Cengiz S, Çavaş L, Yurdakoç K (2012) Bentonite and sepiolite as supporting media: immobilization of catalase. Appl Clay Sci 65:114–120
42. Madhu A, Chakraborty JN (2017) Developments in application of enzymes for textile processing. J Clean Prod 145:114–133
43. Kan CW, Yam LY, Ng SP (2014) Effect of stretching on ultraviolet protection of cotton and cotton/coolmax blended weft knitted fabric in a wet state. Materials 7(1):58–74
44. Kan CW (2015) Washing techniques for denim jeans. In: Denim. Woodhead Publishing, pp 313–356
45. Souza RP, Ambrosio E, Souza MT, Freitas TK, Ferrari-Lima AM, Garcia JC (2017) Solar photocatalytic degradation of textile effluent with TiO 2, ZnO, and Nb 2 O 5 catalysts: assessment of photocatalytic activity and mineralization. Environ Sci Pollut Res 24(14):12691–12699
46. Yu Y, Yuan J, Wang Q, Fan X, Ni X, Wang P, Cui L (2013) Cellulase immobilization onto the reversibly soluble methacrylate copolymer for denim washing. Carbohyd Polym 95(2):675–680
47. Yu Y, Yuan J, Wang Q, Fan X, Wang P, Cui L (2014) A study of surface morphology and structure of cotton fibres with soluble immobilized-cellulase treatment. Fibers Polym 15(8):1609–1615
48. Bhat MK (2000) Cellulases and related enzymes in biotechnology. Biotechnol Adv 18(5):355–383
49. Cavaco-Paulo A, Gubitz G (eds) (2003) Textile processing with enzymes. Elsevier
50. Kuhad RC, Gupta R, Singh A (2011) Microbial cellulases and their industrial applications. Enzym Res
51. Choudhury AKR (2017) Environmental impacts of denim washing. In: Sustainability in denim. Woodhead Publishing, pp 49–81

52. Choudhury AR (2014) Environmental impacts of the textile industry and its assessment through life cycle assessment. In: Roadmap to sustainable textiles and clothing. Springer, Singapore, pp 1–39
53. Kunamneni A, Camarero S, García-Burgos C, Plou FJ, Ballesteros A, Alcalde M (2008) Engineering and applications of fungal laccases for organic synthesis. Microb Cell Fact 7(1):32
54. Samanta KK, Basak S, Chattopadhyay SK (2017) Environmentally friendly denim processing using water-free technologies. In: Sustainability in Denim. Woodhead Publishing, pp 319–348
55. Maryan AS, Montazer M (2013) A cleaner production of denim garment using one step treatment with amylase/cellulase/laccase. J Clean Prod 57:320–326
56. Maryan AS, Montazer M, Harifi T, Rad MM (2013) Aged-look vat dyed cotton with anti-bacterial/anti-fungal properties by treatment with nano clay and enzymes. Carbohyd Polym 95(1):338–347
57. Vigneswaran C, Jayapriya J (2010) Effect on physical characteristics of jute fibres with cellulase and specific mixed enzyme systems. J Text Inst 101(6):506–513
58. Vigneswaran C (2011) Biovision in textile wet processing industry-technological challenges. J Text Appar Technol Manag 7(1)
59. Mir S, Hossain M, Biswas P, Hossain A, Idris MA (2014) Evaluation of mechanical properties of denim garments after enzymatic bio-washing. World Appl Sci J 31(9):1661–1665
60. Ahuja SK, Ferreira GM, Moreira AR (2004) Utilization of enzymes for environmental applications. Crit Rev Biotechnol 24(2–3):125–154
61. Schwermann B, Pfau K, Liliensiek B, Schleyer M, Fischer T, Bakker EP (1994) Purification, properties and structural aspects of a thermoacidophilic a-amylase from *Alicyclobacillus acidocaldarius* ATCC 27009. Insight into acidostability of proteins. Eur J Biochem 226:981–991
62. Rehman AMM, Imran MA (2014) Revolution of biotechnology in finishing sector of textile. Chem Mater Res 6(2):92–103
63. Noreen H, Zia MA, Ali S, Hussain T (2014) Optimization of bio-polishing of polyester/cotton blended fabrics with cellulases prepared from *Aspergillus niger*
64. Penava Ž, Šimić-Penava D, Knezic Ž (2014) Determination of the elastic constants of plain woven fabrics by a tensile test in various directions. Fibres Text Eastern Eur
65. Nielsen PH, Kuilderd H, Zhou W, Lu X (2009) Enzyme biotechnology for sustainable textiles. In: Sustainable textiles. Woodhead Publishing, pp 113–138
66. Kan Z, Yang MB, Yang W, Liu ZY, Xie BH (2015) Investigation on the reactive processing of textile-ramie fiber reinforced anionic polyamide-6 composites. Compos Sci Technol 110:188–195
67. Saravanan D, Vasanthi NS, Ramachandran T (2009) A review on influential behaviour of biopolishing on dyeability and certain physico-mechanical properties of cotton fabrics. Carbohydr Polym 76(1):1–7
68. Duran N, Duran M (2000) Enzyme applications in the textile industry. Rev Prog Color Relat Top 30:41–44
69. Kan CW, Au CH (2014) Effect of biopolishing and UV absorber treatment on the UV protection properties of cotton knitted fabrics. Carbohyd Polym 101:451–456
70. Lee I, Evans BR, Woodward J (2000) The mechanism of cellulase action on cotton fibers: evidence from atomic force microscopy. Ultramicroscopy 82(1–4):213–221
71. Mengal N, Arbab AA, Sahito IA, Memon AA, Sun KC, Jeong SH (2017) An electrocatalytic active lyocell fabric cathode based on cationically functionalized and charcoal decorated graphite composite for quasi-solid state dye sensitized solar cell. Sol Energy 155:110–120
72. Mosier N, Hall P, Ladisch CM, Ladisch MR (1999) Reaction kinetics, molecular action and mechanisms of cellulolytic proteins. Adv Biochem Eng Biotechnol 65:23–39
73. Albers A, Weber NF, Cirauqui M (2017) On weaving: new expanded edition. Princeton University Press
74. Naikwade M, Liu F, Wen S, Cai Y, Navik R (2017) Combined use of cationization and mercerization as pretreatment for the deep dyeing of ramie fibre. Fibers Polym 18(9):1734–1740

75. Zheng D, Zhou J, Zhong L, Zhang F, Zhang G (2016) A novel durable and high-phosphorous-containing flame retardant for cotton fabrics. Cellulose 23(3):2211–2220
76. Uddin MG (2016) Effect of biopolishing on dye ability of cotton fabric–a review. Trends Green Chem 2:1–5
77. Brahma S, Dina MRIRB (2018) Role of mercerizing condition on physical and dyeing properties of cotton knit fabric dyed with reactive dyes
78. Jordanov I, Mangovska B (2009) Characterization on surface of mercerized and enzymatic scoured cotton after different temperature of drying. Open Text J 2:39–47
79. Simic K, Soljačić I, Pušić T (2015) Application of cellulases in the process of finishing Up or a bacelulazv process up lemenitenja
80. Meksi N, Haddar W, Hammami S, Mhenni MF (2012) Olive mill wastewater: a potential source of natural dyes for textile dyeing. Ind Crops Prod 40:103–109
81. Singh L (2017) Biodegradation of synthetic dyes: a mycoremediation approach for degradation/decolourization of textile dyes and effluents. J Appl Biotechnol Bioeng 3:1–7
82. Imran M, Crowley DE, Khalid A, Hussain S, Mumtaz MW, Arshad M (2015) Microbial biotechnology for decolorization of textile wastewaters. Rev Environ Sci Bio/Technol 14(1):73–92
83. Balapure K, Jain K, Bhatt N, Madamwar D (2016) Exploring bioremediation strategies to enhance the mineralization of textile industrial wastewater through sequential anaerobic-microaerophilic process. Int biodeterior biodegradation 106:97–105
84. Singh SA, Rao AGA (2002) A simple fractionation protocol for, and a comprehensive study of the molecular properties of two major endopolygalacturonases from *Aspergillus niger*. Biotechnol Appl Biochem 35:115–123
85. Zhang H, Zhang J, Zhang X, Geng A (2018) Purification and characterization of a novel manganese peroxidase from white-rot fungus Cerrena unicolor BBP6 and its application in dye decolorization and denim bleaching. Process Biochem 66:222–229
86. Irshad M (2011) Characterization of Ligninolytic enzymes produced by schyzophyllum commune in solid state cultures for industrial applications (Doctoral dissertation, University Of Agriculture Faisalabad)
87. Chatha SAS, Asgher M, Ali S, Hussain AI (2012) Biological color stripping: a novel technology for removal of dye from cellulose fibers. Carbohyd Polym 87(2):1476–1481
88. Chatha SA, Mallhi AI, Hussain AI, Asgher M, Nigam PS (2014) A biological approach for color-stripping of cotton fabric dyed with CI reactive black 5 using fungal enzymes from solid state fermentation. Curr Biotechnol 3(2):166–173
89. Sisti L, Totaro G, Vannini M, Celli A (2018) Retting process as a pretreatment of natural fibers for the development of polymer composites. In; Lignocellulosic composite materials. Springer, Cham, pp 97–135
90. Mohammed L, Ansari MN, Pua G, Jawaid M, Islam MS (2015) A review on natural fiber reinforced polymer composite and its applications. Int J Polym Sci
91. Liang D, Hsiao BS, Chu B (2007) Functional electrospun nanofibrous scaffolds for biomedical applications. Adv Drug Deliv Rev 59(14):1392–1412
92. Liu M, Thygesen A, Summerscales J, Meyer AS (2017) Targeted pre-treatment of hemp bast fibres for optimal performance in biocomposite materials: a review. Ind Crops Prod 108:660–683
93. Akin DE (2012) Linen most useful: perspectives on structure, chemistry, and enzymes for retting flax. ISRN Biotechnol
94. Paridah MT, Basher AB, SaifulAzry S, Ahmed Z (2011) Retting process of some bast plant fibres and its effect on fibre quality: a review. BioResources 6(4):5260–5281
95. Oosterhuis F (2006) Substitution of hazardous chemicals: a case study in the framework of the project, assessing innovation dynamics induced by environment policy. Institute for Environmental Studies, Amsterdam, Netherlands
96. Osuji AC, Eze SOO, Osayi EE, Chilaka FC (2014) Biobleaching of industrial important dyes with peroxidase partially purified from garlic. Sci World J

97. Budak E, Ozturk E, Tunc LT (2009) Modeling and simulation of 5-axis milling processes. CIRP Ann 58(1):347–350
98. Zhu HY, Jiang R, Xiao L (2010) Adsorption of an anionic azo dye by chitosan/kaolin/γ-Fe2O3 composites. Appl Clay Sci 48(3):522–526
99. Callewaert C, De Maeseneire E, Kerckhof FM, Verliefde A, Van de Wiele T, Boon N (2014) Microbial odor profile of polyester and cotton clothes after a fitness session. Appl Environ Microbiol 80(21):6611–6619
100. Vollrath F, Knight DP (2001) Liquid crystalline spinning of spider silk. Nature 410(6828):541
101. Brandt L, Rawski TG, Sutton J (2008) China's industrial development. China's Great Econ Transf 569–632
102. Chauhan S, Sharma AK, Jain RK (2013) Enzymatic retting: a revolution in the handmade papermaking from Calotropisprocera. In: Biotechnology for environmental management and resource recovery. Springer, India, pp 77–88
103. Hoondal G, Tiwari R, Tewari R, Dahiya NBQK, Beg Q (2002) Microbial alkaline pectinases and their industrial applications: a review. Appl Microbiol Biotechnol 59(4–5):409–418
104. Rebello S, Anju M, Aneesh EM, Sindhu R, Binod P, Pandey A (2017) Recent advancements in the production and application of microbial pectinases: an overview. Rev Environ Sci Bio/Technol 16(3):381–394
105. Meng C, Zhong N, Hu J, Yu C, Saddler JN (2019) The effects of metal elements on ramie fiber oxidation degumming and the potential of using spherical bacterial cellulose for metal removal. J Clean Prod 206:498–507
106. Bacci L, Di Lonardo S, Albanese L, Mastromei G, Perito B (2011) Effect of different extraction methods on fiber quality of nettle (*Urticadioica* L.). Text Res J 81(8):827–837
107. Jacob N, Niladevi KN, Anisha GS, Prema P (2008) Hydrolysis of pectin: an enzymatic approach and its application in banana fiber processing. Microbiol Res 163(5):538–544
108. Yadav D, Yadav S, Dwivedi R, Anand G, Yadav P (2016) Potential of microbial enzymes in retting of natural fibers: a review. Curr Biochem Eng 3(2):89–99
109. De Prez J, Van Vuure AW, Ivens J, Aerts G, Van de Voorde I (2018) Enzymatic treatment of flax for use in composites. Biotechnol Rep e00294
110. Bajpai P (2018) Bioretting. In: Biotechnology for pulp and paper processing. Springer, Singapore, pp 97–111
111. Morvan C, Jauneau A, Flaman A, Millet J, Demarty M (1990) Degradation of flax polysaccharides with purified endo-polygalacturonase. Carbohyd Polym 13(2):149–163
112. Somashekarappa H, Annadurai V, Subramanya G, Somashekar R (2002) Structure–property relation in varieties of acid dye processed silk fibers. Mater Lett 53(6):415–420
113. Yang G, Zhang L, Cao X, Liu Y (2002) Structure and microporous formation of cellulose/silk fibroin blend membranes: Part II. Effect of post-treatment by alkali. J Membr Sci 210(2):379–387
114. Koch R, Spreinat A, Lemke K, Antranikian G (1991) Purification and properties of a hyperthermoactive a-amylase from the archaeobacterium *Pyrococcus woesei*. Arch Microbiol 155:572–578
115. Liu M, Silva DAS, Fernando D, Meyer AS, Madsen B, Daniel G, Thygesen A (2016) Controlled retting of hemp fibres: effect of hydrothermal pre-treatment and enzymatic retting on the mechanical properties of unidirectional hemp/epoxy composites. Compos A Appl Sci Manuf 88:253–262
116. Pray CE, Huang J, Hu R, Rozelle S (2002) Five years of Bt cotton in China–the benefits continue. Plant J 31(4):423–430
117. Cheung HY, Ho MP, Lau KT, Cardona F, Hui D (2009) Natural fibre-reinforced composites for bioengineering and environmental engineering applications. Compos B Eng 40(7):655–663
118. Das S (1992) The preparation and processing of tussah silk. J Soc Dyers Colour 108(11):481–486
119. Jin HJ, Kaplan DL (2003) Mechanism of silk processing in insects and spiders. Nature 424(6952):1057

120. Barclay S, Buckley C (2000) A waste minimisation guide for the textile industry. The Pollution Research Group, Water Research Commission, WRC Report No.TT 139/00
121. Barrozo A, Borstnar R, Marloie G, Kamerlin SCL (2012) Computational protein engineering: bridging the gap between rational design and laboratory evolution. Int J Mol Sci 13(10):12428–12460
122. Sivalokanathan S, Vijayababu MR, Balasubramanian MP (2006) Effects of Terminalia arjuna bark extract on apoptosis of human hepatoma cell line HepG2. World J Gastroenterol: WJG 12(7):1018
123. Basra AS, Saha S (1999) Growth regulation of cotton fibers. Basra, AS Food Products Press, Binghamton, NY
124. Yang YM, Wang JW, Tan RX (2004) Immobilization of glucose oxidase on chitosan–SiO2 gel. Enzym Microbial Technol 34(2):126–131
125. Chang SF, Chang SW, Yen YH, Shieh CJ (2007) Optimum immobilization of *Candida rugosa* lipase on Celite by RSM. Appl Clay Sci 37(1–2):67–73
126. Eiben CB, Siegel JB, Bale JB, Cooper S, Khatib F, Shen BW et al (2012) Increased Diels-Alderase activity through backbone remodeling guided by Foldit players. Nat Biotechnol 30(2):190
127. Hossain KMG, González MD, Juan AR, Tzanov T (2010) Enzyme-mediated coupling of a bi-functional phenolic compound onto wool to enhance its physical, mechanical and functional properties. Enzym Microbial Technol 46(3–4):326–330
128. Forbes P (2008) Self-cleaning materials: lotus leaf-inspired nanotechnology. Sci Am 299(2):88
129. Freddi G, Mossotti R, Innocenti R (2003) Degumming of silk fabric with several proteases. J Biotechnol 106(1):101–112
130. Coco WM, Levinson WE, Crist MJ, Hektor HJ, Darzins A, Pienkos PT et al (2001) DNA shuffling method for generating highly recombined genes and evolved enzymes. Nat Biotechnol 19(4):354
131. Singh S, Bajaj BK (2017) Potential application spectrum of microbial proteases for clean and green industrial production. Energy Ecol Environ 2(6):370–386
132. Fu Chengjie, Shao Zhengzhong, Fritz Vollrath (2009) Animal silks: their structures, properties and artificial production. Chem Commun 43:6515–6529
133. Gedik G, Avinc O (2018) Bleaching of hemp (*Cannabis sativa* L.) fibers with peracetic acid for textiles industry purposes. Fibers Polym 19(1):82–93
134. Holland GP et al (2008) Quantifying the fraction of glycine and alanine in β-sheet and helical conformations in spider dragline silk using solid-state NMR. Chem Commun 43:5568–5570
135. Pandey AK, Vishwakarma SK, Srivastava AK, Pandey VK (2012) Extracellular xylanase production by Pleurotus species on lignocellulosic wastes under in vivo condition using novel pretreatment. Cell Mol Biol 58(1):170–173
136. Kapoor M, Beg QK, Bhushan B, Singh K, Dadhich KS, Hoondal GS (2001) Application of an alkaline and thermostable polygalacturonase from *Bacillus* sp. MG-cp-2 in degumming of ramie (*Boehmeria nivea*) and sunn hemp (*Crotalaria juncea*) bast fibres. Process Biochem 36(8–9):803–807
137. James C (2001) Global review of commercialized transgenic crops: 2000
138. Kashyap DR, Vohra PK, Chopra S, Tewari R (2001) Applications of pectinases in the commercial sector: a review. Biores Technol 77(3):215–227
139. Losonczi AK (2004) Bioscouring of cotton fabrics. Ph.D. dissertation. Budapest university of Technology and Economics. Department of Plastics and Rubber technology
140. Wang WM, Yu B, Zhong CJ (2012) Use of ultrasonic energy in the enzymatic desizing of cotton fabric. J Clean Prod 33:179–182
141. Rossbach V, Patanathabutr P, Wichitwechkarn J (2003) Copying and manipulating nature: innovation for textile materials. Fibers Polym 4(1):8–14
142. Qiu XM (2000) Advances in natural colored cotton. China Cotton 27(5):5–7
143. Nikolaou A, Meric S, Fatta D (2007) Occurrence patterns of pharmaceuticals in water and wastewater environments. Anal Bioanal Chem 387(4):1225–1234

144. Xue R, Chen H, Cui L, Cao G, Zhou W, Zheng X, Gong C (2012) Expression of hGM-CSF in silk glands of transgenic silkworms using gene targeting vector. Transgenic Res 21(1):101–111
145. Pandey R (2016) Fiber extraction from dual-purpose flax. J Nat Fibers 13(5):565–577
146. Holmquist H, Schellenberger S, van Der Veen I, Peters GM, Leonards PEG, Cousins IT (2016) Properties, performance and associated hazards of state-of-the-art durable water repellent (DWR) chemistry for textile finishing. Environ Int 91:251–264
147. Wool R, Sun XS (2011) Bio-based polymers and composites. Elsevier
148. Truong LV, Tuyen H, Helmke E, Binh LT, Schweder T (2001) Cloning of two pectatelyase genes from the marine Antarctic bacterium *Pseudoaltero monashalo planktis* strain ANT/505 and characterization of the enzymes. Extremophiles 5:35–44
149. Pasternack R, Dorsch S, Otterbach JT, Robenek IR, Wolf S, Fuchsbauer HL (1998) Bacterial pro-transglutaminase from Strep to verticillium mobaraense-purification, characterisation and sequence of the zymogen. Eur J Biochem 257:570–576
150. Ruscio JZ, Kohn JE, Ball KA, Head-Gordon T (2009) The influence of protein dynamics on the success of computational enzyme design. J Am Chem Soc 131(39):14111–14115
151. Ahuja SK, Ferreira GM, Moreira AR (2004) Utilization of enzymes for environmental applications. Crit Rev Biotechnol 24(2–3):125–154
152. Zhao Y, Chen X, Peng WP, Dong L, Huang JT, Lu CD (2001) Altering fibroin heavy chain gene of silkworm bombyx mori by homologous recombination. Sheng wu hua xue yu sheng wu wu li xue bao Acta biochimica et biophysica Sinica 33(1):112–116
153. Tamura T, Thibert C, Royer C, Kanda T, Eappen A, Kamba M et al (2000) Germline transformation of the silkworm *Bombyx mori* L. using a piggyBac transposon-derived vector. Nat Biotechnol 18(1):81
154. De Fries W, Altenhofen U, Föhles J, Zahn H (1983) A protein chemical investigation of the chlorine-hercosett process. J Soc Dyers Colour 99(1):13–16
155. Gouveia IC, Fiadeiro JM, Queiroz JA (2008) Combined bio-carbonization and dyeing of wool: a possibility using cell wall-degrading enzymes and 1: 1 metal-complex dyes. Eng Life Sci 8(3):250–259
156. Duan S, Liu Z, Feng X, Zheng K, Cheng L, Zheng X (2012) Diversity and characterization of ramie-degumming strains. Sci Agricola 69(2):119–125
157. Duffield PA, Lewis DM (1985) The yellowing and bleaching of wool. Rev Prog Color Relat Top 15(1):38–51
158. Gubitz GM, Paulo AC (2003) New substrates for reliable enzymes: enzymatic modification of polymers. Curr Opin Biotechnol 14(6):577–582
159. Cecchini MP, Turek VA, Paget J, Kornyshev AA, Edel JB (2013) Self-assembled nanoparticle arrays for multiphase trace analyte detection. Nat Mater 12(2):165
160. Hussain F, Kamal S, Rehman S, Azeem M, Bibi I, Ahmed T, Iqbal HMN (2017) Alkaline protease production using response surface methodology, characterization and industrial exploitation of alkaline protease of *Bacillus subtilis* sp. Catal Lett 147(2017):1204–12013
161. Johannes TW, Zhao H (2006) Directed evolution of enzymes and biosynthetic pathways. Curr Opin Microbiol 9(3):261–267
162. Naranjo SE (2005) Long-term assessment of the effects of transgenic Bt cotton on the abundance of nontarget arthropod natural enemies. Environ Entomol 34(5):1193–1210
163. Porter D, Vollrath F (2009) Silk as a biomimetic ideal for structural polymers. Adv Mater 21(4):487–492; [184] Termonia Y (1994) Molecular modeling of spider silk elasticity. Macromolecules 27(25):7378–7381
164. Shaheen T, Tabbasam N, Iqbal MA, Ashraf M, Zafar Y, Paterson AH (2012) Cotton genetic resources. A review. Agron Sustain Dev 32(2):419–432
165. Bornscheuer UT, Pohl M (2001) Improved biocatalysts by directed evolution and rational protein design. Curr Opin Chem Biol 5(2):137–143
166. Çetinus ŞA, Öztop HN (2003) Immobilization of catalase into chemically crosslinked chitosan beads. Enzym Microbial Technol 32(7):889–894
167. Zaouali R, Msahli S, Sakli F (2016) Energy modelling of fabrics wrinkle recovery behaviour. J Text Inst 107(11):1434–1441

168. Chen J, Wang Q, Hua Z, Du G (2007) Research and application of biotechnology in textile industries in China. Enzym Microbial Technol 40(7):1651–1655
169. Bajwa KS, Shahid AA, Rao AQ, Kiani MS, Ashraf MA, Dahab AA et al (2013) Expression of *Calotropis procera* expansin gene CpEXPA3 enhances cotton fibre strength. Aust J Crop Sci 7(2):206
170. Chakravarthy VS, Reddy TP, Reddy VD, Rao KV (2014) Current status of genetic engineering in cotton (*Gossypium hirsutum* L): an assessment. Crit Rev Biotechnol 34(2):144–160
171. Liu JF, Zhao CY, Ma J, Zhang GY, Li MG, Yan GJ et al (2011) Agrobacterium-mediated transformation of cotton (*Gossypium hirsutum* L.) with a fungal phytase gene improves phosphorus acquisition. Euphytica 181(1):31–40
172. Liu YD, Yin ZJ, Yu JW, Li J, Wei HL, Han XL, Shen FF (2012) Improved salt tolerance and delayed leaf senescence in transgenic cotton expressing the agrobacterium IPT gene. Biol Plant 56(2):237–246
173. Naranjo SE (2005) Long-term assessment of the effects of transgenic Bt cotton on the function of the natural enemy community. Environ Entomol 34(5):1211–1223
174. Pant HR, Bajgai MP, Nam KT, Seo YA, Pandeya DR, Hong ST, Kim HY (2011) Electrospun nylon-6 spider-net like nanofiber mat containing TiO2 nanoparticles: a multifunctional nanocomposite textile material. J Hazard Mater 185(1):124–130
175. Hinman MB, Jones JA, Lewis RV (2000) Synthetic spider silk: a modular fiber. Trends Biotechnol 18(9):374–379
176. Zhang M, Yi TT, Zhang YM, Zhang L, Wu W, Zhang AL, Pan ZJ (2011) Ornithoctonus huwenna spider silk protein attenuating diameter and enhancing strength of the electrospun PLLA fiber. Polym Adv Technol 22(1):151–157
177. Saravanan D (2006) Spider silk-structure, properties and spinning. J Text Appar Technol Manag 5(1):1–20
178. Mao YB, Tao XY, Xue XY, Wang LJ, Chen XY (2011) Cotton plants expressing CYP6AE14 double-stranded RNA show enhanced resistance to bollworms. Transgenic Res 20(3):665–673
179. Ganesan M, Jayabalan N (2006) Isolation of disease-tolerant cotton (*Gossypium hirsutum* L. cv. SVPR 2) plants by screening somatic embryos with fungal culture filtrate. Plant cell, Tissue Organ Cult 87(3):273–284

Application of Enzymes in Textile Functional Finishing

Shrabana Sarkar, Karuna Soren, Priyanka Chakraborty and Rajib Bandopadhyay

Abstract Textile is one of the strongest economic pillars of any developing country. There is increased demand of eco-friendly and comfortable wearer with improving quality, which comes due to advanced and environment friendly finishing. Enzymes plays important role in bio-processing of textile. Enzymes are there from nineteenth century but less than only two decades ago it was first used in textile processing. Enzymes increase the functionality by minimising the strength and effort. Chemical textile processing is already at hand from very first day, but that cause hazard. In place of that, enzymatic processing is bio-friendly, non-toxic, and trimming down the pollution rate generating due to textile processing. In addition with that, enzyme technology reduces the use of vast amount of water and causes the savings of most important part of ecosystem. Therefore, day by day enzymatic textile progression is getting worldwide appreciation to the manufacturers. Finishing is the process in textile manufacture which converts knitted clothes into functional materials. Predominant enzymes used in textile processing are cellulase, amylase, pectinase, catalase, laccase and many more. Enzymatic finishing includes UV shielding properties on clothes. Use of immobilized enzymes, enzymatic nanoparticles may cause a revolutionary change in textile trade. Enzyme assisted bio-processing has an immense role on increment of textile quality.

Keywords Enzymes · Finishing · Microorganism · Textile · Wool

1 Introduction

Like food, clothing is another fundamental human necessity for surviving. Finishing is the final mechanical and chemical process of textile manufacture which converts knitted cloth into functional and softer one. Both dyeing and bleaching are the part of final textile finishing. This process helps to improve the performance of final fine clothings. For example, it makes the cotton clothes washfree and wrinklefree which

S. Sarkar · K. Soren · P. Chakraborty · R. Bandopadhyay (✉)
UGC-Center of Advanced Study, Department of Botany, The University of Burdwan, Golapbag, Burdwan, West Bengal 713104, India
e-mail: rajibindia@gmail.com

M. Shahid and R. Adivarekar (eds.), *Advances in Functional Finishing of Textiles*, Textile Science and Clothing Technology, https://doi.org/10.1007/978-981-15-3669-4_5

are very necessary. Finishing procedure makes cotton cloth wrinkle free and suitable for wearing.

In textile finishing process different chemicals have been used which cause environmental hazards. According to estimation impact of water pollution on individual's healthiness and environment is nearly 41.73 billion yuan per year [1]. As per increasing awareness on eco-toxic effects, enzymatic treatment has been introduced in textile finishing process [2]. Many harmful side products are being released in water bodies and air after chemical finishing process and cause dangerous pollution. Hence to reduce the environmental challenges related to pollution, eco-friendly enzymatic treatment has been incorporated in textile production. Normally textile industry requires a huge amount of water in each step of its processing, which causes relatively high water pollution. Enzymatic textile processing is a wonderful approach for conserving water by using relatively low amount of water [3].

From past four decades, enzymes are being used in different industries like food, beverage including textile in its finishing stage [4]. Different enzymes used in textile manufacturing and processing, viz., bleaching termination, finishing of denim, softening of cotton, scouring including removal of excess dyes also [4]. Amylase is widely used in clothes desizing and strengthen procedure for its thermostable nature and drastic industrial application. So, this enzyme has a very important role in fining of woven cotton clothes [5]. Control enzymatic treatment on woollen component change surface prototyping of final clothing materials [6]. Being an environment friendly implement enzymatic treatment is also a biotechnology based technique which makes textile products more decorative and attractive by changing the surface patterning than traditional one [6]. Besides this, it also has economic benefits. Enzymes are used in very minute concentration, for which low amount of money is required and production become well.

2 Background on Textile Industry and Production

Textile is a very old industry in the world and second largest just after the agriculture. This is because of cloth is a very basic need of human being in their daily life. Therefore, textile is one of the most primitive human craft which has always been elementary part of economy [7]. This textile history was started back somewhat 30,000 years ago. According to archaeological and anthropological evidences, worn textile and wove clothing were started at prehistoric age. Now a days, spinning, reeling, sizing, processing and printing all are done in order to improve the quality. Unlike ancient age, in recent era, in addition with cotton, silk, nylon, rayon, polyester, man-made synthetic fibers are being used as raw materials for textile industry. In early eighteenth century, Great Britain was being thought as leader in textile trade due to presence of some modern technologies like spinning, power loom in their manufacturing process [8]. In the year of 1786, U.S. Government encouraged with huge amount of money for designing British spinning technology, but production was unsatisfactory and irregular due to use of huge electricity. Up to the year 1790,

U.S. was failed to develop even a single successful spinner for textile manufacture [8]. In the year of 1831, first ready-made cloth production was started on large scale in New York City of USA.

Textile history has an important role in innovation of design on clothes. In ancient era, basic colours were used mostly than complex colours. In fact, colour was a fundamental part for gender bias. In archaeological era like other countries, Greek clothes were not ornamented highly [9]. Designing on clothes is an important part of finishing, because it makes the clothes better looking. In old age time, only geometric figures were used for designing purpose. In recent time stars, flowers, dots, grids are being used for designing purpose. First, Indian mill was established in fort gloster near Calcutta in the year of 1818 [10]. First early hand held spindle followed by hand looms and then today's highly automated spindle along with spinning mechanism has been evolved from time to time. Previously, fibers were extracted only from vegetable plants but in addition with those now a days, complex synthetic fibers are being produced. Till now, the basic mechanism of woven cotton and fibers are same as it was in million years ago [11]. Nearly thirty years ago, the basic principle of weaving was discovered by ancient people. The technique is known as finger weaving and widely practiced till date. In Neolithic period (c. 9000–4000) woven fabric was first created for household and personal uses, which is popular for millennia. From this history, it can be well understood that weaving is the oldest finishing technique exploded in ancient age. After that in middle age, chain sequences of textile machineries and techniques were discovered, viz., dying, spinning, weaving, and tailoring. Though weaving was first started in Europe, but up gradations was introduced from China and other global entity. In India, between 1000–500 CE spinning technique was first introduced. Later explosive advancement was occurred through industrial revolution, which was occurred after 1774. In this era of huge revolutionary development, different advanced machineries were developed in order to improve the textile production [12]. Previously, all the process of looms were done in different locations but after introduction of Robert's power loom in the year of 1812, all phases of converting cotton into threads was done in single factory [12]. This was the result of rapid and glamorous industrial revolution which also added huge capital for more production as in modern age (Fig. 1).

Long historic story of introduction and development of different important textile technique supports the commercial and fashionable textile industry today. In recent era, textile is dominating the trade of any country in addition with economic base of that country too. From previously described history, it is well understood that weaving, an important stage of textile procedures, is the most primordial skill which was practiced by human.

3 Microbial-Origin Enzymes in Textile Industry

In textile industry for finishing purpose as well as in biodegradation purpose different microbial enzymes are used in different stages. Global market of enzyme in textile

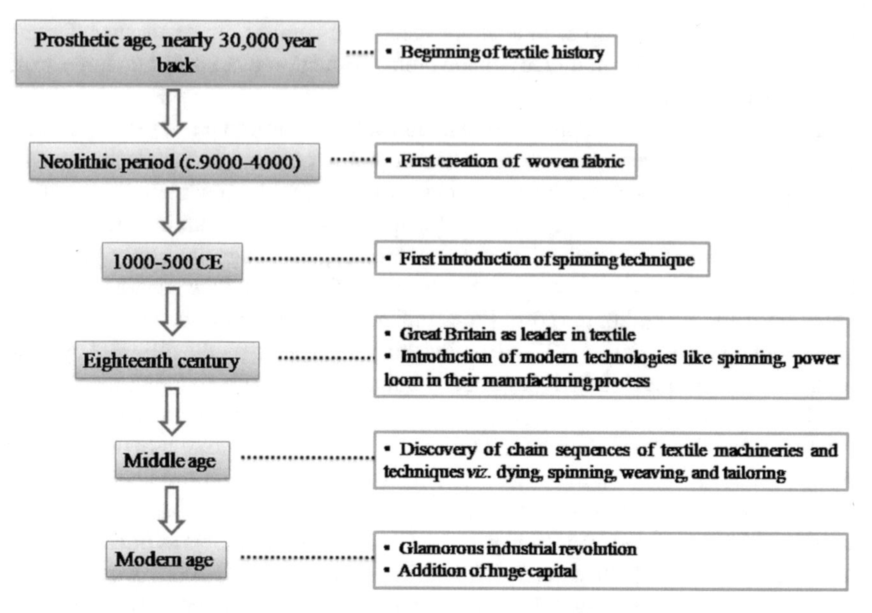

Fig. 1 History of textile revolution

is accounting a huge profit due to its less polluting technology and produces very less waste. For enzyme production, microorganisms are used as because they can produce a diverse range of enzymes due to its quick adaptability with any situation. Microorganism can be easily biotechnologically engineered for more production of enzyme, which is another important reason behind using of microbial enzymes. Oxido-reductase and hydrolyase are two main class of enzymes used in pretreatment of cotton [13]. There are many more enzymes like amylase, catalase, cellulase, pectinase, lipase, protease etc. which have many uses in different steps of textile finishing for different types of raw materials (Table 1). Felt-free finishing of wool, cotton finishing, biostoning of denim, biopolishing of fabrics are popular useful techniques done by different microbial origin [13]. Other than these for decolourisation purpose laccase is used widely. There are many common enzymes which have role in textile processing. Pectinase is an important enzyme used for increasing scouring effect of cotton fibers, produced by very common bacteria, *Bacillus subtilis* [14]. Catalase from *Micrococcus luteus* is used in cotton processing [15]. *Bacillus* seems to be very common microorganisms to produce different types of enzymes which have huge application in textile finishing procedure (Table 1).

Table 1 Microorganism mediated enzymes used in textile finishing process

Name of the enzyme	Name of the microorganism	Used in specific step of finishing	References
Cellulase	*Bacillus* sp. *Aspergillus oryzae*	Biostoning of denim; biopolishing and softening of denim	[15, 17]
Catalase	*Micrococcus luteus*	Used after bleaching for cotton processing	[15, 16]
Amylase	*Bacillus* sp.	Remove starchy layer from	[13, 15]
Protease	*Bacillus licheniformis*	prevent decolourisation of denim; antifelting finishing treatment on wool and silk materials	[15, 16]
Pectinase	*Bacillus subtilis*	Hydrolysis of pection in cotton fiber preparation, scouring of cotton	[14]

4 Enzymes in Finishing Process and Its Role

Recent ongoing trends are application of different types of biological treatments in order to prevent environmental pollution. Enzymatic application is ecofriendly and cheap in nature. Now, in textile wet processing microbial enzymes are being used regularly and successfully because they can be used in very minute condition and provide promising result due to high specificity [16]. Normally, textile industry uses excessive amount of water. Use of enzyme in finishing process reduce the usage of water. Introduction of enzymatic finishing process in textile has improved the product quality than before.

4.1 Cellulase in Bio-finishing and Denim Biostoning

First use of cellulase enzyme was started in the area of agriculture and food, in addition with fermentation of alcohol and brewing industry. In textile dye-house, laundries, use of cellulase was started less than two decades ago due to its special characters. Cellulose has slow kinetics in enzymatic degradation of cellulose [17]. Therefore, bioconversion has inadequate feasibility due to crystalline structure of cellulose. Because of this kinetic property cellulase has controlled action over improvement of cellulosic fibers and has wide use in textile industry. Indeed, commercial textile cellulase production is in third position and one of the fastest growing trades just after starch and detergent industry [17, 18]. In the field of textile finishing, cellulase play an important role in biopolishing. After dying on cotton fabric, biopolishing has shown great report though it can be done before too. One key role of cellulase treatment in case of denim is softening of fibers, which makes the wearer more comfortable [19].

Biostoning is a finishing technique first started long year back in 1989, first in Europe [18]. At that time this enzyme was isolated from a fungal species namely *Trichoderma reesei*, though cellulase can be collected from bacteria and protozoans too. First isolated cellulase had shown its maximum reactivity at pH −5.0 [18]. Day by day it has been distributed through out the world as very popular finishing technique mainly for denim. Till date, *Trichoderma* cellulase is favourite for trade as because it takes shortest time of reaction [17]. As it is biological enzymatic process, so, cellulase is used widely on surface of fabric to make denim more comfortable. In order to make wearers' surface smooth, enzymatic biostoning remove waxy and starchy materials from fabric after dying. Therefore, beside making denim more comfortable to wear, this mechanism also help to keep the colour stable [20]. Due to satisfactory result, only in textile industry global sale of cellulase enzyme get reached nearly $165 million within just 10 years of discovery [21]. Before this new approach, denim polishing was done by using one mechanism widely known as 'stone washing'. This process sometime used to create problems, which was known as 'back staining'. Due to loosening of fibers dye molecules get redeposited in the back surface instead of washing, which cause decolorisation. Biostoning is the process uses the same mechanism as 'stone washing', which is environment friendly and cost-effective. This is because, enzyme is eco-friendly, used in minute quantity and can be recycled. Whereas, previously stone was used in large quantity and technique was time consuming. Enzymatic reaction under moderate pH and temperature has removed the problem of older stone washing technique.

In textile non-cellulasic fibers like polyester or nylon etc., are used besides cellulosic one. Cellulosic treatment on these materials make colour more vibrant, as per example print on buttons, colorful logo on denim or other dresses etc. Biopolishing of materials other than denim is done by cellulase in pilot scale for making the colour more promising and attracting [17]. Cellulase treatment prevents the denim from tearing, decolourisation after washing, making them comfortable to wear and also workplace friendly. For these advantages and wide use in denim washing many of textile industries have made patent of cellulase isolated from different promising species [17].

4.2 Protease in Felt-Free Woollen Fibre Processing

Wool normally gets felted and shrinked under wet condition due to the presence of cuticle scale on woollen fibers [22]. Many chemical methods like chemical oxidation, coating of woollen fibers have been done in order to eradicate the problem of shrinkage, but the result was not so satisfactory [23]. Other than these processes produce environmental hazards [24]. Enzymatic treatment is an eco-friendly procedure which makes the textile product safer and comfortable to human. Protease has important role in production of shrink proof wool. Partial hydrolysis by protease,

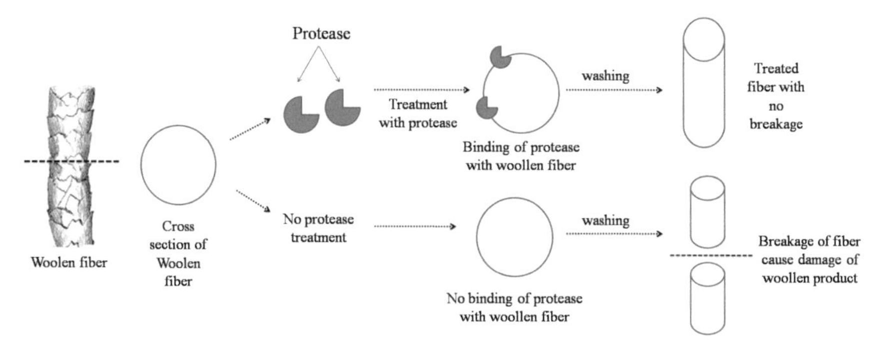

Fig. 2 Role of protease in production of felt free woollen clothes (diagrammatic representation)

papine makes the woollen silky and increased its market value. Pre-treatment with papine helps to produce shrinkage free woollen tops. This enzymatic finishing treatment is known as Zero absorbable organohalogens (AOX) which is a shrink proofing practice [25]. Proteolytic treatment helps to hydrolyse the cuticle layer which makes it smooth and shrink proof [23]. On the other hand, protease action also increases and strength of woollen fiber nearly 30% more than untreated one [23]. Cellulase treatment increases the activity of proteolytic effect on cotton. Enzymatic treatment of woollen fiber modifies the cuticular scale and reduces the tearing strength [26]. Shrinkage free woollen fiber has increased dying capability and it also makes the woollen wearer more comfortable [22]. Actually enzymatic treatment diminishes the resistance against dye diffusion for woollen fiber [26]. Overall, protease treatment increases the comfort of woollen wearer. Protease enzyme from bacteria *Streptomyces fradiae* is used industrially for production of felt free woollen [25]. Sometime shrinkage free treatment by protease, loose the strength of woollen fiber. This can be improved by transglutaminase treatment [25, 27]. Therefore, biological enzymatic treatment increases softness of woollen garments, makes it felt free (Fig. 2). Other than this, it also reduces chemical allergen reaction with skin and do mild or no environmental pollution [28].

Protease has another role in bioscouring and contraction of cellulase. At the time of biostoning sometime stain molecule get redeposited on the back surface which cause the wearer unstrained or roughly stained. Counter action of protease, isolated from microbes, can prevent this decolourisation of fabric of denim. Addition of protease prevent cellulase to bind with specific dye molecule on undesired position on back surface. In addition, protease does not show any adverse effect on biopolishing of fabrics by cellulase, because of their synergistic effect [17].

4.3 Catalase and Lipase in Cotton Processing

Catalase and lipase are though very common enzymes but play an important use in finishing of cotton. Catalase has two important usages in two different stages, one in hydrogen peroxide treatment just before bleaching and another is in waste water treatment [15]. In order to reduce environmental pollution catalase treatment remove residual hydrogen peroxide from cotton. Textile companies mainly consume high amount of water which is one of the main cause behind the water shortage. This process also limits the usage of water. This is an appropriate green usage of enzyme in textile finishing. It is used in washing to remove knits from it and commercial finishing after bleaching [29]. There is one drawback of using catalase in removal of residual hydrogen peroxide. At the time of reaction it produces high temperature and high alkaline condition. As enzyme shows optimal activity under ambient temperature and neutral pH, so catalase fails to withstand in adverse condition. Extremophilic microorganism's enzyme are being tried to use in order to mitigate this problem [15]. Sometime catalase activity reduces dye uptake capability. Use of immobilised microbes can alleviate this trouble.

Lipase is an enzyme of hydrolyase section of classification. For useful processing of cotton, microorganism mediated lipase is used widely in textile industry. Hydrolytic enzymes has effective role in pre-treatment of cotton. In textile lipase removes lubricant from cotton fiber in order to improve absorbance efficiency of dye [30]. This will increase the brightness and quality of colour on cotton wearer. Lipase treatment also decreases the regularity of crack and streak in cotton denim. It mainly removes wax and oil from raw cotton material at the time of bioscouring, a step of cotton processing. Lipase is responsible for proteinolytic activity too. Lipase activity in addition with pectinase activity helps to lowering down the water absorbance of cotton which increases the scouring effect [14].

5 Immobilized Enzymes and Enzyme Nanoparticles in Textile Finishing Process: A Paradigm Shift in Eco-friendly Textile Processing

In recent era with increased demand of products new technologies have been evolved, immobilisation and nanoparticle formation is one of them. Immobilisation of enzymes increases the productivity of enzyme. Immobilisation of amylase is done by chemical cross linking of enzymes with functional magnetite nanoparticles. It also increases the thermal stability and can be stored for long time [31]. This process shows increased activity in desizing of cotton. Laccase immobilisation is done with modified green coconut fibres and 3-glycidoxypropyl tri-methyloxysilane which increase substrate specificity [31]. This has prolonged usage in textile waste water treatment. Immobilised protease shows increased thermal stability. This is mainly used in production of shrink free finished woollen clothes [31]. All these immobilised enzymes have shown increase activity in different stages of textile finishing. These will open a new path towards future application in textile industry for production of comfortable clothes. Enzymes are being used in order to increase the rate and effectivity of reaction. Nanoparticle is a way to increase the rate of reaction. Formation of enzyme-nanoconjugates with other chemical increases the efficacy of reaction. Nano-conjugates of citric acid and Fe_3O_4 with protease has been developed which is used in pretreatment of wool. This process will increase tensile strength of woollen fiber and also make it resistance to alkaline condition [32]. Enzyme immobilised nanoparticle shows increased catalytic activity which is very much preferable for increasing production rate [33]. ZnO- and TiO_2 nanoparticle along with citric acid/sodium is used for treatment of denim in eco-friendly way after enzymatic treatment of denim fabric [34]. It is well understood that in addition with textile finishing process above mentioned advanced techniques are also being used in bioremediation of waste water.

6 Laccase in Decolorisation of Textile Effluent and Mitigating Important Environmental Issue

In textile, mainly synthetic azo dyes are used in order to make colour variation in wearers. Dying is a main part of finishing process. Dyes are water soluble and bind with fibers to make them colourful. Nearly, 10–15% dyes get released through effluent without binding with fibers [35]. These dyes are very harmful to the environment and human beings because of being a xenobiotic compound. Traditional physical and chemical methods can remove dyes partially and produce aromatic amine as secondary pollutant, but enzymatic treatment has the ability to remove dyes efficiently without producing any secondary pollutants [35].

Multicopper oxidase enzyme Laccase is a very good source of enzyme in order to remove unused dyes from waste water before release. Laccase has ability to degrade azo dyes like methyl orange effectively [36]. This special enzyme isolated from microorganisms is reported extensively for its capacity in azo dye degradation. It has non-specific oxidation capacity which is imperative for bioremediation process. On the other hand laccase doesn't necessitate any cofactor or oxygen for reaction [37]. Mainly dye molecule with low molecular weight is being used as redox mediator in the course of decolourisation. Aromatic azo dyes, phenolic compound, etc. in addition with pesticides is degraded by laccase by redox reaction [38]. Now, pH stable and thermo-stable laccase are being isolated which also exhibit tolerance to organic solvent [39]. In the path of reaction, laccase uses Cu^{2+} as a mediator to oxidise aromatic amine. Recently, in order to increase the degradation capacity laccase is being immobilised. This immobilised and insolubilised laccase has more efficiency to remove micro-pollutants in addition with dyes [40]. Immobilized laccase from fungal source has been used with the support of SiO_2 on magnetic separators to remove azo dyes from waste water efficiently [41]. Nanocomposites of enzymes are also being used in recent research for remediation of dye containing waste water. Textile finishing process is based on enzymatic treatment which is environment friendly. The releasing dyes and aromatic compounds are also being treated by eco-friendly enzymatic process. This is more acceptable to environment due to its less pollution capacity.

7 Patenting Trends on Use of Microbial Enzymes on Textile Finishing

Patenting trends of enzymes in textile finishing are also interesting. Cellulase is mostly used enzyme in textile. As per this perspective, it has been found that there are a number of petants on cellulase. Europe, United Kingdom, United State have several patent from 1990s. Previously, patents were done based on application as detergent or starch. Later, due to some trade legal issues, laundries increased their use of cellulase [17]. In fact for antimicrobial finishing different types of enzymes from microbial origin are being used and patented commercially from diverse countries (Table 2).

Table 2 Patents of microorganism mediated enzymes used in different steps of textile finishing

Patent number	Year	Publication on patent	Mechanism of action	References
EP2064385B1	2016	Enzymatic treatment of textiles using a pectate lyase from *Bacillus subtilis*	Microbial enzyme used in textile finishing processing	[42]
CN101864406B	2016	Endo-β-1,4-glucanase	Enzyme is used as detergent in bleaching process of textile finishing	[43]
EP3088502B1	2016	Method of treating a fabric	Use in textile finishing step	[44]
CN105544198A	2016	Antibacterial and anti-felting wool fabric finishing method	Anti felting finishing of woollen fabric	[45]
CN106758261A	2017	Biological enzyme method protein fiber product flame retardant finishing method	Biological enzyme used in pre-treatment and textile finishing method	[46]
CN105951434B	2018	A biological enzymatic silk fabric water repellent finishing methods	Act as water repellent in silk finishing process	[47]

(continued)

Table 2 (continued)

Patent number	Year	Publication on patent	Mechanism of action	References
CN108251319A	2018	Cutinase fermentation and cotton fabric enzyme refining process	For the refining process of cotton by the help of microbial enzyme	[48]
CN109457468A	2019	A kind of antibiotic finishing method of cellulosic fabrics	Antibiotic finishing of cellulosic fibers.	[49]

8 Conclusion

Microorganism mediated enzymes have major application in textile finishing. Enzyme mediated finishing procedure is a way to establishment of greener textile. From history it has been found that though in low scale but enzymes were used in different stages of textile finishing. In this chapter use of different enzymes have been discussed, like lipase and catalase in cotton processing, cellulase in bio-finishing and biostoning of denim, protease in production of felt free woollen wear. In addition with textile finishing in case of remediation of textile waste water even microorganism mediated enzymes are being used in large scale. Functional textile is future of global apparel industry which faces both new challenges and opportunities. In order to mitigate challenges researchers are using opportunities of enzyme immobilisation and nano-conjugates formation. This process has ability to increase the activity. Patenting trends have also been discussed from which it is understood that enzymes are used widely throughout the world though patents are mainly done from china, then Europe followed by other countries. This result proves the extensive use of enzymes in textile industry.

Acknowledgements Authors are thankful to UGC-Center of Advanced Study, Department of Botany, The University of Burdwan for pursuing research activities. SS is thankful to SVMCM-Non Net fellowship for the financial assistance.

References

1. Lu X, Liu L, Liu R, Chen J (2010) Textile wastewater reuse as an alternative water source for dyeing and finishing processes: a case study. Desalination 258:229–232
2. Omrane M, Moujehed E, Elhoul MB, Mechri S, Bejar S, Zouari R, Jaouadi B (2018) A novel detergent-stable protease from *Penicillium chrysogenium* X5 and its utility in textile fibres processing. In: Proceedings of MOL2NET 2018, International Conference on Multidisciplinary Sciences 4 MDPI AG
3. Samanta KK, Pandit P, Samanta P, Basak S (2019) Water consumption in textile processing and sustainable approaches for its conservation. Water in Textiles and Fashion. Woodhead Publishing (pp 41–59)
4. Kirk O, Borchert TV, Fuglsang CC (2002) Industrial enzyme applications. Curr Opin Biotechnol 13:345–351

5. Zafar A, Aftab MN, Iqbal I, ud Din Z, Saleem MA (2019) Pilot-scale production of a highly thermostable α-amylase enzyme from *Thermotoga petrophila* cloned into *E. coli* and its application as a desizer in textile industry. RSC Advances 9:984–992
6. Prajapati CD, Smith E, Kane F, Shen J (2019) Selective enzymatic modification of wool/polyester blended fabrics for surface patterning. J Clean Prod 211:909–921
7. Gleba M, Mannering U (2012) Textiles and textile production in Europe from prehistory to AD 400. Oxbow Books
8. https://www.thoughtco.com/textile-revolution-britains-role-1991935. Accessed on 30 Jan 2019
9. Perivoliotis MC (2005) The role of textile history in design innovation: a case study using hellenic textile history. Text Hist 36:1–19
10. https://www.fibre2fashion.com/industry-article/543/indian-textile-industry?page=1. Accessed on 30 Jan 2019
11. https://www.thoughtco.com/history-of-textile-production-1991659. Accessed on 12 Feb 2019
12. https://thescrubba.com/blogs/news/the-history-of-weaving-and-the-textile-industry. Accessed on 14 Feb 2019
13. Singh R, Kumar M, Mittal A, Mehta PK (2016) Microbial enzymes: industrial progress in 21st century. 3 Biotech 6:174
14. Ahlawat S, Dhiman SS, Battan B, Mandhan RP, Sharma J (2009) Pectinase production by *Bacillus subtilis* and its potential application in biopreparation of cotton and micropoly fabric. Process Biochem 44:521–526
15. Shahid M, Mohammad F, Chen G, Tang RC, Xing T (2016) Enzymatic processing of natural fibres: white biotechnology for sustainable development. Green Chem 18:2256–2281
16. Vankar PS, Shanker R (2008) Ecofriendly ultrasonic natural dyeing of cotton fabric with enzyme pretreatments. Desalination 230:62–69
17. Galante YM, De Conti A, Monteverdi R (2014) Application of *Trichoderma* enzymes in the textile industry. Trichoderma and Gliocladium 2:311–325
18. Cavaco-Paulo A (1998) Mechanism of cellulase action in textile processes. Carbohyd Polym 37:273–277
19. Uddin MG (2016) Effect of biopolishing on dye ability of cotton fabric–a review. Trends Green Chem 2:1–5
20. Belghith H, Ellouz-Chaabouni S, Gargouri A (2001) Biostoning of denims by *Penicillium occitanis* (Pol6) cellulases. J Biotechnol 89:257–262
21. https://www.biology.iupui.edu/biocourses/N100/2k3goodfor2.html. Accessed on 04 April 2019
22. Ammayappan L (2013) Application of enzyme on woolen products for its value addition: an overview. J Text Appar, Technol Manag 8
23. Udakhe J, Honade S, Shrivastava N (2011) Recent advances in shrink proofing of wool. J Text Assoc 72:171–176
24. Lenting HBM, Schröder M, Gübitz GM, Cavaco-Paulo A, Shen J (2006) New enzyme-based process direction to prevent wool shrinking without substantial tensile strength loss. Biotech Lett 28:711–716
25. Allam OG (2011) Improving functional characteristics of wool and some synthetic fibres. Open J Org Polym Mater 3:8–19
26. Ammayappan L, Moses JJ, Senthil KA, La JK (2011) Effect of alkaline and neutral protease enzyme pretreatment followed by finishing treatments on performance properties of wool/cotton union fabric: a comparative study. J Nat Fibers 8:272–288
27. Cortez J, Anghieri A, Bonner PL, Griffin M, Freddi G (2007) Transglutaminase mediated grafting of silk proteins onto wool fabrics leading to improved physical and mechanical properties. Enzym Microb Technol 40:1698–1704
28. Xiao-Wei Y, Guan WJ, Yong-Quan L, Ting-Jing G, Ji-Dong Z (2005) A biological treatment technique for wool textile. Braz Arch Biol Technol 48:675–680
29. Shai-xiang LENG, Guo-di QIAN, Zhao-zhe HUA, Guo-cheng DU, Jian CHEN (2006) Catalase for H_2O_2 removal after cotton knits bleaching. Dye Finish 19

30. Hasan F, Shah AA (2006) A hameed industrial applications of microbial lipases. Enzym Microb Technol 39:235–251
31. Silva C, Martins M, Jing S, Fu J, Cavaco-Paulo A (2018) Practical insights on enzyme stabilization. Crit Rev Biotechnol 38:335–350
32. Nazari A (2017) Treatment of enzymatic wool with Fe_3O_4 nanoparticles and citric acid to enhance mechanical properties using RSM. J Text Inst 108:1572–1583
33. Yi S, Dai F, Zhao C, Si Y (2017) A reverse micelle strategy for fabricating magnetic lipase-immobilized nanoparticles with robust enzymatic activity. Sci Rep 7:9806
34. Ibrahim NA, Eid BM, Aziz MSA, Hamdy SM, Allah SEA (2018) Green surface modification and nano-multifunctionalization of denim fabric. Cellulose 25:6207–6220
35. Sarkar S, Banerjee A, Halder U, Biswas R, Bandopadhyay R (2017) Degradation of synthetic azo dyes of textile industry: a sustainable approach using microbial enzymes. Water Conserv Sci Eng 2:121–131
36. Akansha K, Chakraborty D, Sachan SG (2019) Decolorization and degradation of methyl orange by *Bacillus stratosphericus* SCA1007. Biocatal Agric Biotechnol 18:101044
37. Singh RL, Singh PK (2015) RP Singh Enzymatic decolorization and degradation of azo dyes–a review. Int Biodeterior Biodegrad 104:21–31
38. Kupski L, Salcedo GM, Caldas SS, de Souza TD, Furlong EB (2019) EG Primel Optimization of a laccase-mediator system with natural redox-mediating compounds for pesticide removal. Environ Sci Pollut Res 26:5131–5139
39. Guan ZB, Song CM, Zhang N, Zhou W, Xu CW, Zhou LX, Zhao H, Cai YJ, Liao XR (2014) Overexpression, characterization, and dye-decolorizing ability of a thermostable, pH-stable, and organic solvent-tolerant laccase from *Bacillus pumilus* W3. J Mol Catal B Enzym 101:1–6
40. Ba S, Arsenault A, Hassani T, Jones JP (2013) H Cabana Laccase immobilization and insolubilization: from fundamentals to applications for the elimination of emerging contaminants in wastewater treatment. Crit Rev Biotechnol 33:404–418
41. Arun J, Felix V, Monica MJ, Gopinath KP (2019) Application of nano-photocatalysts for degradation and disinfection of wastewater. Advanced Research in Nanosciences for Water Technology. Springer, Cham, pp 249–261
42. EP2064385B1.https://patents.google.com/patent/EP2064385B1/en?q=microbial&q=enzyme&q=textile+finishing&oq=microbial+enzyme+in+textile+finishing
43. CN101864406B.https://patents.google.com/patent/CN101864406B/en?q=microbial&q=enzyme&q=textile+finishing&oq=microbial+enzyme+in+textile+finishing
44. EP3088502B1.https://patents.google.com/patent/EP3088502B1/en?q=microbial&q=enzyme&q=textile+finishing&oq=microbial+enzyme+in+textile+finishing
45. CN105544198A.https://patents.google.com/patent/CN105544198A/en?q=microbial&q=enzyme&q=textile+finishing&before=priority:20190101&after=priority:20150101
46. CN106758261A.https://patents.google.com/patent/CN106758261A/en?q=microbial&q=enzyme&q=textile+finishing&before=priority:20190101&after=priority:20150101
47. CN105951434B.https://patents.google.com/patent/CN105951434B/en?q=microbial&q=enzyme&q=textile+finishing&before=priority:20190101&after=priority:20150101
48. CN108251319A.https://patents.google.com/patent/CN108251319A/en?q=microbial&q=enzyme&q=textile+finishing&before=priority:20190101&after=priority:20150101
49. CN109457468A.https://patents.google.com/patent/CN109457468A/en?q=microbial&q=enzyme&q=textile+finishing&before=priority:20190101&after=priority:20150101

Recent Advances in Development of Antimicrobial Textiles

Shagufta Riaz and Munir Ashraf

Abstract Cellulosic textiles mostly facilitate the growth of microorganism due to their larger surface area and capability to hold moisture, which is responsible for several devastating effects not only to consumer but also to the textile itself. Health related consciousness and hygienic perception of products stimulated the intensive research to mitigate the pathogenic effects and development of microbe free materials. Textile researchers have also been paying rigorous attention for developing different antibacterial agents to fulfill the increased demands of ecofriendly antibacterial textiles. The main emphasis of this chapter was to give an overview of advancements in development of different organic and inorganic antibacterial agents, their inclusion during fiber formation or at textile finishing stage, mode of action of different antibacterial agents, durability and the environmental impact of leaching and bound antibacterial textiles.

Keywords Antibacterial textiles · Nanoparticles · Antibacterial agents · Natural antibacterial agents · Leaching agents

1 Introduction

The clothing concept is shifting towards the innovations in the multifunctional textiles as the emergence of new technologies has revolutionized the approach of inducing improved performances and characteristics. Fibers and fibrous materials have been used for different application like home furnishing, sports goods, medical textiles, water purification and food packaging besides traditional clothing applications. In last few decades, with the increased awareness of environmental and human concerns, trend is moving towards antibacterial products because of microbe full environment [1].

Cotton textiles, having promising comfort properties, can easily be attacked by microorganisms because of superior absorbency and moisture retaining characteristics [2], thus causing different unwanted effects like unpleasant odor, discoloration,

---

S. Riaz · M. Ashraf (✉)
Functional Textiles Research Group, National Textile University, Faisalabad 37610, Pakistan
e-mail: munir.ashraf01@gmail.com

M. Shahid and R. Adivarekar (eds.), *Advances in Functional Finishing of Textiles*, Textile Science and Clothing Technology, https://doi.org/10.1007/978-981-15-3669-4_6

loss in mechanical properties and sometimes undesired staining. Therefore, there is a great need to develop products that offer protection against microbes. Antibacterial textiles have become the fastest growing sector of textile industry due to its inflated production over the years [3, 4]. Antibacterial textiles are of great concern in medical, sportswear, food industry, and water purification systems etc. due to health, fitness and hygienic concerns.

Natural and synthetic fibres can be made antibacterial through application of various natural and synthetic compounds such as plant extracts (Aloe vera, Neem, lotus), chitosan, curcumin, triclosan, *N*-halamines, polybiguanides, quaternary ammonium based compounds and nanomaterials [5, 6]. In particular nanoparticles of inorganic metals and metal oxides (Ag, Cu, CuO, ZnO) are extensively used to functionalize natural, synthetic and regenerated polymers because of their effective antibacterial activity against different bacterial strains [4, 7–13]. Nanoengineered textile fabrics are of huge interest owing to their superior functionality and performance without affecting inherent characteristics of textiles [14].

To give an overview of recent advances in production and application of antibacterial agents to make antimicrobial textiles was the prime objective for this chapter. Later, the particular focus was on inorganic antibacterial agents (metal oxide nanoparticles) and herbal antibacterial agents based textiles because they have the potential for bringing advancements in textile to make it microbe resistant.

2 Mode of Action of Antibacterial Agents

Killing of bacteria is referred as biological self-cleaning. Biological self-cleaning textile is capable to destroy or inhibit the growth of bacteria attached to them. Action of antibacterial agent to kill or stop the bacterial growth depends upon the type of bacteria. Antibacterials now identified as agents have been used to eliminate or disinfect the surfaces from potentially harmful bacteria. According to the speed of action of antibacterial agents or the residue production these agents are divided into two groups i.e. bacteriostatics, and bactericides.

On attachment of bacteria the surfaces become contaminated, therefore, the bacteria should be killed or their growth should be inhibited on their contact with any surface. Mostly bacteria are killed by antiseptics and disinfectants immediately because of bacterial cell explosion which is called bacterial conjugation due to consumption of bacterial resources and preventing multiplication of bacteria. Due to carboxylic peripheral groups in bacterial cell wall, the total charge on bacterial strain is negative between pH 3–9. When exposed to some aqueous solutions, the cell wall gets negative charge because of dissociation of these negative functional groups. When cationic agents like silver, zinc, copper, manganese, chitosan etc. and bacteria are in close proximity, the electrostatic force of attraction between oppositely charged bacteria and cationic agent shows first interaction and strong adhesion causes strong binding of bacteria with antibacterial agent. Then antibacterial agents can either kill bacteria or they may inhibit the growth of bacteria. There are many ways of agents to

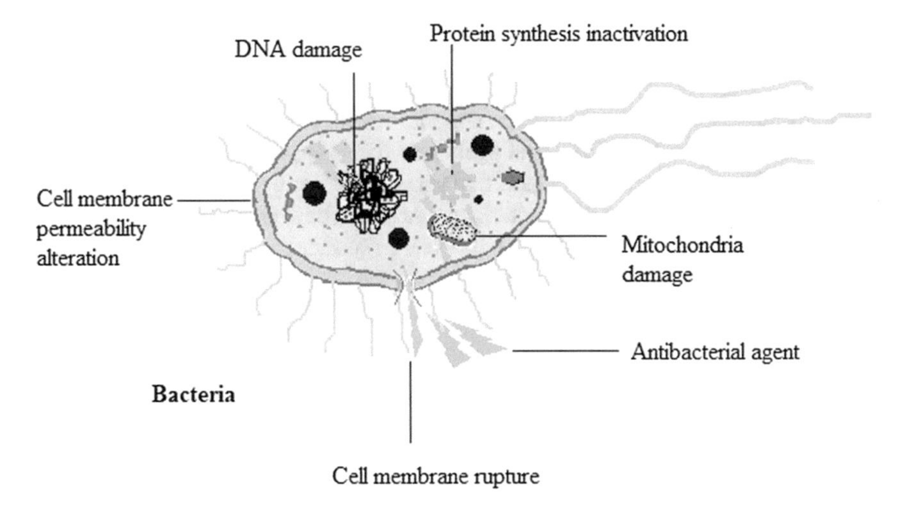

Fig. 1 Mechanism of action of antibacterial agent

exhibit antibacterial activity [15] depending upon the structure and attraction level of target site inside the bacterial cell. They may obstruct, (1) Cell-wall synthesis, (2) Protein synthesis, (3) Nucleic acid synthesis, (4) Enzymatic activity, (5) Folate metabolism, (6) Damage cell membrane (Fig. 1).

3 Why Antimicrobial Textile?

Textile materials/clothing play important role in providing media for growth of microorganisms such as bacteria, fungi, etc. According to recent reports, microorganisms could survive on fabric materials for more than 90 days in a hospital environment. Such a high survival rate of pathogens on medically used textiles may contribute to transmissions of diseases in hospitals. As a mean to reduce bacterial population in healthcare settings and possibly to cut pathogenic infections caused by the textile materials, the use of antimicrobial textile become a vital importance to be used [16].

In recent years, antimicrobial finishing of textiles has become enormously important in the production of protective, decorative and technical textile products. This has provided chances to expand the use of such textiles to different applications in the textile, pharmaceutical, medical, engineering, agricultural, and food industries [17].

Owing to their importance, the number of different antimicrobial agents suitable for textile application on the market has increased noticeably. These antimicrobial agents differ in their chemical structure, effectiveness, method of application, and influence on lives and the environment as well as cost [18].

4 Antimicrobial Textiles Classification

Presently, all disinfectant or antiseptic surfaces resist bacteria mainly by two mechanisms: (i) diffusion of antibacterial agent into the contacting moist surface, or (ii) chemically bound antibacterial agent that show its action upon direct meeting with microorganisms. Therefore, antibacterial surfaces can be divided into two main categories; biocide-releasing or leaching antibacterial surface and contact-active or non-leaching antimicrobial surface depending upon the efficiency and mod of action of antibacterial agents [18, 19].

4.1 *Leaching Antimicrobial Textiles*

The antibacterial textiles that operate on controlled release mechanism are called leaching antibacterial textiles because they release biocides slowly into its surrounding environment to attack the microbes (Fig. 2a). They act as venom for both bacteria

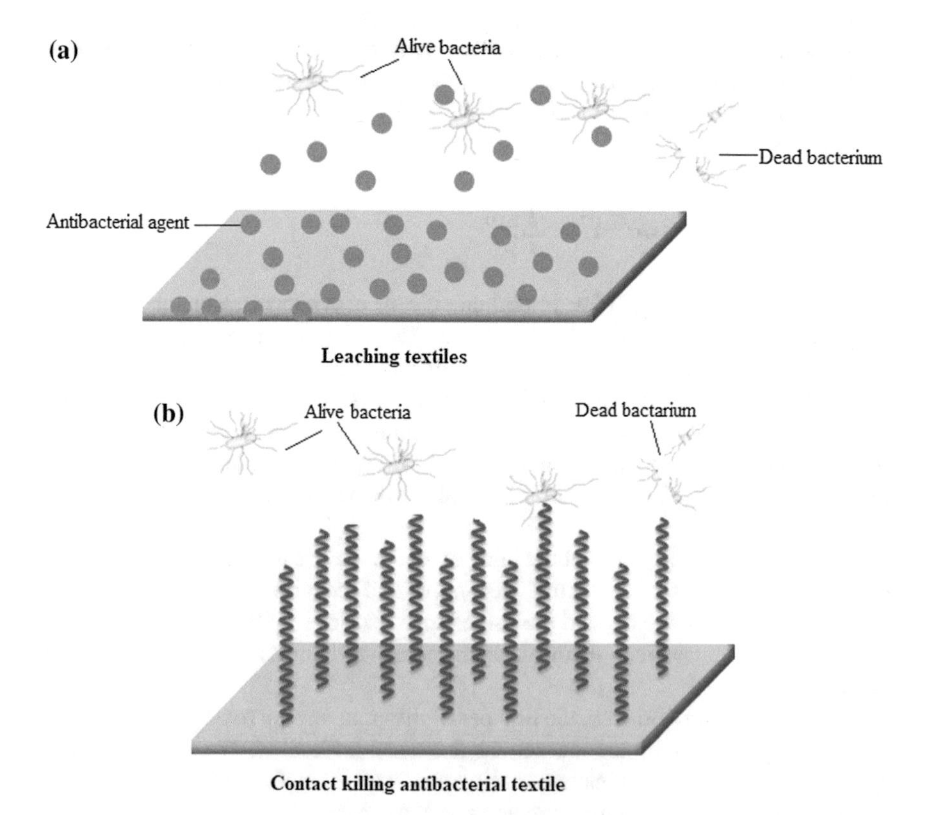

Fig. 2 Schematic showing leaching and non-leaching textile

attached with the surface or present in its surrounding by releasing the agent regularly and constantly that is not chemically bound to textiles. Therefore, the main limitation for their use is due to environmental concern. Because they could be hazardous to human as well as other microbes essential for the effluent treatment [20].

The concentration of antimicrobial agent called minimum inhibitory concentration (MIC) is a significant factor for its proper functioning. Below this level the biocide would not be able to kill or inhibit the growth of bacteria [17]. Another drawback of their use is depletion of reservoir (textile surface containing the antibacterial agent) due to continuous release that eventually would no longer be effective for killing the bacteria. Along with negative effects of leaching antimicrobial agents used for textiles, they also are not durable because they can be wiped out from textile surface during different lauding cycles.

4.2 *Non-leaching or Bounded Antimicrobial Textiles*

Non-leaching or contact killing antibacterial textiles are capable of killing the microbes only on their contact with the substrate because such textiles have chemically bounded organic polycationic materials through covalent bonding directly or via some cross-linker and do not discharge any biocide into its surrounding environment (Fig. 2b). Such textiles show long term functionality because they antibacterial agents do not scrap off due to their bonding with them. They are also human and environmental friendly because they cannot provoke health issues by their release into environment.

When the bacteria come in contact with an antibacterial surface, the organic material spears the outer lipid bilayer of the bacterial cell membrane and preferentially moves into proteins in the cellular organelles to increase the permeability of unwanted substances into microorganism causing its death. Such type antibacterial action offers greater surface activity with considerably low leaching into environment [21]. Though, they are ecofriendly but less effective in action against bacteria because no active substance is released from the textile surface. They show their efficacy only against the bacteria that directly meet the surface. These finishes show no efficiency in their surrounding environment.

5 Application Methods of Antibacterial Agents on Textiles

Antibacterial agents can be imparted to textile using different chemical, physical or physio-chemical approaches at fiber forming stage or attextile finishing stage, depending upon the composition, type of fiber, chemical structure, fiber composition, and fiber surface. Various antimicrobial agents have been applied by direct pad-dry-and cure method, spraying, coating and foaming method. Herbal antibacterial agents are mostly applied after fabric processing and at finishing stage to get more evident outcomes. The methods are as follow:

5.1 Direct Application

Natural antibacterial agents are applied to textile by direct application method. Pad-dry-cure method is conventional direct application method in which textile is dipped in aqueous finishing agent for some specified time. After immersion the fabric is padded between the nip of rollers at particular pressure range for certain pick-up of antibacterial finish. Further, the fabric is dried and cured for a particular time at specified temperature to attach the finish with the textile. Though this is an easy approach to apply finishing agent but to achieve enough durability is difficult if some kind of chemical bonding does not exist between finishing agent and the textile. Multiple finishes could be applied on textile by this technique and the effect would be the same regardless of number of finishing agents applied. This technique is fundamental for all other applied finishes.

5.2 Nanotechnology

As the herbal antibacterial agents are considered less hazardous to human and environment therefore, their use is growing for textile applications as well. But, the main issue associated with these plant based antibacterial agents is there durability. Different methodologies have been adopted to make them durable and nanotechnology is one of them. Different nanoparticles by using plant extracts have been developed for biomedical application and drug release [22, 23] but, there use in textile is very limited and still a space is present to work on them for their application as At nanoscale the adhesion between fibers and natural antibacterial agents could be enhanced without compromising the appearance, handle and comfort of textile [24]. Based on neem extract herbal nanoparticles were synthesized and applied on cotton fabric by conventional pad-dry-cure method increasing the adhesion up to 25 washing cycles showing antibacterial activity against *S. aureus* and *E. coli* both qualitatively and quantitatively [25]. Similarly, chitosan and herbal nanocomposite on cellulosic fabric gave enhanced antibacterial, comfort without affecting physical properties with greater wash durability [26].

5.3 Microencapsulation Method

Functionalization of textiles by nanoparticle encapsulation is micropacking approach to improve the properties as well as to impart the new characteristics in textile product [27]. Nano/microcapsules can be used as phase change materials (PCMs) encapsulation, fire retardants, antimicrobial, fragrance finish, polychromic/thermochromics etc. but here, only drug release (insect repellent, antibacterial, moisturizer, slimming agents) and thermal comfort are stated. During encapsulation, tiny particles of solid, liquid or gaseous components are entrapped as core material inside a shell material [28]. Different active materials like drugs, enzymes, vitamins, pesticides, flavours, skin softeners, PCM and catalysts have been successfully been encapsulated inside nano balloons or nanocapsules made from a variety of polymeric and non-polymeric materials [29]. In this technique one material is packed in a second material as shown in Fig. 3 for the purpose of shielding the active ingredient from surrounding environment [30].

There are many techniques for nanoencapsulation in general divided into three main categories, listed in Table 1.

But, the most frequently used technique is oil-in-water emulsion followed by solvent evaporation method. Durability is an issue after application of microspheres on textile and microencapsulation is an expensive technique therefore, different cross linkers or binders have been used to apply microcapsules, but, it can affect the handle of fabric. To resolve this issue to some extent, application of microencapsulated

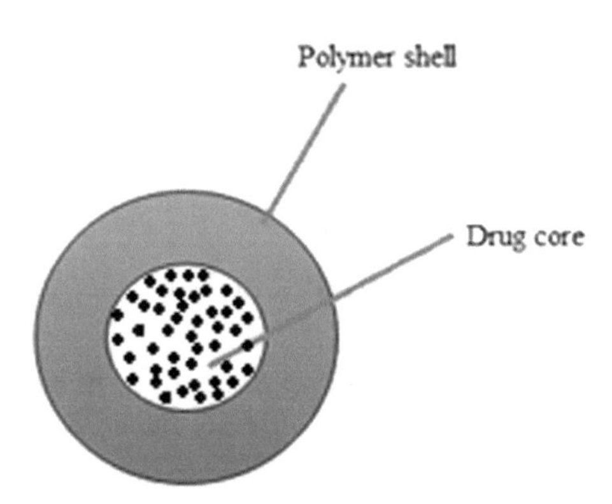

Fig. 3 Microencapsulation: core material surrounded by polymeric shell

Table 1 Different techniques for microencapsulation

Chemical processes	Physico-chemical processes	Physico-mechanical process
Interfacial polymerization	Coacervation and phase separation	Spray drying and congealing
In situ polymerization	Sol-gel encapsulation	Fluid bed coating
Poly condensation	Supercritical CO_2 assisted microencapsulation	Pan coating solvent evaporation

PCMs and effect of plasma surface modification on adhesion of microcapsules on wool fabric was investigated [31]. Research is done on the encapsulation of various essential oils in chitosan and their application on cotton fabric. It was found that by application of such capsules on desized, bleached and mercerized cotton fabric by using pad-dry-cure method enhanced antimicrobial activity of treated fabric. A modified dihydroxy ethylene urea was used as a cross-linking agent. Antimicrobial activity was increased as the concentration of essential oil and chitosan increased [32].

5.4 Crosslinking Method

The chemical bonding between polymeric chains and applied finish is mostly happened by using cross linker for durable functionality. Therefore, different physical, chemical and irradiation techniques have been used [33, 34]. In case of chemical cross-linkers, the introduction of chemical bridges is done between polymeric chains. For irradiation cross linkers no heat and other catalyst is needed, but, for physical crosslinking which is environmental friendly process the durability is less because ionic bond is established between polymeric chains with efficient antibacterial activity and good thermal properties [35, 36]. The physical crosslinking method is most widely used method in which herbal extract or oil is applied on textile by conventional pad-dry-cure method and by using cross linker in presence of some catalyst [37]. Many researchers applied Neem extract and Aloe vera extract by using 1,2,3,4-butanetetracarboxylic acid (BTCA) and citric acid (CA) cross linker to make the textile antimicrobial with enhanced durability [38–40] against Gram-positive and Gram-negative bacteria also some fungi.

6 Assessment of Antibacterial Activity

To evaluate the antibacterial activity of treated textiles, various different standards have been developed. Mainly inhibition zone method also known as qualitative method and bacterial colonies counting method also known as quantitative method have adopted depending upon the leaching and non-leaning antibacterial agents applied on textile. Qualitative test methods 147 (Fig. 4b) include: AATCC 147, JIS L 1902-Halo method, ISO/DIS 20645 and EN ISO 20645 in which antimicrobial testing has been carried out based on agar diffusion plate method. AATCC and JIS L 1902-Halo methods are effective for both non-leaching and leaching textiles while ISO/DIS 20645 and EN ISO 20645 are effective for only leaching or diffusible antibacterial agents. On the other hand, quantitative test methods (Fig. 4a) include AATCC 100 and JIS L 1902-Absorption method), the amount of antibacterial agent applied on textile is more accountable for JIS L 1902 test method rather than AATCC 147 [41]. To perform this testing special training, attention, and careful study of test

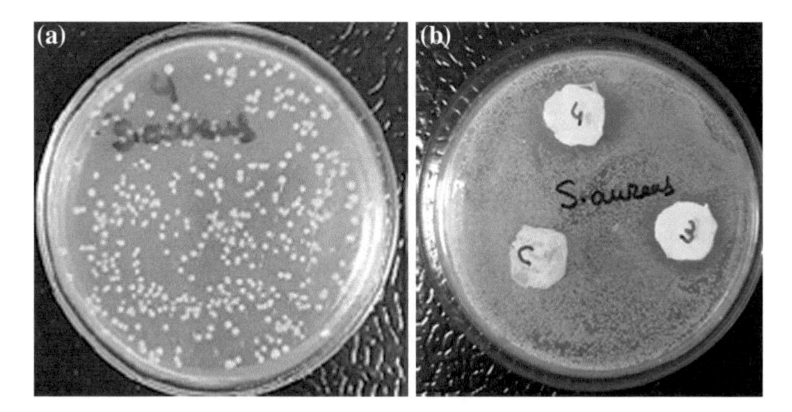

Fig. 4 Antibacterial actvity. **a** quantitative, **b** qualitative

methods, and laboratory equipment for testing is needed to get most accurate results with reproducibility and without any contamination [42].

"British Textile Technology Group (BTTG)" developed a more vigorous antibacterial test method based of adenosine triphosphate luminescence at the end of 1980s in which antibacterial activity was evaluated by analyzing the adenosine triphosphate and bioluminescent recognition.

7 Antimicrobial Compounds

7.1 Organic Antibacterial Agents for Textiles

7.1.1 *N*-Halamine

Organic antibacterial agents have been considered to be the best biocides against wide spectrum of microbes. Among these *N*-halamine is heterocyclic organic antibacterial agent containing one or more covalent bonds between nitrogen and halogen formed by the halogenation of amine, imide or amide groups (Fig. 5) [43]. On its contact with the microorganism the halogen exchange reaction takes place causing the cell

Fig. 5 Structure of *N*-halamine

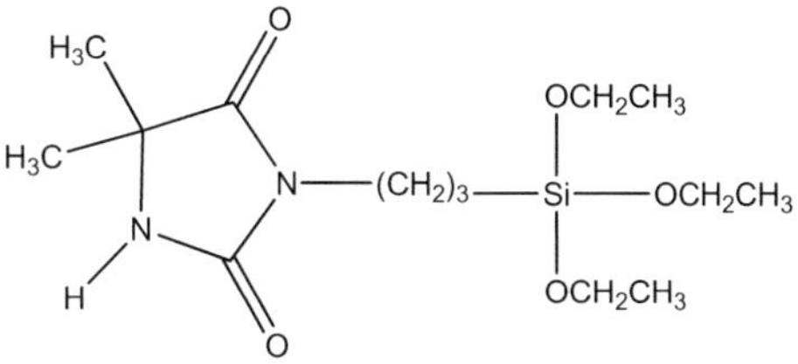

Fig. 6 Basic structure of quaternary ammonium salt

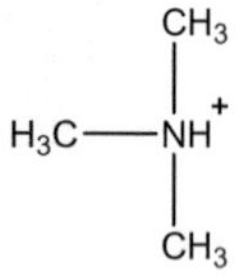

$$H_3C-NH^{+}(CH_3)_2$$

death due to inhibition of enzymatic and metabolic activities [44–46]. *N*-halamine is nonhazardous, more stable with less inclination towards the hydrocarbon generation compared with inorganic antibacterial compounds. Different synthetic and natural textiles could be coated with this organic antibacterial compound to be used for medical and protective purposes with maximum functionality for a longer period of time due to chemical bonding with the substrate [47]. For higher biocide functionality the *N*-halamine coated material should be in contact with pathogens for a longer period of time.

The *N*-halamine based coatings or finishes provide antibacterial activity [48, 49]. They could be functionalized with different silanes and depending upon the carbon chain length the hydrophobicity of textile could also be enhanced [50].

7.1.2 Quaternary Ammonium Compounds (QACs)

The quaternary ammonium compounds are the positively charged compounds. Usually these cationic agents are attached to anionic fabric due to electrostatic force of attraction. QACs represents to a group with approximately 191 compounds which mainly composed of long chain hydrocarbons (hydrophobic part) and other hydrophilic counterpart. The basic structure is shown in Fig. 6. These QACs are applied on natural and synthetic textiles for making them antibacterial and superhydrophobic [51–53]. The electrostatic force of attraction is basic mechanism of antibacterial activity, due to which negatively charged bacterial come closer to the QACs and then long chain hydrocarbons could penetrate into cell membrane causing the cell leakage and ultimately death of bacteria. So, antibacterial action depends upon the chain length of hydrocarbons and perfluorinated group.

These QACs have been applied on textile to make them effective against *Staphylococcus aureus* and *Escherichia coli*. These organic silanes have been used along with metal or metal oxide nanoparticles to make them superhydophobic and antibacterial [54, 55].

7.1.3 Chitosan

Due to extensive research on properties of chitosan it has been considered as interesting biopolymer with biological properties like: Non toxicity, biodegradability, biocompatibility, cytocompatibility, antimicrobial activity, antioxidant activity, haemostatic action and anti-inflammatory action. Chitin and chitosan is a polysaccharide

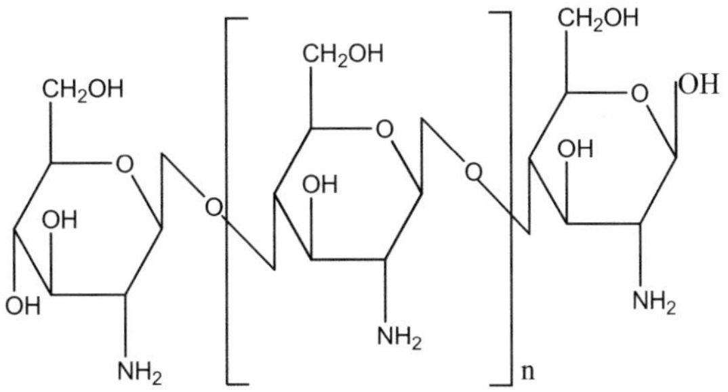

Fig. 7 Structure of chitosan

which is consisting of different amount of β-(1 → 4)-linked 2-amino-2-deoxy-β-D-glucopyranose and 2-acetamido-2-deoxy-β-D-glucopyranose residues [56, 57]. The element composition of the chitosan polymer is carbon (44.11%), hydrogen (6.84%) and nitrogen (7.97%). Chitosan with chemical structure shown in Fig. 7 is soluble in acidic solution and free amino group on its chain are formed which generate positive charge [58]. The antibacterial activity of chitosan is due to amino group present in its structure [59].

Chitosan has been used as durable press finishing [60], antistatic finishing [61] and functional properties like antibacterial activity could be imparted along with durable dyeing [62] and the increased concentration of chitosan could be responsible for more hydrophilic nature of textile [60] degree of acetylation of chitosan also have the effect on microbe inhibition [63]. Anionic and cationic modification in some studies have been done for enhanced corporation of functionality. pH sensitive chitosan hydrogel have been applied for antibacterial activity against *S. aureus* and *E. coli* [60]. Nanotechnology have been playing an important role due to its extraordinary properties therefore, research has been conducted on modification of textiles by nano-scaled chitosan [64]. To increase effectiveness it has been used with zinc and silver in the form of film and nanoparticles on the textiles [65, 66].

7.1.4 Halogenated Phenols (Triclosan)

The antibacterial agents also include bis-phenols the organic compounds exhibiting antibacterial activity against a broad spectrum of bacterial strains. Triclosan and hexachlorophene are the maximally used antibacterial members, but, Triclosan (2,4,4′-trichloro-2′-hydroxydiphenyl ether, Fig. 8) is most prevalent in toxicity against bacteria and fungi. Over last thirty years, the chlorinated bisphenol (Triclosan) has been the most effective organic compound that has been used in health care products, cosmetic, plastics and protective textiles. The antibacterial mode of action is mainly due to biosynthesis inhibition of fatty acids because it could constrain the lipid biosynthesis thus affecting the cell membrane [67]. Triclosan have been applied by using cross-linkers like as (BTCA, CA on textile to make it antibacterial by pad-dry-cure method. Very small concentration of it gives the effective results against different

Fig. 8 Structure of triclosan

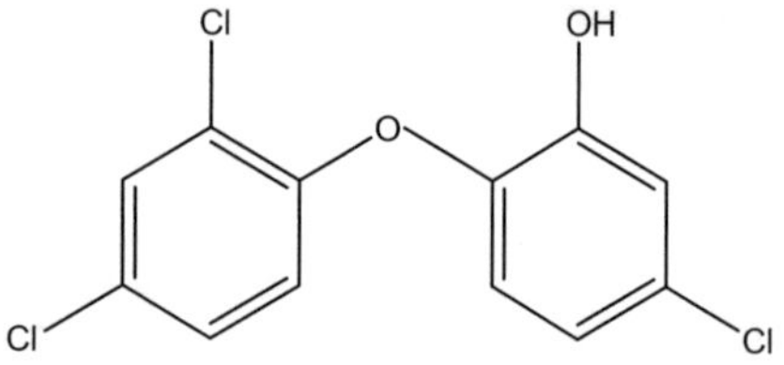

bacteria and fungi with maximum durability upto 50 washing cycles [68]. The activity of triclosan could be affected by various acidic, basic and urine conditions [69]. It could be used in combination with other biopolymers for enhanced functionality like antibacterial, durable press, antistatic property etc. [70]. Among silver, chitosan and triclosan that were applied on textile by pad-dry-cure method separately. The concentration of applied dispersion and pH affected the antibacterial action. And the repeated laundering could reduce the action of textile. The silver is highly effective then triclosan and chitosan is least effective against Gram positive bacteria (*S. aureus*) and Gram negative bacteria (*E. coli*) [70].

7.1.5 L-Cysteine

L-cysteine is an uncharged polar amino acid where the polarity is due to "R" groups which are polar due to sulfhydryl groups and hydrogen of these sulfhydryl groups in L-cysteine structure (Fig. 9) could be easily replaced by any radical for covalent linkage with other molecular structure. It could damage the cells because it acts as free radicals scavenger. Thus, this reactive compound is highly renowned for its devastating effect against several metabolic functions occurring inside cells of large number of microbes because it could affect various enzymatic reactions making it potential organic antibacterial agent. However it's antibacterial action is not clear and still a room is available to find its mod of action against different microorganisms [71].

L-cysteine have been used for textile functionalization to impart antibacterial activity with greater durability due to covalent bonding with prospective biomedical, pediatric and geriatric textile applications [72]. To increase the immobilization and to enhance the antibacterial activity nanoparticles have been applied on L-cysteine modified cotton fabric. Copper and silver nanoparticles have been used for different researches and then durability and antibacterial efficacy of treated textiles was

Fig. 9 Basic chemical structure on L-cystein

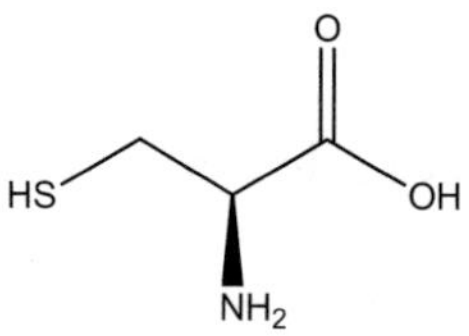

analyzed [73, 74]. The L-cysteine could be covalently attached onto textile substrate via esterification of OH-groups present at the surface of cellulosic material. On the other hand, due to coordinate covalent bond formation between nanoparticles and L-cysteine made it possible to adhere nanoparticles tightly with the textile [75]. The L-cysteine have also been applied on wool fabrics to preserve its quality and to make it antibacterial for different medical and healthcare applications [76].

7.2 *Inorganic Antibacterial Bacterial Agents for Textiles*

Killing of bacteria is referred as biological self-cleaning. Biological self-cleaning textile is capable of killing bacteria attached to them. In last few decades trend is moving towards antibacterial products because the environment is full of microbe. There is a great need to develop products that offer protection against microbes. Nanomaterials in particular nanoparticles of inorganic metals and metal oxides are extensively used to functionalize natural, synthetic and regenerated polymers because of their effective antibacterial activity against different bacterial strains. Researchers worked on different polymers and different antibacterial agents to give antibacterial property to textiles.

The antibacterial activity of contact active surfaces is mainly due to its topography. If the surface has nanoroughness like presence of nanorods, nanoparticles or any other nanostructure, the attaching bacteria on this surface will be killed due to penetration of these structures into membranes which leads to its disintegration. This mechanism is valid for all type of nanostructures whether they are inherently antibacterial or not. Like surface structures of many butterflies are not inherently antibacterial but their morphologies make them antibacterial [77, 78].

Nanoparticles can be applied by two methods to modify fabric as antibacterial. Generally, first method involves direct deposition of metal NPs on the textile surface by some electrostatic force. In this methods colloidal metal nanoparticles are prepared using some reducing agent to reduce metallic salt to metal nanoparticles. Then polymeric textiles are directly immersed in colloidal solution containing metal NPs and are deposited on fibres as shown in Fig. 10b [79]. Sometimes surface modification of textile is done to enhance surface energy, to make surface rough so that adhesion/affinity between fibre surface and metal nanoparticles can be enhanced to a greater level. Plasma treatment [80], corona treatment [81], and enzyme treatments [82] are done for this purpose. In second method, metal ions are adsorbed at the surface, then, these ions are converted to metal NPs by UV radiation [83], heat treatment [84] or by some chemical reduction method.

Currently NPs of silver [85], titanium dioxide [86, 87], silver bromide, zinc oxide [88], gallium, gold, carbon nanotubes and copper oxide are being used as biocide releasing antibacterial agents (Fig. 11) that kill bacteria before their access to the surface.

Fig. 10 SEM image. **a** Untreated textile, **b** textile treated with NPs

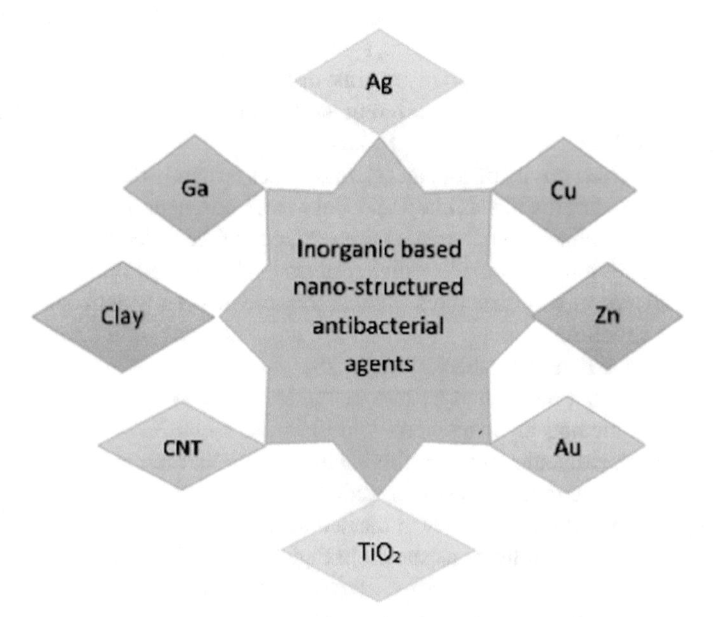

Fig. 11 Category of inorganic based antibacterial agents applied on textile for desired efficacy

7.2.1 Silver Nanoparticles

Silver ions and nanoparticles are most frequently used antibacterial agents that are integrated into fibers and onto fiber surface and have been reported for their most effective antibacterial activity against bacteria, viruses, and eukaryotic cells [89, 90]. NPs showed more bactericidal activity compared to ionic or bulk material [91, 92] because of larger surface area of NPs that provides more interaction of antibacterial agent with microbes per surface unit and due to penetration of nanostructures in bacterial membranes. Silver has capability to attack the cell membrane and penetrate through it, causing malfunctioning of bacterial cell respiratory system, thus, responsible for the cells death [93].

Shateri et al. studied the antibacterial activity of silver nanoparticles deposited on cationized cotton. Cationic agent (3-chloro-2-hydroxy propyl trimethyl ammonium chloride) in presence of alkali (NaOH) was used to make ordinary surface of cotton fibers cationic and increased affinity of cotton fibers for AgNPs that showed outstanding antibacterial activity [94]. Yeo and Jeong [95] prepared bicomponent fibers using polypropylene (PP) and AgNPs by melt spinning method. AgNPs were in core and PP was spun as sheath of bicomponent fiber. But fibers didn't show antibacterial activity. The fibers in which silver was added in sheath showed superb antibacterial property.

Dubas et al. [96] worked on layer by layer deposition of Ag NPs which were stabilized with capping agent poly (methacrylic acid) PMA. The layer of anionic poly (methacrylic acid) capped AgNPs was immobilized by cationic poly(diallyldimethylammonium chloride) when silk and nylon fibers were coated with these layers. Thus, rendering these fibers as highly antibacterial for different applications. Wool fibres were also investigated for their antibacterial property by direct deposition of AgNPs, at fibre surface reduction of metal salt was done by reducing agent and trisodium citrate (TSC) was used as linker that bound AgNPs with amino acid of keratin that is wool fibres protein. Along with antibacterial activity, AgNPs also gave surface plasmon resonance optical effect, enhanced electrical conductivity and antistatic property to fibres [97]. In another approach AgNps were applied on surface of silk fibres by hydrogen bonding [86].

AgNPs were fabricated by in situ direct metallization technique. The cellulose membranes were treated with Aqueous $AgNO_3$ solution. After washing, Ag^+ ions containing membranes were treated with solution of reducing agents and some colloid protector under certain process conditions. These composite fibers showed superb antibacterial activity [87]. To enhance the antibacterial effect, combination of different antibacterial agents have also been used. Silver loaded chitosan nanoparticles when applied to textile showed enhanced antibacterial activity compared with only chitosan nanoparticles applied textile because synergistic antimicrobial effect against *S. aureus* bacteria was showed by silver loaded chitosan nanoparticles [98].

Surface modification of textile substrates with AgNPs and antibacterial activity was explored by using sodium hydroxide at different concentrations. Alkali treated cotton fabric was immersed in silver nitrate solution ($AgNO_3$) and then chemical reduction caused in situ AgNPs on fabric surface. Higher the concentration of NaOH more was silver content at surface. Homogeneously distributed NPs at treated fabric surface exhibited excellent antibacterial activity against *Escherichia coli* and *Staphylococcus aureus*. That procedure gave durable, superficial, cost effective method for higher silver content on textile surface [89].

In most recent studies electrospinning is fascinating technique for nanofiber fabrication. Researchers are developing electrospun nanofibers for different biomedical, filtration applications. Solution of cellulose acetate (CA) was direct electrospun containing $AgNO_3$ in small amount. Nanofibers containing Ag+ ions. AgNPs were synthesized by direct UV irradiation of ultrafine CA fibers and stabilized by carbonyl oxygen in CA, having strong antibacterial activity because of silver NPs and unreduced ions of silver [83]. In same way fibers chitosan/gelatin nanofibers containing AgNPs were fabricated, only difference was that, instead of γ or UV irradiation and heat treatment, AgNPs were synthesized by reducing agent like chitosan that also act as stabilizer for silver NPs [99, 100]. Nano AgZ (Silver-loaded zirconium phosphate nanoparticles) were added to PCL Poly(ε-caprolactone) and biocompatible fibers were electrospun for medical applications with enhanced bacteriostatic activities [90].

Carbon nanotubes Ag-coated CNTs were prepared by ultrasonic irradiation of dimethyl formamide (DMF) solution containing multi-walled carbon nanotubes, silver acetate solution. Dry mixing of nylon-6 powder and silver coated CNTs was done, through melt spinning prepared fibers had enhanced bactericidal property [91], because single-walled and multi-walled carbon nanotubes are well known for their antibacterial activity [101, 102]. Hence, CNTs have great potential for being used as antibacterial agent, in textile industry.

Detrimental effect of AgNPs, when they are used as antibacterial agent is of particular concern as they come in contact with human skin. Size and surface area have direct effect on cytotoxicity of silver NPs in relation to concentration. If size of AgNPs is 2–3 nm and concentration of colloidal silver is 100 ppm then, it has devastating effect on skin [92]. But, if the colloidal concentration is less i.e. 10, 20 ppm, etc. then, it is less toxic compared to 100 ppm.

In vitro cytotoxicity of AgNPs having diameter of about 15 nm, was studied in male mammalian mouse germline stem cells. This study showed that if AgNPs are used in concentration more than 5 μg/ml, it reduced the cell viability and function of mitochondria by increasing lactase dehydrogenase (LDH) [103]. Therefore, the trend is moving towards less toxic nanoparticles for antibacterial characteristics of fibers. ZnO, TiO_2 and CuO nanoparticles are being used as a substitute of silver for antibacterial characteristics. Photocatalysis is the main reason of bacterial inactivation in case of all these nanoparticles. The schematic illustration of the whole photocatlytic process in given in Fig. 12.

7.2.2 Copper Oxide Nanoparticles

In one study CuO NPs in crystalline form of monoclinic phase were developed and adsorbed directly onto raw cotton fibers surface by ultrasonic irradiation of metal hydroxide, which could be used as highly antibacterial raw material for protective clothing, medical textiles, sportswear etc. [104]. Thin film formation of nanoscale of metal by direct sputtering had also been done on textile to inactivate bacteria. When Cu react with air O_2, CuO formation occurred. Semiconductor CuO (p-type) have

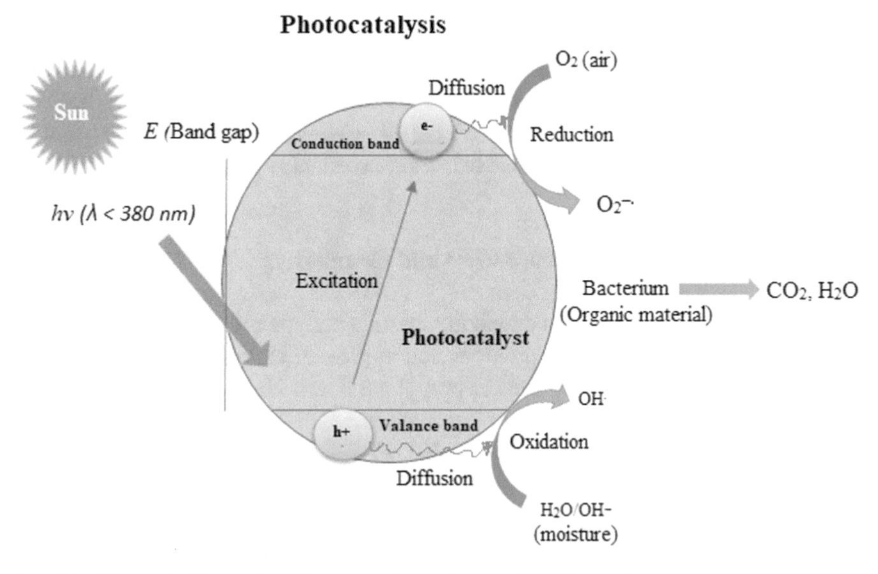

Fig. 12 Schematic illustration of photocatalytic degradation of bacterium

1.7 eV band-gap energy. Electron hole pair is formed with photon energies exceeding the band-gap of CuO under irradiation. The excited e^-s react with O_2 forming $O_2^{\bullet -}$ radical [105], that is responsible for bacterial inactivation.

$$CuO + h\nu(<660\ nm) \rightarrow CuO\left(e_{cb}^{-},\ h_{vb}^{+}\right) \tag{1}$$

CuO produced electron and hole pair (e_{cb}^{-}, h_{vb}^{+}) with photon energies greater than band-gap energy of CuO as given in Eq. (1). Thus produced excited electron (e_{cb}^{-}) can either react with (1) directly the atmospheric O_2 forming $O^{\bullet -2}$ (Eq. 2) or (2) by reducing the Cu^{2+} to Cu^{+} as shown in Eq. (3) that on reacting with O_2 converted into CuO (Cu^{2+}) [105] shown in Eq. (4).

$$CuO\left(e_{cb}^{-}\right) + O_2 \rightarrow CuO + O^{\bullet -2} \tag{2}$$

$$CuO\left(e_{cb}^{-}\right) \rightarrow CuO\left(Cu^{+}\right) \tag{3}$$

$$CuO\left(Cu^{+}\right) + O_2 \rightarrow CuO\left(Cu^{2+}\right) \tag{4}$$

Overall reaction is given in Eq. (5). And generated $O^{\bullet -2}$ radical is responsible for antibacterial activity.

$$CuO\left(Cu^{+}\right) + O_2 \rightarrow CuO\left(Cu^{2+}\right) + O^{\bullet -2} \tag{5}$$

In another study chitosan attached cellulose fibers were prepared and then soaked in copper aqueous solution to make copper bound chitosan attached cellulose fibers, reduction of copper into copper nanoparticles was done by using sodium borohydride solution. These CuNP's attached fibers have great prospective to be used as antibacterial wound dressing because of having excellent antibacterial activity [106].

7.2.3 Titanium Dioxide Nanoparticle and Composites

TiO_2 nanomaterials are focused due to their photocatalytic property [107, 108], this photocatalytic activity of TiO_2 provides active protection against bacteria. Therefore, this attribute was investigated for propylene fibers. Here, composite Ag/ TiO_2 nanocomposite filler were used for melt spinning of polypropylene (PP). Biostatic effectiveness was excellent for Ag/TiO_2/PP nanocomposite filaments [109]. In another antibacterial property investigation, nano-sol was prepared in which cellulosic fibers were pressed through nano-sol to coat nano TiO_2 on substrate. Fibers showed substantial antibacterial activity in UV light, however, the activity was not very effective in dark [110]. Synergistic effect of TiO_2 and Cu nanocomposite film made by direct current pulsed magnetron sputtering onto polyester and cotton fabric was investigated [111]. When sputtering by direct current TiO_2 sputtering of textile material for 10 min was followed by direct current pulsed sputtering of Cu-layers for 40 s. Due to band gap and band position the hole injection from CuO to TiO_2 occurred, preventing e^--hole recombination of CuO that allowed holes in TiO_2 to inactivate *Escherichia coli*. Because of large difference between to valance band (vb) a significant induced force was responsible for interfacial charge transfer from CuO to TiO_2. TiO_2 holes (+) react with OH^- group at surface of TiO_2 releasing $OH^\bullet$ radical that inactivated bacteria (Fig. 13).

TiO_2 nanowires alone and Ag-doped TiO_2 nanowires were also applied on textile substrate to compare their antibacterial properties due to photocatalysis. TiO_2 nanowires were fabricated by hydrothermal process using TiO_2 NPs. Ag-doped TiO_2 NPs and nanowires were synthesized by photo-reduction of Ag^+ ions to Ag metal on the surface of TiO_2 NPs and nanowires. Different concentrations of PVP were coated on bleached cotton fabric was dip-pad-dry method. These treated and untreated fabric samples were then applied with nano-sol containing TiO_2 NPs and nanowires and Ag-doped TiO_2 NPS and nanowires by pad-dry-cure method. PVP acted as dispersant and stabilizer for NPs and nanowires. The finished cotton fabrics were investigated for their antibacterial properties. Sample pre-treated with highest concentration of PVP that was finished with highest concentration of Ag-doped TiO_2 nanowires exhibited highest activity against different Gram-positive and Gram-negative bacterial strains and fungi. Thus, a potential finished textile for medical and industrial applications [112].

Ultrasound energy is used is sonochemistry for induction of some physical and chemical change in medium via acoustic cavitation. TiO_2 were deposited by ultrasonic mechanism onto cotton substrate. Use of ultrasonic irradiation for coating on textile substrate is economic, simple, fast and "green" approach that does not involve

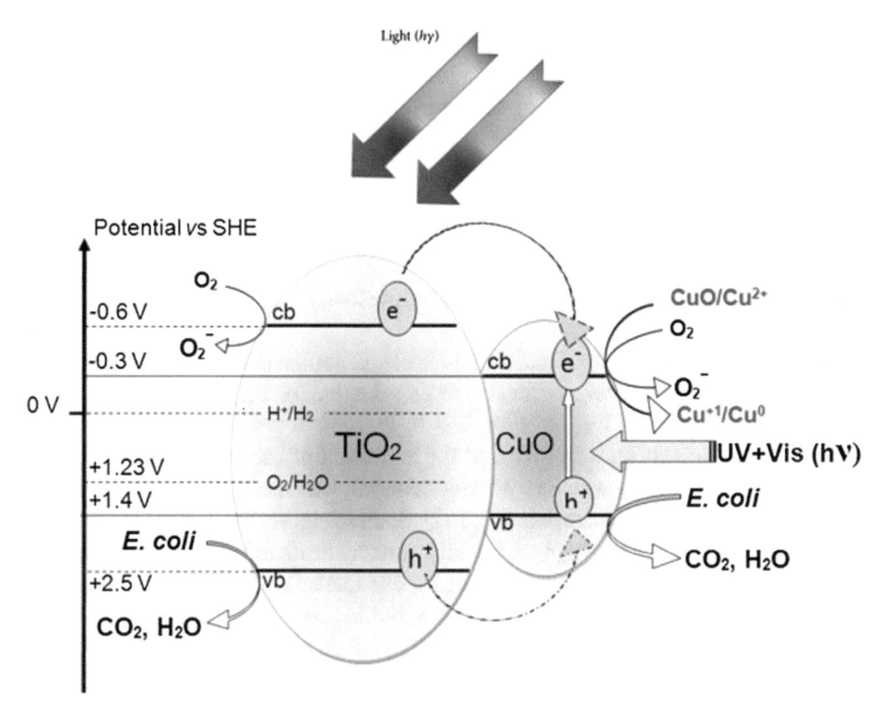

Fig. 13 Schematic diagram of injection of charge during interfacial charge transfer between Cu and TiO_2 under visible simulated light and bacterial inactivation due to generated radical species [111]

toxic materials. It was reported that reactive oxygen species due to photocatalytic activity by visible light, are responsible for antibacterial activity against *S. aureus* and *E. coli*. Optimization of minimal effective concentration of the deposited TiO_2 NPs on the fabrics was done, that showed antibacterial activity [99]. The effect of enzymatic pre-treatment on textile functionalities was investigated [100]. Due to hydrolysis of textile substrate and introduction of new reactive groups, the adsorption of nano-TiO_2 on textile substrate increased. Proteases and lipases enzymes were used for this surface activity of wool and polyester respectively. Pre-treated wool/polyester was then immersed into ultrasound bath that contained TiO_2 NPs and cross linking agent butane tetracarboxylic acid (BTCA). Treated and untreated samples were investigated for their activity against *E. coli*. It was shown that the treated sample have superb activity because bacterial reduction was 100% when 0.75% TiO_2 was used and 99% when 0.25% concentration of TiO_2 was used. This study clarified that photocatalytic activity of TiO_2 enhanced self-cleaning (chemical and biological).

To prepare nanocomposite for enhanced functionality of textile, chitosan was used along with nano-TiO_2. Novel chitosan/nano-TiO_2 composite suspension was prepared by inverse suspension technique, in which chitosan was dissolved in acetic acid (2.5%) solution, then epichlorohydrin was added in solution. TiO_2 was dispersed for 15–20 min in chitosan emulsion by ultrasonic vibrator. Three gauze samples were prepared and treated with chitosan, nano-TiO_2 and nano-TiO_2/chitosan emulsion hen their antibacterial assessment showed highest antibacterial activity for nano-TiO_2/chitosan composite emulsion. against *E. coli*, *A. niger* and *C. albicans* [113]. That was a new development of antibacterial green textile, to be used for medical application especially for wound care.

7.2.4 Zinc Oxide Nanoparticles and Composites

Recently, ZnO has found applications in textile as antibacterial agent. It appears to be promising functionalizing agent due to its selective toxicity towards prokaryotic and eukaryotic [114]. It has been applied to textiles to render them antibacterial. Different mechanisms may be involved for inhibition of bacterial growth or killing of bacteria. First, due to structure of nanomaterial, because various nanostructures can be grown on different materials [88, 101]. Such structure disintegrate the bacterial cell membrane by damaging it mechanically, causing death of bacteria [102]. Second, due to generation of oxygen reactive species like $OH^•$, H_2O_2, ${O_2}^{•-2}$ that penetrates into cell membrane, inhibiting metabolic activities and later death of cell [115]. Another possibility is leaching of Zn^{++} ions that attach to the bacterial surface due to electrostatic attraction as bacterial cell wall carries negative charge, inhibiting further growth [116].

ZnO nanomaterials are applied on different textile substrates like polyester, cotton, their blend etc. to impart multifunctional attributes. ZnO NPs of different size were synthesized, polyester knitted fabrics was dip coated in nano-ZnO aqueous suspension containing binder to enhance adhesion of NPs with hydrophobic surface. The treated polyester fabric showed antibacterial activity up to 10 washes against *S. aureus* and *K. pneumoniae* [37]. Instead of ZnO NPs Ashraf et al. [117] functionalized polyester fabric by the growth of ZnO NRs using hydrothermal fabrication process. ZnO NPs were coated on plasma modified textile substrate by pad-dry-cure method, to make seeded fabric. Fabric was dried at 120 °C for 2 min and then cured at 170 °C for about 8 min. Then, ZnO NRs were grown on seeded fabric, by placing fabric in reagents solution at 90 °C for 4 h. Qualitative and quantitative antibacterial assessment showed high efficacy of ZnO NRs against bacterial strains *S. aureus* and *E. coli*. To make multi-functional textile that was used as biosensor or pesticides, ZnO NRs were grown on conductive textile substrate consisting of silver (55%) and nylon (45%). Zinc acetate and potassium hydroxide (KOH) solution was made in methanol at 60 °C temperature for 2 h by continuous stirring. Textile sample was heated at 100 °C in prepared solution to make seed over which the ZnO NRs were grown in an oven at 90 °C for about 6 h, then, cleaning and drying of finished conductive textile was done. ZnO NRs based functional textile exhibited excellent photocatalytic and antibacterial activity along with sensing properties. Such concept of smart textile for wearable sensing without any odor and bacterial growth was given first time by Hatamie et al. [118], after the commercially available silver based product that is considered hazardous to health because of its toxicity [107]. For biomedical application

NPs are frequently used, by incorporating them in dope for wet spun fibers. ZnO NPs were electrospun with sodium alginate as absorbent polysaccharide and polyvinyl alcohol as fiber forming additive agent to produce highly antibacterial fiber. Such produced fibers were operative for wound dressing against *Staphylococcus aureus* and *Escherichia coli*. ZnO being effective agent against these bacterial strains provided least toxicity [108]. Nanofibers incorporated with nanoparticles are made to use for biomedical applications. Mostly silver and ZnO NPs have been used for antibacterial efficacy against different strains of bacteria [119, 120]. These nanofibers are superabsorbent, better mechanical and antibacterial properties because of their novel and differentiating properties due to higher surface to volume ratio and extremely small size.

Chitosan is abundantly available natural polysaccharide known for its activity against various microorganisms. Therefore, hybrid metal nanoparticles are prepared with this polysaccharide to increase antibacterial activity symbiotically. AbdElhady [121] synthesized ZnO/chitosan nanoparticles at different temperatures and different concentrations. These ZnO/chitosan NPs were applied on cotton fabric by pad-dry-cure method. After padding dried for 10 min at 100 °C then cured for 5 min at 170 °C. Washed and finished cotton fabric presented exceptional antibacterial activity against Gram-negative and Gram-positive bacteria along with significant enhancement in UV-protection. In another study ZnO/carboxymethyl chitosan bionanocomposite were applied as finishing agent to the cotton fabric by pad-dry-cure method. Dipping of fabric was done for 15 min and dried at 100 °C for 15 min and cured at 160 °C for 3 min. Thorough washing with water and drying of fabric was done [122]. One step sonochemical coating process was used for deposition of ZnO/chitosan hybrid NPs on cotton fabric to evaluate their antimicrobial properties against *Staphylococcus aureus* and *Escherichia coli*. Combination of chitosan and ZnO resulted in increased activity compared to individual chitosan and ZnO NPs application of textile. Biopolymer like chitosan not only increased antibacterial efficiency but durability of applied finishing agent was augmented. Such finished textile could be useful in hospital environment to prevent microorganism growth and spread of infections [123]. Ag:ZnO/chitosan nanocomposite finish was prepared by sol-gel process using 3-glycidyloxypropyltrimethoxysilane (GPTMS) and tetraethoxysilane (TEOS) in two steps, in another study. Solution of chitosan in acetic acid (1%) was prepared then ZnO, Ag doped ZnO NPs or $AgNO_3$ were added to the prepared solution with functionalization agents GPTMS and TEOS. Due to hydrolysis and condensation of ethoxy and methoxy groups of GPTMS and TEOS in chitosan solution organic/inorganic hybrid sol formation occured. Scoured and bleached blend (cotton/polyester) fabric was applied with this hybrid sol by pad-dry-cure method. Significant bacterial reduction was observed against *M. Luteus* and *E. coli*. Such treated fabric was strongly recommended for potential applications in medical for infection prevention during wound healing and burns [124].

Similarly, different antibacterial agents have been used along with ZnO to fabricate different nanomaterials. To get synergistic activity against microorganisms and to reduce toxicity level of individual compound combination therapy is practiced [101].

In the study used chalcones (E)-3-(4-hydroxyphenyl)-1-(4-methoxyphenyl) prop-2-en-1-one are antibacterial agents that along with ZnO were used to prepare flowers like nanorods. *Acacia arabica* was used as binder. Cotton fabric was immersed in solution containing chalcones, ZnO and *acacia arabica* for 5 min to coat them on textile by dip-pad-dry method. 99% bacterial reduction against *S. aureus*, *E. coli*, *P. aeruginosa* of treated cotton was observed. That was first study reported on bacterial reduction by using chalcones, ZnO and *acacia arabica* by Sivakumar et al. [101].

Natural plant based antimicrobial agents, due to the safety and their functionality are gaining more attention of researchers than synthetic or inorganic antimicrobial agents.

7.3 Plant Extract Based Organic Antibacterial Textile

Plant have been a rich source of antibacterial agents and usually they have been used for the treatment of many human infections [125]. In different cultures and regions about 60–90% of diseases are cured by using different plants derived medicines and their crude extract [126]. Different phytochemicals like phenolic compounds (tannins, flavonoids, stilbenes, quinines, lignans and coumarins), terpenoids and nitrogen compounds (betalains, alkaloids and amines) [127] contain substantial antioxidant activity providing significant antibacterial, antifungal, anti-inflammatory, anticarcinogenic [125, 126] and antiviral action have been sourced by medicinal plants. Instead of synthetic drugs now, world has been moving towards the naturally occurring antimicrobial drugs to heal the infections, because they can be easily extracted from the plants and their further purification can produce benign and certainly active medications. That's why natural drugs have been replacing now the synthetic drugs to mitigate the adverse side effects of synthetic medication. The plant extracts and essential oils that exhibit antibacterial activity, mainly come from different parts of plants like leaves, trunk, roots, flowers, seeds, bark etc. Different plant extracts have been used to treat the textiles to resist microbes in the field of health care and medical products because of their ecofriendly, human and skin friendly, non-toxic nature.

7.3.1 Aloe Vera

Aloe vera plant of *genus Aloe* belongs to family *Asphodelaceae* originated from "Arabian Peninsula" has been used in many human useable products like cosmetics, beverages, ointments and skin lotions etc. because of being safe and effective against different microbes. Due to presence of about 75 many active compounds [128], including minerals, essential amino acids and different vitamins, enzymes, anthraquinones, hormones, it has been used in many skin care products, traditional health care products for burn wounds and infectious wounds since last 2000 years. Aloe vera comprises of 6 disinfectant and antiseptic ingredients that are responsible of antifungal, antibacterial, and antiviral properties i.e. phenols, urea nitrogen, Lupeol,

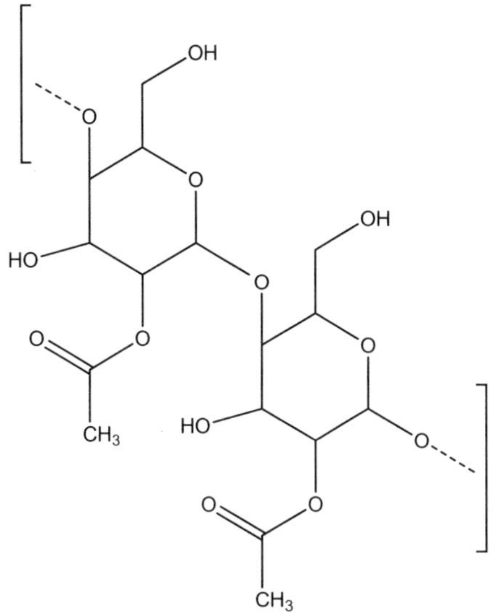

Fig. 14 Chemical structure of acemannan

cinnamonic acid sulfur and salicylic acid. Aloe vera contains different polysaccharides including acetylated mannan (β(1,4)-acetylated mannan, Fig. 14) which is also known as acemannan has antifungal, anti-inflammatory, immunostimulant and anti-tumor attributes [129]. Due to extravagant antiseptic characteristics it could be used for medical textiles like, surgical gowns, absorbable and non-adorable sutures.

Different researches have made attempts to apply the Aloe vera gel and its extract on textile to make it antimicrobial due its awesome antiseptic properties and their application has been done by coating and pad-dry-cure method by using different cross linkers to get maximum efficiency with certain durability. Due to application of Aloe vera gel by pad-dry-cure method the whitening index could be less affected as compared to coating of gel. Also, along with antibacterial properties softness can be enhanced due to inherent softening attribute of Aloe vera. Highest the concentration of applied gel higher would be the antibacterial efficacy as investigated by Ibrahim et al. [130]. When a combination of Aloe vera, chitosan, and curcumin was applied by exhaust technology, on cotton, wool and rabbit hair to investigate the antimicrobial activity. Results showed that peroxide treated cotton fabric and formic acid treated wool/rabbit hair substrates demonstrated better antibacterial activity. The efficacy of Aloe vera increased when used in combination with chitosan and curcumin as compared when applied alone. The treated textiles showed durability up to twenty washings [131].

Ali et al. [38] developed antibacterial cotton fabric by application of Aloe vera gel using 1,2,3,4-butanetetracarboxlic acid (BTCA) crosslinker. Application was done by pad-dry-cure method. The treated fabric showed antibacterial activity against both Gram-positive bacteria (*S. aureus*) and Gram-negative bacteria (*E. coli*) (2014). Different extraction methods have been used for separation of antibacterial agents by using organic solvents. The efficiency of antibacterial agent is also dependent upon the solvent used and the extraction method and functional textile fabric could be developed with excellent performance against bacteria to be used for medical textiles [132]. Ayyub et al. [133] studied the synergistic effect of different extracts by developing the durable antibacterial textiles. Aloe vera gel and Neem extract were separately applied and their functionality was examined antifungal property against *Aspergillus niger* and antibacterial activity against strains i.e. *E. coli* and *S. aureus*. And the composite finish showed better results than separate application of each agent.

Inherent and mechanical properties of treated textile are affected by the application of Aloe vera extract. This was studied by Hein et al. [134] and antibacterial testing was performed for Gram-positive bacteria (*S. aureus*, *B. subtilis*, and *B. pumilus*) and Gram-negative bacteria (*Pseudomonas*, *Candida*, and *E. coli*). Best results were obtained against *Pseudomonas* and *E-coli*, while, finished fabric have shown fair enough activity *B. subtilis* and *B. pumilus*. But, in case of *S. aureus* and *Candi* bacterial strain, the treated fabric could not resist the bacterial growth. Similar to cotton textiles, the silk textiles are also prone to microbes due to their hygroscopic nature. Therefore, antibacterial treatment of silk textiles is also needed. Combination of organic and inorganic antibacterial agents could be used for enhanced activity against microbes. A study was conducted in which Silver NPs and Aloe vera gel have been used along with BTCA crosslinker to treat silk fabric. Excellent antibacterial activity was shown against *S. aureus* and *K. pneumoniae* bacteria with durability up to 5 washing cycles [40].

7.3.2 Neem

Pant from Genus *zadirachta,* which is known as Neem (A*zadirachta indica*) or *margosa* belongs to family (*Meliaceae*), Mohagny. It is mostly found in Pakistan, Maldives, India, Sri-lanka, Bangladesh and Nepal, as well as in some parts of America and Africa, is well appreciated as a medicinal plant to be used in folk medicines, cosmetics, and organic farm applications (pest and insect control) due to its antiseptic, antifungal, antibacterial, antiviral, antidiabetic, sedative, anthelmintic, and contraceptive properties [119, 120]. The active compounds are found mostly in all parts of plant including leaves, fruit, seed, roots, flowers, twig, gum and bark. From where its extract or oil has been obtained to cure different human diseases from ancient times to the modern age. More than 700 Neem based products have been used in Amchi, Siddha, Ayurveda, Unani, and many other traditional health care remedies especially as skin related medicaments. Because of being organic material it is safe with its extraordinary attributes but, its long-term use can cause some side effects. Neem tree consists of more than 250 active ingredients but, the most active compounds obtained from different parts of tree are nimbin, azadirachtin, and salannin [135] and nimbin with chemical structure shown in Fig. 15 is most prevalent due to its antibacterial activity.

Different classes of antimicrobial compounds and their extract found their application in textile industry [136]. Neem extract have been used widely to inhibit the

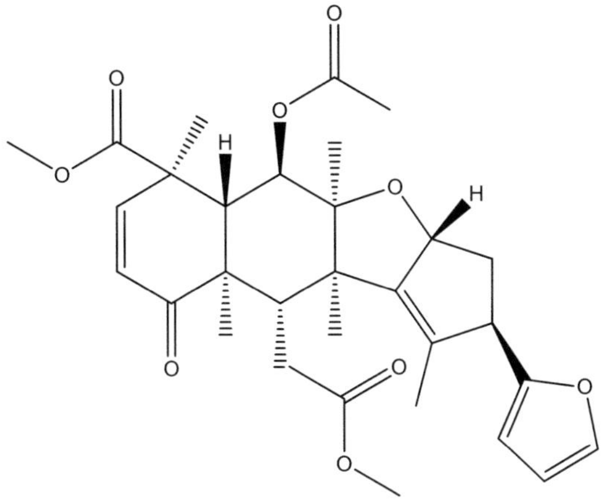

Fig. 15 Chemical structure of nimbin (active limonoids present in neem giving antimicrobial property)

growth of different bacterial strains, and currently many researches have been trying to impart antibacterial characteristic to textiles by using it alone or in combination with other organic and inorganic materials. Direct application of Neem oil and extract or in the form of microencapsulation it has been applied onto textile by pad-dry-cure method and by exhaust method. It was investigated by Ganesan and Vardhini [137] that the microencapsulation technique is more durable than direct method of application of an extract. And the functionality of fabric against bacteria (*S. aureus* and *E. coli*) would be more for longer time period in case of microencapsulation up to 15 washing cycles.

Blend of cotton/polyester textiles was applied with Neem seed extract by using cross-linker glyoxal/glycol to investigate its efficiency against *Bacillus subtilis* and *Proteus vulgaris* bacteria by Joshi et al. [39]. Quantitative antibacterial analysis showed that antibacterial activity was higher for *Bacillus subtilis* than *Proteus vulgaris*. 90% growth inhibition was observed for treated textiles. They conducted similar study, in which Neem bark and seed extract was applied on textile with a crosslinker using microwave cure method. It was investigated that seed extract treated textile was more durable than bark extract treated textile without significant change in mechanical properties [138].

The materials at nano scale show better results because of their larger surface area and higher surface energy leading to better adhesion with the textile substrate increasing the durability for longer functionality. In another study, nanocomposite of Neem and chitosan was developed by emulsion and solvent evaporation method and applied on cotton fabric for protective clothing. The spherical shaped nanocomposite of Neem and chitosan with size 50–200 nm exhibited excellent antibacterial activity against bacteria which was higher than alone Neem and chitosan [139]. Another researcher condensed extracted Neem oil into micro and nano Neem oil by using different emulsifier and then applied on textile. The developed textiles showed tremendous antimicrobial activity against bacteria image [140]. The antibacterial activity of could be enhanced by developing composite finish of different organic antibacterial agents. Neem and Aloe vera could be a good combination of antibacterial agents to inhibit the growth of bacteria. A research work was reported in which different concentrations of Neem and Aloe vera were used to coat the fabric separately. The hybrid combination was found very active against fungi and bacteria compared to alone Aloe vera and Neem with moderate washing durability [141].

Along with antimicrobial activity the Neem extract could be used as natural dye. Various green shades could be attained with the use of different mordant. Polyurethane based fibers were dyed by Patel [142] and then K/S values were determined digitally. The dyed fabrics showed moderate to hood fastness to rubbing, light and washing along with encouraging results of bacterial inhibition. Similar work is presented by Inprasit et al. [143]. In which tannin-rich natural dye was extracted at 65 °C by Soxhlet method, from Neem tree bark and antibacterial agent from leaves using organic solvent. The hemp fabric was dyed with this extracted dye giving high color strength (K/S value) at high temperature and higher concentration for a longer period of dyeing time. The reddish brown fabric reduced 99.9% of the *S. aureus* bacteria with dye fastness to washing, water and perspiration. Patel and Pratibha [144] found that the natural antimicrobial agents are better than hazardous synthetic antibacterial agents. For this purpose natural dyes and antimicrobial agents extracted from bioactive plants were used for finishing of textile for medical applications. Herbal extract application made woven and nonwoven fabrics resistant against bacteria. Effect of organic and inorganic antibacterial agent was also compared by Rajput et al. [145] for baby diapers. ZnO NPs, TiO_2 NPs, and Ag NPs were applied on baby diapers and compared with the diapers coated with natural extracts of Turmeric, Tulsi, Aloe vera and Neem. It was reported that the natural ingredients showed the results of antimicrobial effects similar to inorganic agents showing the cost effective, and human-friendly diapers.

7.3.3 Tulsi Leaves

Ocimum tenuiflorum belong to family Lamiaceae (mint family) is commonly known as Tulsi or holy basil with aromatic leaves, which is known for its religious and medicinal importance. It contains many phytochemicals like, oleanolic acid, ursolic acid, rosmarinic acid, eugenol, carvacrol, linalool, β-caryophyllene (about 8%), and the essential oils of Tulsi contains eugenol (70%) β-elemene (~11.0%), β-caryophyllene (~8%), 2-methoxy 4-(1-propyl) phenol (2.65%), 1,4-diethyl benzene (1.03%), germacrene (~2%), (2.01%) essential oil of Tulsi and vanillin (1.27%). Tulsi leaves and essential oils have properties like insecticide, antiviral, expectorant, antiprotozoal, antifungal and antibacterial. The antimicrobial activity of Tulsi is most prevalently due to eugenol (Fig. 16a), and also due to alpha terpinol (Fig. 16b) and gamaterpine (Fig. 16c) compounds presents in it.

Some physicians have been applying different plants after grinding and forming their smooth paste to treat skin diseases. Cotton, bamboo and soybean fabrics have been coated with different natural extracts such as lemon grass, Aloe vera, Tulsi, kadukai, karpuravalli and vettukayathalai to make textile good antibacterial [146].

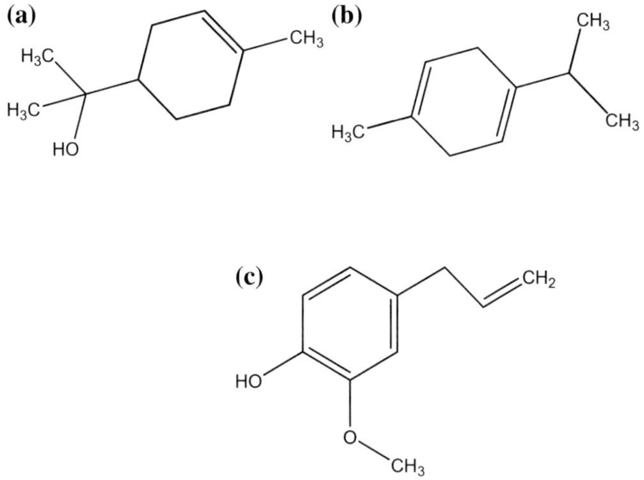

Fig. 16 Chemical structure. **a** Eugenol, **b** alpha terpinol, **c** gamaterpine

Tulsi leaf extract is an effective bioactive mediator that is being applied on cotton and polyester/cotton fabric to make antibacterial. To enhance the affinity with textile, cross linking agents have been used. A study was conducted in which different concentrations of leaf extract by using cross linking agent glutaraldehyde catalyzed by sodium hypophosphite were applied on polyester/cotton blend fabric. The treated fabric showed antibacterial activity against *S. aureus* and *E. coli*. Bending and mechanical properties were decreased while improvement of crease recovery was observed after application of natural herbal extract [147].

Some researchers have been applying softeners to improve the physical properties of textile along with imparting the antibacterial activity. In this way the effect of herbal extract on textile has been minimized. A study was conducted in which cotton fabric was modified with cationizing agent and then modified cotton was treated with different concentration of Tulsi and Neem extract in presence of silicon based softener. Excellent antibacterial textile was developed with minimum effect on properties like tensile strength, bending, elongation and roughness [148].

For controlled release of extracts from textile they have been applied in the form of microcapsules. Numerous extracts like Tulsi, Neem, and Turmeric have been encapsulated in some polymeric shell and then application of these microcapsule have been done by conventional pad-dry-cure method using binder to make silk textile effective against different bacterial strains like *Pseudomonas aeruginosa*, *Escherichia coli*, and *Staphylococcus aureus* [8].

7.3.4 Turmeric (Curcumin)

Tumeric have been used since last many years as natural coloring material because of possessing properties such as biodegradable, non-allergenic, non-toxic and remarkable antibacterial activity. Turmeric with botanical name Curcuma longa L. belongs to family *Zingiberaceae*. It is originated from India but now found in most area like southern Asia and throughout the tropics. Curcumin is the major ingredient (Fig. 17) of Turmeric with other minor ingredients like bisdemethoxycurcumin and demethoxycurcumin having biological attributes like antibacterial, anti-fungal, anticancer, anti-inflammatory, and healing [149, 150]. In recent years, due to these characteristics throughout the world they have grabbed a significant attention of medical and textile researchers. *Curcuma longa* L. with core phytochemicals (1,7-bis(4-hydroxy-3-methoxyphenyl)-1,6-heptadiene-3,5-dione) have been applied on textile for dyeing and antimicrobial characteristics.

Several studies have been conducted in which curcumin have been proven as antibacterial, antiviral, antifungal, and insect repellent. Also, it has been widely accepted suitable antimicrobial agent for textiles [151, 152]. Silk fabric was died with Turmeric and in pre-metallization process the Copper sulfate, ferrous sulfate and potassium aluminium sulfate were used as mordant. Antibacterial silk textile was prepared in which direct relation was found between degree of antimicrobial activity and metal ion concentration. The mordant dyed textiles showed excellent antibacterial activity against *E. coli* and *S. aureus* an desired color fastness was obtained [153].

Cotton fabric was modified with chitosan and enzyme [152] and then dyed with curcumin. The dyed cotton showed enhanced antibacterial activity and desirable color. The study was reported to show enhanced durability of curcumin on cotton fabric along with more dye uptake. Silk fabric was died in the same way with various concentrations of curcumin and using different mordants (ferric sulfate, cupric sulfate, and potassium aluminum sulfate). Then the developed silk fabric was assessed for bacteriostatic property against Gram-positive and Gram-negative bacteria. The results showed that Turmeric died fabric have excellent antibacterial activity and enough durability to washing [151]. Researchers have been finding the ways to improve the functionality of textiles by combining the curcumin with other natural

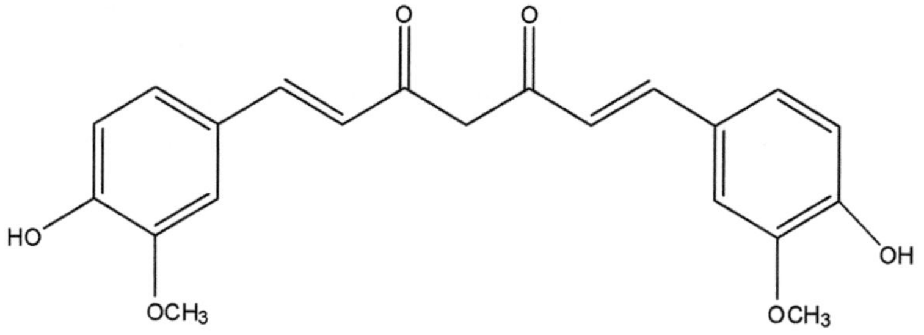

O
O
HO
OCH3
OH
OCH3

Fig. 17 Chemical structure of curcumin

antibacterial agents. Combination of chitosan and curcumin was applied on viscose fabric and then the adsorption/desorption of compounds was analyzed. The adsorbed curcumin enhanced the antibacterial efficiency of already established antibacterial performance of chitosan [154].

At nanoscale the properties of material could be enhanced too many folds. Therefore, nanomaterials and nano-emulsions have been applied to textile for greater functionality. Curcumin nano-emulsions have also been applied by exhaust method using crosslinkers i.e. β-cyclodextrin and polycarboxylic acid and antibacterial activity was assessed by using both standards AATCC 147 and AATCC 100. The results were positive for the treated textiles with greater adhesion and potential to be used in wound care management as medial textiles [155]. To improve the antibacterial activity, curcumin could be mixed with other organic and inorganic antibacterial agents for development of antibacterial textiles. In this way dyeing and finishing of fabric could be done in single step by conventional pad-dry-cure method. Turmeric dye, chitosan/TiO_2 composite nanoparticles were applied on textiles by using crosslinking agent (citric acid) in presence of sodium pyrophosphate catalyst. Under optimum conditions of all (w/v) such as TiO_2 (0.75%, w/v), Turmeric dye (10% dye, w/v), chitosan (2.5%, w/v), citric acid (30 g/L), sodium pyrophosphate (4 g/L), drying tem. (70 °C for 5 min.) and curing temp. (180 °C for 2 min), the treated fabric showed significant improvement in self-cleaning, color fastness and antibacterial properties [154].

7.3.5 Tobacco Leaf

The herbaceous plant tobacco (*Nicotiana tabacum*) that is native to tropical and subtropical America has been cultivated in most parts of world due to its properties like narcotic, antiseptic, sedative, emetic and antispasmodic. This plant belongs to family Solanaceae and leaves are rich in several bioactive phytochemicals/phenols. Also, Nicotine has been extracted from the leaves that is zinc associated and showed antibacterial active against many bacterial strains [156]. The flavonoids of *Nicotiana tabacum* contains free radical scavenging capability to ascorbic acid and superoxide ions due to which it possess antioxidant activity [157]. The antioxidant and antibacterial activity of tobacco plant is well known. However, no much research work have yet been reported showing the antibacterial activity of textiles imparted by polyphenols of *Nicotiana tabacum.* Duangsri et al. [158] developed ecofriendly antibacterial finish from extract by tobacco leaves and applied on cotton fabric. The inhibition activity against *E. coli* and *S. aureus* confirmed the antibacterial activity of polyphenol extracted from tobacco leaf. There is still a room to explore the antibacterial properties of tobacco plant after their application on different textile materials.

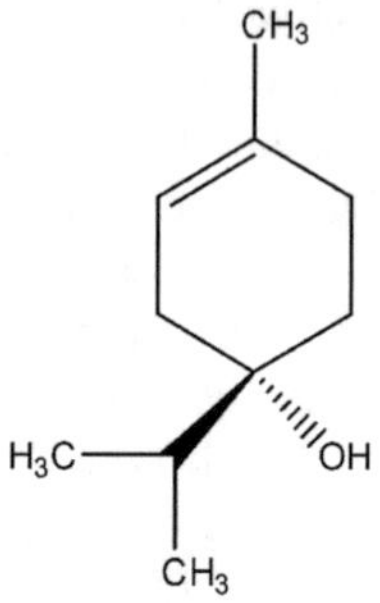

Fig. 18 Chemical structure of terpinen-4-ol

7.3.6 Tea Tree

Melaleuca alternifolia is Australian native tree that is known for its antimicrobial properties and the Tea tree oil is an essential oil that has been obtained from leaves of Tea tree by steam distillation. It has been used as a remedy for many diseases in Europe, North America and Australia. Now it has been widely accepted medicinal plant all over the word due to its antiseptic properties against infections. Tea tree oil is effective against different bacterial strains such as *P. aeruginosa, E. coli*, *Proteus microbilis, S. auresus and P. vulgaris*, etc. Along with many other compounds the terpinen-4-ol (Fig. 18) present in oil is mainly responsible for most of antibacterial activities [159].

Due to antimicrobial and antiseptic properties, it has been used as medicinal plants in cosmetics since last many years [160]. It has also found its application in textiles and have been applied on as nanospheres and microcapsules for controlled release of encapsulated drug. High pressure homogenization technique have been used for making nanoparticle's dispersion and then their application on nonwoven by spray method made them antibacterial that could be used for feminine hygienic products [161]. These essential oils could be encapsulated in some wax or wax/chitosan shell. Then could be applied on material to impart antimicrobial characteristics with controlled release of Tea tree oil against *S. aureus* and *E. coli* [162]. Tea tree oil can also be used for bandages and wound dressings and it could also be used for textile applications alone or in combination with other organic and inorganic antibacterial agents. All the plants described, there antibacterial active agents and the bacterial strains against which they show antibacterial activity are given in Table 2.

8 Conclusion

Most promising organic and inorganic antibacterial agents that have been used for antibacterial textiles are discussed in this chapter. These agents in crude form, as micro or nanoparticles and in the form of microcapsules attached with the textile substrate. They could be strongly or loosely bound depending upon the nature of

Table 2 Plant used for antibacterial textile, active antibacterial agents and the bacterial stains against which plants show antibacterial activity

Sr.#	Botanical name	Antibacterial agent	Bacterial strain and fungi against which plants show activity	Plant
1	Aloe vera	Acemannan	*Aspergillus Niger, S. aureus, B. subtilis, B. pumilus, Pseudomona, K. pneumoniae*	
2	Neem	Nimbin	*Bacillus subtilis, Proteus vulgaris, Staphylococcus aureus, Escherichia coli*	
3	Tulsi	Euginol, alpha Terpinol and Gama terpine	*Staphylococcus aureus, Escherichia coli* and *Pseudomonas K. pneumoniae*, and *Pseudomonas aeruginosa*	

(continued)

Table 2 (continued)

Sr.#	Botanical name	Antibacterial agent	Bacterial strain and fungi against which plants show activity	Plant
4	Tea tree		*Staphylococcus aureus, Escherichia coli, Streptococcus pyogenes, Salmonella Typhimurium, Pseudomonas aeruginosa, Helicobacter pylori, Proteus vulgeris, Proteus mirabilis*	
5	Turmeric	Phenolic compounds (Curcumin)	*Salmonella typhimurium, S. epidermidis, Staphylococcus aureus, Bacillus subtilis, Escherichia coli,, Pseudomonas aeruginosa*, and *E. coli, B. coagulans, B. cereus, A. hydrophila*	
6	Tobacco leaf	Polyphenols (chlorogenic acid and rutin)	*Escherichia coli, Staphylococcus aureus* and *Bacillus subtilis, Micobacteriumphlei, Candida aldicans, Cryptococcus neoformans*	

antibacterial agent, textile substrate, attachment process, and process conditions. As far as the antibacterial finishing concerned their environmental impact is as important as their functionality. Therefore, their health and safety concerns must be taken into account before their efficient application on textiles. Here in this chapter inorganic and plant based (herbal) organic antibacterial agents, and their application methods have been presented in details. Still there is a great need for intensive research on herbal based antibacterial agents to be applied on various natural and synthetic textile substrates.

References

1. Textiles Intelligence LTD (2013) Antimicrobial fibres, fabrics and apparel: innovative weapons against infection. Peform Appa Mark 47:25–57
2. Zhou C-E, Kan C, Matinlinna J, Tsoi J (2017) Regenerable antibacterial cotton fabric by plasma treatment with dimethylhydantoin: antibacterial activity against *S. aureus*. Coatings 7(1):11
3. Windler L, Height M, Nowack B (2013) Comparative evaluation of antimicrobials for textile applications. Environ Int 53:62–73
4. Kyung Wha O, Young Joo N (2014) Antimicrobial activity of cotton fabric treated with extracts from the lotus plant. Text Res J 84(15):1650–1660
5. Morais DS, Guedes RM, Lopes MA (2016) Antimicrobial approaches for textiles: from research to market. Materials 9(6):498
6. Ibrahim NA (2015) Nanomaterials for antibacterial textiles. In: Nanotechnology in diagnosis, treatment and prophylaxis of infectious diseases. Academic Press, pp 191–216
7. Jothi D (2009) Experimental study on antimicrobial activity of cotton fabric treated with Aloe gel extract from *Aloe vera* plant for controlling the *Staphylococcus aureus* (bacterium). Afr J Microbiol Res 3(5):228–232
8. Saraswathi R, Krishnan P, Dilip C (2010) Antimicrobial activity of cotton and silk fabric with herbal extract by micro encapsulation. Asian Pac J Trop Med 3(2):128–132
9. Saif MJ, Zia KM, Rehman F, Ahmad MN, Kiran S, Gulzar T (2015) An eco-friendly, permanent, and non-leaching antimicrobial coating on cotton fabrics. J Text Inst 106:907–911
10. Javid A, Raza ZA, Hussain T, Rehman A (2014) Chitosan microencapsulation of various essential oils to enhance the functional properties of cotton fabric. J Microencapsul 2048:1–8
11. Tang J, Chen Q, Xu L, Zhang S, Feng L, Cheng L, Xu H, Liu Z (2013) Graphene oxide–silver nanocomposite as a highly effective antibacterial agent with species-specific mechanisms. ACS Appl Mater interfaces 5(9):3867–3874
12. Javid A, Kumar M, Yoon S, Lee JH, Han JG (2017) Size-controlled growth and antibacterial mechanism for Cu: C nanocomposite thin films. Phys Chem Chem Phys 19(1):237–244
13. Ashraf M, Campagne C, Perwuelz A, Champagne P, Leriche A, Courtois C (2013) Development of superhydrophilic and superhydrophobic polyester fabric by growing Zinc Oxide nanorods. J Colloid Interface Sci 394(1):545–553
14. Yetisen AK, Qu H, Manbachi A, Butt H, Dokmeci MR, Hinestroza JP, Skorobogatiy M, Khademhosseini A, Yun SH (2016) Nanotechnology in Textiles. ACS Nano 10(3):3042–3068
15. Gebbharadt LD, Bachtold JG (1955) Proceedings of the society for experimental biology and medicine, vol 88. Blackwell Science
16. Qian L, Sun G (2003) Durable and regenerable antimicrobial textiles: synthesis and applications of 3-methylol-2,2,5,5-tetramethylimidazolidin-4-one (MTMIO). J Appl Polym Sci 89(9):2418–2425
17. Curteza A (2011) Sustainable textiles. Radar 2(1):19–21
18. Dring I (2003) Antimicrobial, rot proofing and hygiene finishes. In: Textile finishing. Society of Dyers and Colourists, Bradford, pp 351–371
19. Siedenbiedel F, Tiller JC (2012) Antimicrobial polymers in solution and on surfaces: overview and functional principles. Polymers (Basel) 4(4):46–71
20. Shahidi S, Wiener J (2012) Antibacterial agents in textile industry. In: Antimicrobial agents, pp. 387–406
21. Sawan SP, Shalon T, Subramanyam S, Yurkovetskiy A (1996) Contact-killing non-leaching antimicrobial materials. US5849311A
22. Karthik S, Suriyaprabha R, Vinoth M, Srither SR, Manivasakan P, Rajendran V, Valiyaveettil S (2017) Larvicidal, super hydrophobic and antibacterial properties of herbal nanoparticles from: *Acalypha indica* for biomedical applications. RSC Adv 7(66):41763–41770
23. Mamillapalli V (2016) Nanoparticles for herbal extracts. Asian J Pharm 10(2 Supplement):S54–S60

24. Vigneshwaran N, Ashtaputre NM, Varadarajan PV, Nachane RP, Paralikar KM, Balasubramanya RH (2007) Biological synthesis of silver nanoparticles using the fungus *Aspergillus flavus*. Mater Lett 61(6):1413–1418
25. Ahmed H, Rajendran R, Balakumar C (2012) Nanoherbal coating of cotton fabric to enhance antimicrobial durability. Appl Chem 45(2012):7840–7843
26. Chandrasekar S, Vijayakumar S, Rajendran R (2014) Application of chitosan and herbal nanocomposites to develop antibacterial medical textile. Biomed Aging Pathol 4(1):59–64
27. Singh MN, Hemant KSY, Ram M, Shivakumar HG (2010) Microencapsulation: a promising technique for controlled drug delivery. Res Pharm Sci 5(2):65–77
28. Gupta A, Dey B (2013) Microencapsulation for controlled drug delivery: a comprehensive review. Sunsari Tech Coll J 1(1):48–54
29. Nagavarma BVN, Yadav HKS, Ayaz A, Vasudha LS, Shivakumar HG (2012) Different techniques for preparation of polymeric nanoparticles—a review. Asian J Pharm Clin Res 5:16–23
30. Fievez V, Garinot M, Schneider Y, Préat V (2006) Nanoparticles as potential oral delivery systems of proteins and vaccines: a mechanistic approach. J Controlled Rel 116(1):1–27
31. Oliveira FR, Fernandes M, Carneiro N, Pedro Souto A (2013) Functionalization of wool fabric with phase-change materials microcapsules after plasma surface modification. J Appl Polym Sci 128(5):2638–2647
32. Rehman A, Javed A, Raza ZA, Hussain T (2014) Chitosan micro capsulation of various essential oils to enhance the functional properties of cotton fabric. J Microencapsul 31:461–468
33. Hayashi MA, Bizerra FC, Da Silva PI (2013) Antimicrobial compounds from natural sources. Front Microbiol 4:195
34. Sivakumar V, Vijaeeswarri J, Anna JL (2011) Effective natural dye extraction from different plant materials using ultrasound. Ind Crops Prod 33(1):116–122
35. Syamili E, Elayarajah B, Rajendran R, Venkatraja B, Kumar PA (2012) Antibacterial cotton finish using green tea leaf extracts interacted with copper. Asian J Text 2(1):6–16
36. Straccia MC, Romano I, Oliva A, Santagata G, Laurienzo P (2014) Crosslinker effects on functional properties of alginate/N-succinylChitosan based hydrogels. Carbohydr Polym 108(1):321–330
37. Rode C, Zieger M, Wyrwa R, Thein S, Wiegand C, Weiser M, Ludwig A, Wehner D, Hipler U (2015) Antibacterial zinc oxide nanoparticle coating of polyester fabrics. J Text Sci Technol 1(August):65–74
38. Ali SW, Purwar R, Joshi M, Rajendran S (2014) Antibacterial properties of aloe vera gel-finished cotton fabric. Cellulose 21(3):2063–2072
39. Joshi M, Ali SW, Rajendran S (2007) Antibacterial finishing of polyester/cotton blend fabrics using Neem (*Azadirachta indica*): a natural bioactive agent. J Appl Polym Sci 106(2):793–800
40. Nadiger VG, Shukla SR (2017) Antibacterial properties of silk fabric treated with aloe vera and silver nanoparticles. J Text Inst 108(3):385–396
41. Pinho E, Magalhães L, Henriques M, Oliveira R (2011) Antimicrobial activity assessment of textiles: standard methods comparison. Ann Microbiol 61(3):493–498
42. Schindler WD, Hauser PJ (2004) Chemical finishing of textiles. CRC, The Textile Institute, Manchester
43. Hui F, Debiemme-Chouvy C (2013) Antimicrobial N-halamine polymers and coatings: a review of their synthesis, characterization, and applications. Biomacromol 14(3):585–601
44. Li R, Hu P, Ren X, Worley SD, Huang TS (2013) Antimicrobial N-halamine modified Chitosan films. Carbohydr Polym 92(1):534–539
45. Liang J, Wu R, Wang JW, Barnes K, Worley SD, Cho U, Lee J, Broughton RM, Huang TS (2007) N-halamine biocidal coatings. J Ind Microbiol Biotechnol 34(2):157–163
46. Ren T, Dormitorio TV, Qiao M, Huang T-S, Weese J (2018) N-halamine incorporated antimicrobial nonwoven fabrics for use against avian influenza virus. Vet Microbiol 218:78–83
47. Liu Y, Ren X, Liang J (2015) Antibacterial modification of cellulosic materials. BioResources 10(1):1964–1985

48. Liang J, Chen Y, Ren X, Wu R, Barnes K, Worley SD, Broughton RM, Cho U, Kocer H, Huang TS (2007) Fabric treated with antimicrobial N-halamine epoxides. Ind Eng Chem Res 46(20):6425–6429
49. Demir B, Cerkez I, Worley SD, Broughton RM, Huang TS (2015) N-halamine-modified antimicrobial polypropylene nonwoven fabrics for use against airborne bacteria. ACS Appl Mater Interfaces 7(3):1752–1757
50. Cheng X, Li R, Du J, Sheng J, Ma K, Ren X, Huang TS (2015) Antimicrobial activity of hydrophobic cotton coated with N-halamine. Polym Adv Technol 26(1):99–103
51. Xue Y, Xiao H, Zhang Y (2015) Antimicrobial polymeric materials with quaternary ammonium and phosphonium salts. Int J Mol Sci 16(2):3626–3655
52. Zhu P, Sun G (2004) Antimicrobial finishing of wool fabrics using quaternary ammonium salts. J Appl Polym Sci 93(3):1037–1041
53. Riaz S, Ashraf M, Hussain T, Hussain MT (2019) Modification of silica nanoparticles to develop highly durable superhydrophobic and antibacterial cotton fabrics. Cellulose 26(8):5159–5175
54. Aslanidou D, Karapanagiotis I (2018) Superhydrophobic, superoleophobic and antimicrobial coatings for the protection of silk textiles. Coatings 8(3):101
55. Berendjchi A, Khajavi R, Yazdanshenas ME (2011) Fabrication of superhydrophobic and antibacterial surface on cotton fabric by doped silica-based sols with nanoparticles of copper. Nanoscale Res Lett 6(1):594
56. Aiba S (1992) Studies on chitosan: 4. Lysozymic hydrolysis of partially N-acetylated Chitosans. Int J Macromol 14:225–228
57. Ravi Kumar MNV (2000) A review of chitin and chitosan applications. React Funct Polym 46:1–27
58. Tolimate A, Desrieres J, Rhazia M (2003) Contribution to the preparation of chitins and chitosans with controlled physico-chemical properties. Polymer (Guildf) 44:7939–7952
59. Goy RC, de Britto D, Assis OBG (2009) A review of the antimicrobial activity of Chitosan. Polímeros 19(3):241–247
60. Huang KS, Wu WJ, Chen JB, Lian HS (2008) Application of low-molecular-weight chitosan in durable press finishing. Carbohydr Polym 73(2):254–260
61. Abdel-Halim ES, Abdel-Mohdy FA, Al-Deyab SS, El-Newehy MH (2010) Chitosan and monochlorotriazinyl-β-cyclodextrin finishes improve antistatic properties of cotton/polyester blend and polyester fabrics. Carbohydr Polym 82(1):202–208
62. Moses JJ, Venkataraman VK (2016) Study of chemical treated cotton fabric for functional finishes using chitosan. J Text Apparel Technol Manag 10(1):1–17
63. Zhang Z, Chen L, Ji J, Huang Y, Chen D (2003) Antibacterial properties of cotton fabrics treated with chitosan. Text Res J 73(12):1103–1106
64. Şahan G, Demir A (2016) A green application of nano sized Chitosan in textile finishing. TEKSTİL ve KONFEKSİYON 26(4):414–420
65. Scacchetti FAP, Pinto E, Soares GMB (2017) Preparation and characterization of cotton fabrics with antimicrobial properties through the application of chitosan/silver-zeolite film. Procedia Eng 200:276–282
66. Perelshtein I, Ruderman E, Perkas N, Tzanov T, Beddow J, Joyce E, Mason TJ, Blanes M, Mollá K, Patlolla A, Frenkel AI, Gedanken A (2013) Chitosan and chitosan-ZnO-based complex nanoparticles: formation, characterization, and antibacterial activity. J Mater Chem B 1(14):1968–1976
67. Regös J, Hitz HR (1974) Investigations on the mode of action of triclosan, a broad spectrum antimicrobial agent. Zentralbl Bakteriol Orig A 226(3):390–401
68. Orhan M, Kut D, Gunesoglu C (2009) Improving the antibacterial activity of cotton fabrics finished with triclosan by the use of 1,2,3,4-butanetetracarboxylic acid and citric acid. J Appl Polym Sci 111(3):1344–1352
69. Orhan M, Kut D, Gunesoglu C (2007) Use of triclosan as antibacterial agent in textiles. Indian J Fibre Text Res 32(1):114–118

70. Ranganath AS, Sarkar AK (2014) Evaluation of durability to laundering of triclosan and chitosan on a textile substrate. J Text 2014:1–5
71. Kari C, Nagy Z, Kovacs P, Hernadi F (2009) Mechanism of the growth inhibitory effect of cysteine on *Escherichia coli*. J Gen Microbiol 68(3):349–356
72. Caldeira E, Piskin E, Granadeiro L, Silva F, Gouveia IC (2013) Biofunctionalization of cellulosic fibres with l-cysteine: assessment of antibacterial properties and mechanism of action against *Staphylococcus aureus* and *Klebsiella pneumoniae*. J Biotechnol 168(4):426–435
73. Xu Q, Duan P, Zhang Y, Fu F, Liu X (2018) Double protect copper nanoparticles loaded on L-cysteine modified cotton fabric with durable antibacterial properties. Fibers Polym 19(11):2324–2334
74. Perni S, Hakala V, Prokopovich P (2013) Biogenic synthesis of antimicrobial silver nanoparticles capped with L-cysteine. Colloids Surf A Physicochem Eng Asp 460:219–224
75. Xu QB, Gu JY, Zhao Y, Ke XT, Liu XD (2017) Antibacterial cotton fabric with enhanced durability prepared using L-cysteine and silver nanoparticles. Fibers Polym 18(11):2204–2211
76. Gouveia IC, Sá D, Henriques M (2012) Functionalization of wool with L-cysteine: process characterization and assessment of antimicrobial activity and cytotoxicity. J Appl Polym Sci 124(2):1352–1358
77. Kelleher SM, Habimana O, Lawler J, Reilly BO', Daniels S, Casey E, Cowley A (2015) Cicada wing surface topography: an investigation into the bactericidal properties of nanostructural features. ACS Appl Mater Interfaces, acsami.5b08309
78. Bazaka K, Jacob MV, Chrzanowski W, Ostrikov K (2015) Anti-bacterial surfaces: natural agents, mechanisms of action, and plasma surface modification. RSC Adv 5(60):48739–48759
79. Kim SS, Park JE, Lee J (2011) Properties and antimicrobial efficacy of cellulose fiber coated with silver nanoparticles and 3-mercaptopropyltrimethoxysilane (3-MPTMS). J Appl Polym Sci 119(4):2261–2267
80. Liston EM, Martinu L, Wertheimer MR (1993) Plasma surface modification of polymers for improved adhesion: a critical review. J Adhes Sci Technol 7(10):1091–1127
81. Xu W, Liu X (2003) Surface modification of polyester fabric by corona discharge irradiation. Eur Polym J 39(1):199–202
82. Vertommen MAME, Nierstrasz VA, van der Veer M, Warmoeskerken MMCG (2005) Enzymatic surface modification of poly(ethylene terephthalate). J Biotechnol 120(4):376–386
83. Son WK, Youk JH, Lee TS, Park WH (2004) Preparation of antimicrobial ultrafine cellulose acetate fibers with silver nanoparticles. Macromol Rapid Commun 25(18):1632–1637
84. Ifuku S, Tsuji M, Morimoto M, Saimoto H, Yano H (2009) Synthesis of silver nanoparticles templated by TEMPO-mediated oxidized bacterial cellulose nanofibers. Biomacromol 10(9):2714–2717
85. Gray JE, Norton PR, Alnouno R, Marolda CL, Valvano MA, Griffiths K (2003) Biological efficacy of electroless-deposited silver on plasma activated polyurethane. Biomaterials 24(16):2759–2765
86. Zhang G, Liu Y, Gao X, Chen Y (2014) Synthesis of silver nanoparticles and antibacterial property of silk fabrics treated by silver nanoparticles. Nanoscale Res Lett 9(1):216
87. Maria LCS, Santos ALC, Oliveira PC, Valle ASS, Barud HS, Messaddeq Y, Ribeiro SJL (2010) Preparation and antibacterial activity of silver nanoparticles impregnated in bacterial cellulose. Polímeros 20(1):72–77
88. Tam KH, Djurišić AB, Chan CMN, Xi YY, Tse CW, Leung YH, Chan WK, Leung FCC, Au DWT (2008) Antibacterial activity of ZnO nanorods prepared by a hydrothermal method. Thin Solid Films 516(18):6167–6174
89. Yazdanshenas ME, Shateri-Khalilabad M (2012) In situ synthesis of silver nanoparticles on alkali-treated cotton fabrics. J Ind Text 42(4):459–474
90. Duan Y, Jia J, Wang S, Yan W, Jin L, Wang Z (2007) Preparation of antimicrobial poly(ε-caprolactone) electrospun nanofibers containing silver-loaded zirconium phosphate nanoparticles. J Appl Polym Sci 106(2):1208–1214
91. Rangari VK, Mohammad GM, Jeelani S, Hundley A, Vig K, Singh SR, Pillai S (2010) Synthesis of Ag/CNT hybrid nanoparticles and fabrication of their nylon-6 polymer nanocomposite fibers for antimicrobial applications. Nanotechnology 21(9):095102

92. Ji JH, Jung JH, Kim SS, Yoon J-U, Park JD, Choi BS, Chung YH, Kwon IH, Jeong J, Han BS, Shin JH, Sung JH, Song KS, Yu IJ (2007) Twenty-eight-day inhalation toxicity study of silver nanoparticles in Sprague-Dawley rats. Inhal Toxicol 19(10):857–871
93. Rai MK, Deshmukh SD, Ingle AP, Gade AK (2012) Silver nanoparticles: the powerful nanoweapon against multidrug-resistant bacteria. J Appl Microbiol 112(5):841–852
94. Shateri Khalil-Abad M, Yazdanshenas ME, Nateghi MR (2009) Effect of cationization on adsorption of silver nanoparticles on cotton surfaces and its antibacterial activity. Cellulose 16(6):1147–1157
95. Yeo SY, Jeong SH (2003) Preparation and characterization of polypropylene/silver nanocomposite fibers. Polym Int 52(7):1053–1057
96. Dubas ST, Kumlangdudsana P, Potiyaraj P (2006) Layer-by-layer deposition of antimicrobial silver nanoparticles on textile fibers. Colloids Surf A Physicochem Eng Asp 289(1–3):105–109
97. Kelly FM, Johnston JH (2011) Colored and functional silver nanoparticle-wool fiber composites. ACS Appl Mater Interfaces 3(4):1083–1092
98. Ali SW, Rajendran S, Joshi M (2011) Synthesis and characterization of chitosan and silver loaded chitosan nanoparticles for bioactive polyester. Carbohydr Polym 83(2):438–446
99. Perelshtein I, Applerot G, Perkas N, Grinblat J, Gedanken A (2012) A one-step process for the antimicrobial finishing of textiles with crystalline TiO_2 nanoparticles. Chem Eur J 18(15):4575–4582
100. Montazer M, Seifollahzadeh S (2011) Enhanced self-cleaning, antibacterial and UV protection properties of nano TiO_2 treated textile through enzymatic pretreatment. Photochem Photobiol 87(4):877–883
101. Sivakumar PM, Balaji S, Prabhawathi V, Neelakandan R, Manoharan PT, Doble M (2010) Effective antibacterial adhesive coating on cotton fabric using ZnO nanorods and chalcone. Carbohydr Polym 79(3):717–723
102. Appierot G, Lipovsky A, Dror R, Perkas N, Nitzan Y, Lubart R, Gedanken A (2009) Enhanced antibacterial activity of nanocrystalline ZnO due to increased ROS-mediated cell injury. Adv Funct Mater 19(6):842–852
103. Braydich-Stolle L (2005) In vitro cytotoxicity of nanoparticles in mammalian germline stem cells. Toxicol Sci 88(2):412–419
104. El-Nahhal IM, Zourab SM, Kodeh FS, Semane M, Genois I, Babonneau F (2012) Nanostructured copper oxide-cotton fibers: synthesis, characterization and applications. Int Nano Lett 2(1):14
105. Castro C, Sanjines R, Pulgarin C, Osorio P, Giraldo SA, Kiwi J (2010) Structure–reactivity relations for DC-magnetron sputtered Cu-layers during *E. coli* inactivation in the dark and under light. J Photochem Photobiol A Chem 216(2–3):295–302
106. Mary G, Bajpai SK, Chand N (2009) Copper (II) ions and copper nanoparticles-loaded chemically modified cotton cellulose fibers with fair antibacterial properties. J Appl Polym Sci 113(2):757–766
107. Sambale F, Wagner S, Stahl F, Khaydarov RR, Scheper T, Bahnemann D (2015) Investigations of the toxic effect of silver nanoparticles on mammalian cell lines. J Nanomater 2015:1–9
108. Shalumon KT, Anulekha KH, Nair SV, Nair SV, Chennazhi KP, Jayakumar R (2011) Sodium alginate/poly(vinyl alcohol)/nano ZnO composite nanofibers for antibacterial wound dressings. Int J Biol Macromol 49(3):247–254
109. Dastjerdi R, Mojtahedi MRM, Shoshtari AM, Khosroshahi A (2010) Investigating the production and properties of Ag/TiO_2/PP antibacterial nanocomposite filament yarns. J Text Inst 101(3):204–213
110. Daoud WA, Xin JH, Zhang Y-H (2005) Surface functionalization of cellulose fibers with titanium dioxide nanoparticles and their combined bactericidal activities. Surf Sci 599(1–3):69–75
111. Kiwi J, Rtimi S, Pulgarin C (2013) Cu, Cu/TiO_2 thin films sputtered by up to date methods on non-thermal thin resistant substrates leading to bacterial inactivation. Microb Pathog Strateg Combat Them Sci Technol Educ 74–82

112. Hebeish AA, Abdelhady MM, Youssef AM (2013) TiO_2 nanowire and TiO_2 nanowire doped Ag-PVP nanocomposite for antimicrobial and self-cleaning cotton textile. Carbohydr Polym 91(2):549–559
113. Shi L, Zhao Y, Zhang X, Su H, Tan T (2008) Antibacterial and anti-mildew behavior of chitosan/nano-TiO_2 composite emulsion. Korean J Chem Eng 25(6):1434–1438
114. Reddy KM, Feris K, Bell J, Wingett DG, Hanley C, Punnoose A (2007) Selective toxicity of zinc oxide nanoparticles to prokaryotic and eukaryotic systems. Appl Phys Lett 90(21):2139021–2139023
115. Raghupathi KR, Koodali RT, Manna AC (2011) Size-dependent bacterial growth inhibition and mechanism of antibacterial activity of zinc oxide nanoparticles. Langmuir 27(7):4020–4028
116. Zhang L, Jiang Y, Ding Y, Daskalakis N, Jeuken L, Povey M, O'Neill AJ, York DW (2010) Mechanistic investigation into antibacterial behaviour of suspensions of ZnO nanoparticles against *E. coli*. J Nanoparticle Res 12(5):1625–1636
117. Ashraf M, Dumont F, Campagne C, Champagne P, Perwuelz A, Leriche A, Chihib N-E (2014) Development of antibacterial polyester fabric by growth of ZnO nanorods. J Eng Fiber Fabr 9(1):15–22
118. Hatamie A, Khan A, Golabi M, Turner APF, Beni V, Mak WC, Sadollahkhani A, Alnoor H, Zargar B, Bano S, Nur O, Willander M (2015) Zinc oxide nanostructure-modified textile and its application to biosensing, photocatalysis, and as antibacterial material. Langmuir 31(39):10913–10921
119. Agrawal DP (2002) Medicinal properties of neem: new findings, pp 1–5
120. Mao LMJW (2008) Durable freshness through antimicrobial finishes. Ext Mag 30(4):13–16
121. AbdElhady MM (2012) Preparation and characterization of chitosan/zinc oxide nanoparticles for imparting antimicrobial and UV protection to cotton fabric. Int J Carbohydr Chem 2012:1–6
122. Shafei AE, Abou-Okeil A (2011) ZnO/carboxymethyl chitosan bionano-composite to impart antibacterial and UV protection for cotton fabric. Carbohydr Polym 83(2):920–925
123. Petkova P, Francesko A, Fernandes MM, Mendoza E, Perelshtein I, Gedanken A, Tzanov T (2014) Sonochemical coating of textiles with hybrid ZnO/Chitosan antimicrobial nanoparticles. ACS Appl Mater Interfaces 6(2):1164–1172
124. Buşilă M, Muşat V, Textor T, Mahltig B (2015) Synthesis and characterization of antimicrobial textile finishing based on Ag:ZnO nanoparticles/Chitosan biocomposites. RSC Adv 5(28):21562–21571
125. Alviano D, Alviano C (2009) Plant extracts: search for new alternatives to treat microbial diseases. Curr Pharm Biotechnol 10(1):106–121
126. Cai Y, Luo Q, Sun M, Corke H (2004) Antioxidant activity and phenolic compounds of 112 traditional Chinese medicinal plants associated with anticancer. Life Sci 74(17):2157–2184
127. Handique PJ (2013) Antibacterial properties of leaf extracts of *Strobilanthes cusia* (Nees) Kuntze, a rare ethno-medicinal plant of Manipur, India. Int J PharmTech Res 5(3):1281–1285
128. Shelton RM (1991) Aloe vera: its chemical and therapeutic properties. Int J Dermatol 30(10):679–683
129. Day MJ (2008) Immunomodulatory therapy. In: Small animal clinical pharmacology. W.B. Saunders, pp 270–286
130. Ibrahim W, Sarwar Z, Abid S, Munir U, Azeem A (2017) Aloe vera leaf gel extract for antibacterial and softness properties of cotton. J Text Sci Eng 07(03):1–6
131. Ammayappan L, Jeyakodi Moses J (2009) Study of antimicrobial activity of aloe vera, Chitosan, and curcumin on cotton, wool, and rabbit hair. Fibers Polym 10(2):161–166
132. Vastrad JV, Byadgi SA (2018) Eco-friendly antimicrobial finishing of cotton fabric using plant extracts. Int J Curr Microbiol Appl Sci 7(2):284–292
133. Ayyoob M, Khurshid MF, Asad M, Shah SNH (2015) Assessment of eco-friendly natural antimicrobial textile finish extracted from aloe vera and Neem plants. Fibres Text East Eur 23(6):120–123

134. Hein NT, Hnin SS, Htay DH (2008) A study on the effect of antimicrobial agent from aloe vera gel on bleached cotton fabric. Certif J 4(2):7–11
135. Mitchell MJ, Smith SL, Johnson S, Morgan ED (2002) Effects of the Neem tree compounds azadirachtin, salannin, nimbin, and 6-desacetylnimbin on ecdysone 20-monooxygenase activity. Arch Insect Biochem Physiol 35(12):199–209
136. Joshi M, Ali SW, Purwar R, Rajendran S (2009) Ecofriendly antimicrobial finishing of textiles using bioactive agents based on natural products. Indian J Fibre Text Res 34(3):295–304
137. Ganesan P, Vardhini KJ (2015) Herbal treated microbial resistant fabrics for healthcare textiles. Indian J Nat Prod Resour 6(3):227–230
138. Purwar R, Mishra P, Joshi M (2008) Antibacterial finishing of cotton textiles using Neem extract. AATCC Rev 8(2):36–43
139. Rajendran R, Radhai R, Balakumar C, Ahamed HAM, Vigneswaran C, Vaideki K (2018) Synthesis and characterization of neem chitosan nanocomposites for development of antimicrobial cotton textiles. J Eng Fiber Fabr 7(1):155892501200700
140. Sayed U (2017) Application of essential oils for finishing of textile substrates. J Text Eng Fash Technol 1(2):42–47
141. Khurshid MF, Ayyoob M, Asad M (2015) Assessment of eco-friendly natural antimicrobial textile finish extracted from aloe vera and neem plants. Fibres Text East Europe 6(114):120–123
142. Patel BH (2009) Dyeing and antimicrobial finishing of polyurethane fibre with neem leaves extract. Man-Made Text India 52(4):112–116
143. Inprasit T, Motina K, Pisitsak P, Chitichotpanya P (2018) Dyeability and antibacterial finishing of hemp fabric using natural bioactive neem extract. Fibers Polym 19(10):2121–2126
144. Patel MH, Pratibha D (2014) Grafting of medical textile using neem leaf extract for production of antimicrobial textile. Res J Recent Sci 3(IVC-2014):24–29
145. Rajput A, Ramachandran M, Gotmare VD, Raichurkar PP (2017) Recent bioactive materials for development of eco-friendly dippers: an overview. J Pharm Sci Res 9(10):1844–1848
146. Thangamani K, Periasamy R (2017) Study on antimicrobial activity of cotton, bamboo, and soybean fabrics with herbal finishing. Int Res J Pharm 8(5):115–119
147. Ravindra KB, Murugesh Babu K (2016) Study of antimicrobial properties of fabrics treated with *Ocimum sanctum* L. (tulsi) extract as a natural active agent. J Nat Fibers 13(5):619–627
148. El-Shafei A, El-Bisi MK, Zaghloul S, Refai R (2017) Herbal textile finishes—natural antibacterial finishes for cotton fabric. Egypt J Chem 60(2):161–180
149. Prasad S, Aggarwal BB (2011) Turmeric, the golden spice: from traditional medicine to modern medicine. CRC Press/Taylor & Francis
150. Nasri H, Sahinfard N, Rafieian M, Rafieian S, Shirzad M, Rafieian-Kopaei M (2014) Turmeric: A spice with multifunctional medicinal properties. J HerbMed Pharmacol J 3(1):5–8
151. Mirjalili M, Karimi L (2013) Antibacterial dyeing of polyamide using turmeric as a natural dye. Autex Res J 13(2):51–56
152. Reddy N, Han S, Zhao Y, Yang Y (2013) Antimicrobial activity of cotton fabrics treated with curcumin. J Appl Polym Sci 127(4):2698–2702
153. Ghoreishian SM, Maleknia L, Mirzapour H, Norouzi M (2013) Antibacterial properties and color fastness of silk fabric dyed with turmeric extract. Fibers Polym 14(2):201–207
154. Al Sarhan TM, Salem AA (2018) Turmeric dyeing and chitosan/titanium dioxide nanoparticle colloid finishing of cotton fabric. Indian J Fibre Text Res (IJFTR)
155. Gotmare VD, Kole SS, Athawale RB (2018) Sustainable approach for development of antimicrobial textile material using nanoemulsion for wound care applications. Fash Text 5(1):25
156. Sharma Y, Dua D, Srivastava N (2016) Antibacterial activity, phytochemical screening and antioxidant activity of stem of nicotiana tabacum. Int J Pharm Sci Res 7(3):1156–1167
157. Ru QM, Wang LJ, Li WM, Wang JL, Ding YT (2012) In vitro antioxidant properties of flavonoids and polysaccharides extract from tobacco (*Nicotiana tabacum* L.) leaves. Molecules 17(9):11281–11291

158. Duangsri P, Juntarapun K, Satirapipathkul C (2012) The tobacco leaf extract and antibacterial activity in textile. In: RMUTP international conference: textiles & fashion, pp 3–8
159. Carson CF, Hammer KA, Riley TV (2006) *Melaleuca alternifolia* (tea tree) oil: a review of antimicrobial and other medicinal properties. Clin Microbiol Rev 19(1):50–62
160. Kunicka-Styczyńska A, Sikora M, Kalemba D (2009) Antimicrobial activity of lavender, tea tree and lemon oils in cosmetic preservative systems. J Appl Microbiol 107(6):1903–1911
161. Pohlmann M, Paese K, Frank LA, Guterres SS (2018) Production, characterization and application of nanotechnology-based vegetable multi-component theospheres in nonwovens: a women's intimate hygiene approach. Text Res J 88(20):2292–2302
162. Cerempei A, Guguianu E, Muresan EI, Horhogea C, Rîmbu C, Borhan O (2015) Antimicrobial controlled release systems for the knitted cotton fabrics based on natural substances. Fibers Polym 16(8):1688–1695

Advances of Textiles in Tissue Engineering Scaffolds

Pallavi Madiwale, Girendra Pal Singh, Santosh Biranje and Ravindra Adivarekar

Abstract Tissue engineering is the union of engineering and life science principles in the development of biological substitutes for restoration, maintenance or improvement of tissue function or a whole organ. Scaffold provides a location for cells to attach, to proliferate in three dimensions, distinguish and secrete an extra-cellular matrix, ultimately leading to tissue formation. Textile technologies have recently attracted great attention as potential bio-fabrication tools for engineering tissue constructs. As varied the field of tissue engineering is so is the variation required for the preparation of these scaffolds. In this chapter, the use of current textile technologies, advanced designing of textile structures are discussed to attain the required properties that are demanded by different tissue engineering applications. It covers advances in fibre construction, surface properties and materials which are further used in synergy for development of scaffold and tissue engineering and their properties. It covers in depth the use of the primary textile fibres like wool, silk, cellulose and polyester in this emerging field of tissue scaffolds. It also emphasizes on their production for commercial usage and their future challenges and further opportunities.

Keywords Medical textiles · Tissue engineering · Scaffolds · Bio-fabrication · Textile fibres

1 Textiles in Medical Field

The world of technology has become boundary-less and the merger of various disciplines of technology has given astounding results for benefit of human kind. One such union is of medical field and textiles. Textiles are used since stone-age in the form of wound dressings. Surgical papyrus has the first recorded use of fibres as a medicine as long back as 4000 years.

P. Madiwale · G. P. Singh · S. Biranje · R. Adivarekar (✉)
Department of Fibres and Textile Processing Technology, Institute of Chemical Technology, Mumbai 400019, India
e-mail: rv.adivarekar@ictmumbai.edu.in

M. Shahid and R. Adivarekar (eds.), *Advances in Functional Finishing of Textiles*, Textile Science and Clothing Technology, https://doi.org/10.1007/978-981-15-3669-4_7

Textile materials like leather strips, horse hair, cotton fibre, animal sinew and tree bark have been used since 2500 years as medicine as reported in "Susanta Sambita". Variety of medical applications of textile has been ventured with the time [1].

Textile has a wide range of applications in medicine ranging from non-implantable materials like wound care materials, prosthetic socks, bandages, pressure garments adhesive tapes, orthopaedic belts, medical gowns, sanitary pads and eye pads. The market medical textiles in developed countries due to age factor and in developing countries are due to socio-economic changes and technical advancement in implantable materials, tissue engineering and spinal implants are contributing factors for exponentially increasing growth [2]. The advancement and sophistication in the usage of textiles has only evolved for the betterment of the medical world. The presence of very few market players forces the global medical textile market to be consolidated in nature. The global completion in the medical textile industry is very high with Germany ranking first of the global export followed by India, China and U.S. The global demand of medical textile is increasing with the combination of increased demand in developed and developing economies across the globe. However production of medical textile products is geographically concentrated while very few market players are driving the market. Skilled labour, lower manufacturing cost, availability of raw material are the driving forces which can create opportunity in new markets with increase in competition of existing market as well [3].

2 Textile Forms Used in Medical Field

Crucial properties of textile material such as tensile strength, flexibility, air and moisture permeability and wicking render a textile material suitable for the medical application. The end application of the material decides the material to be used to produce medical textile product. However, all the textile fibres cannot be used as medical textile products. Significant desirable properties required in the fibres to be used in medical textiles are non-toxicity, sterilizability, biocompatibility, biodegradability, absorbing ability, softness and it should be additive and contaminants free [4, 5]. The field of medical textiles is wide spread [6]. Medical technology is witnessing developments in terms of the use of fibres. Many textiles are been utilised having properties such as high absorbability or, non-absorbable, biodegradable, or non-biodegradable, or resorbable in the body. The medicinal products used are in various forms like implantable and non-implantable materials, extracorporeal devices, health care and hygiene. The use of textile and textile scaffolds in surgery and medicine has made the medical field inseparable from textiles. The global market of medical fabric can be divided into non-woven, woven, knitted and others (braided) based on manufacturing technique. The non-woven fabrics dominate the medical textile field contributing 64.29% of global market in 2018 in terms of volume. It is projects to grow at 5% CAGR, fastest among all, over the forecast period. The non-woven fabric is having superior properties in terms of disposability, less contamination, cost effectiveness and better performance. Woven fabric has the second largest contribution in total volume of medical textiles i.e. 15.38% globally in 2018 and expected to reach 491.8 k tons by 2015. The typical applications of woven fabrics are hospital and surgical clothing, wound contact layer, bandages and artificial tendons. The manufacturing process of woven fabric is economical and requires less trained labour as compare to other fabric manufacturing processes thus expecting growth in demand. The knitted fabric was valued USD 2130.1 millions in 2018 in terms of revenue, expecting good rate of growth over the forecast period. Application of knitted fabric in medical fields are in implantable devices, surgical mesh, hernia repair prolapsed devices, surgery meshes in reconstructive or cosmetic surgery [2].

3 Tissue Engineering Scaffolds and Textiles

The advancement in this technological world has enabled us to engineer the tissue, cartilages, bones, nerves, skin which are damaged or not reusable. These damages can be recovered by replacing them with artificially engineered tissues. The engineering of these tissues is a compound process. The process requires the contribution of engineering field along with biology and medicine. The textile discipline plays very important role in the tissue engineering.

Tissue engineering can be explained as the generation of three dimensional (3D) artificial tissues. This 3D material help in the regeneration of body tissue and the development of cell based substitute in order to restore or reconstruct the damaged or improve tissue functions in human body. The replacement is done today predominantly by organs or tissues of donating person. The donors available for the required organs are very limited. Also the response of the immune system of the patient to the donor's organ is often very critical and the chance of rejection of the donated organ or tissue by the immune system of the patient is predominant. For these reasons, the rate of successful implantation of the organ or tissue is very low. The engineering of the tissues has made it possible to overcome immune system response by using the cells grown from the host or patient tissue. Also the shortcoming of availability of donor is addressed by engineering the required tissue.

Engineering of tissues consist of three sections- regenerating the tissues; repairing the damaged tissues by stimulating the tissue at a cell or molecular level and; replacement of the tissue by a biological substitute that is created in the lab and implanted to replace the tissue or organ of interest. This engineering includes an elemental step that is growth of the required tissue in external environment. For the growth of the required tissue there is need of a structure to provide a support for growth. This support structure is called as scaffold. The scaffold is known to be as support for tissue growth or load bearing structures or temporary substrate for cell adherence. Some structures are also used as permanent reinforcement in the tissue where they form permanent part of body. These structures provide assistance and support to cell/tissue in growth till they acquire their original shape and strength.

The materials used in tissue engineering as scaffolds form a crucial part of tissue engineering. The materials to be used depend on the need of the scaffold and the part of the body in which the scaffold is to be deployed [7].

Scaffold is a three dimensional material which act as supporting structure for tissue regeneration. An accurate scaffold should have microstructures to enhance cellular attachment, proliferation and differentiation and proper surface chemistry. The scaffold should have mechanical strength and biodegradability without producing any undesirable by products [8].

The scaffold thus is a structure which acts as a skeleton for the growth of seeded cell to attach, proliferate and differentiate. The scaffold is also known as a carrier for cells, growth factors or other bio-molecular signals.

Cells of which the tissue or organ is to be regenerated are expanded in culture and seeded onto a scaffold that will slowly degrade and resorb, as the tissue structures grow in vitro and/or in vivo. It is of utmost importance to develop a scaffold as to imitate exact properties of the desired human tissue and thus make available environment for macroscopic process of tissue formation.

The functions of scaffolds in vitro or in vivo are to allow cell attachment, proliferation and differentiation; deliver and retain cells and growth factors. Scaffolds permit diffusion of cell nutrients and oxygen to the tissue. They provide appropriate mechanical and biological environment for tissue regeneration in appropriate manner [9]. Overall, bio-scaffold can be stated as a structure used to substitute an organ either permanently or temporarily to restore functionality.

4 History

To restore the functionality of damaged organ and thus improve the quality of living, the tissues engineering or scaffold engineering was introduced [10]. In seventeenth century the use of artificial implants with the Romans had started. Romans used legs made out of wood to replace damaged legs/limbs and used it to restore the functionality. Until 1962 the developments in the area of artificial implants were very slow. Charnley replaced the damaged joints using low friction arthoplasty along with polytetraflouroethylene (PTFE). The advancement in artificial implants is very significant in last six decades. Orthopaedic implants designed artificially are done using range of materials like metallic, ceramic or polymers. Due to corrosion resistance stainless steel (surgical grade) was widely used in orthopaedics and dentistry applications. In later development Co–Cr and Ti alloys are used due to their biocompatibility and bio inertness. The metallic implants can be replaced by ceramic but ceramic has its own limitations in its use. The biocompatibility of the material and immune rejection by the host cell are two measure concerns in the development of artificial implants. This forced researchers to engineer and find suitable material which can be used as regenerative medicine in growth of damaged tissues or organs. In 1980 with use of autologuous (use of grafts from same species) skin grafts the research in the field of implantable material got a head start [11]. The field of tissue engineering thereafter

Table 1 Tissue engineered biological substitute and corresponding textile structure

Tissue engineered biological substitute	Scaffold structures
Bladder	Nonwovens [16]
Blood vessel	Woven, knitted, braided, nonwoven
Bone	Nonwovens, foam [17, 18]
Cartilage	Nonwovens [19]
Dental	Foam (porous membrane), nonwoven [20]
Heart valve	Woven, nonwoven [21, 22]
Tendon	Woven, nonwoven [23]
Ligament	Yarn, braided, nonwoven [24]
Liver	Foam, nonwoven, 3D printed [25, 26]
Nerve	Foam, nonwoven [27]
Skin	Foam, woven [28]

has seen rapid developments from the use of synthetic materials to naturally derived materials that include use of autografts (patient's own tissue), allografts (human donor) or xenograft (other animal tissue) for repair or regeneration of tissues. However, in the present scenario use of scaffolds made from degradable bio-materials of synthetic or natural origin has become the most commonly used approach for the researchers. This approach represents the efforts of researchers in the area of tissue engineering and development of various types of biomaterials using different types of fabrication technologies [12, 13]. Because of inherent properties of textile materials such as porosity, cellular matrix these materials are suitable for application in the field of scaffolds and tissue engineering. Scaffolds can be categorise in three types including foams, 3D printed substrate and textile structures. Textile structures are used in porous scaffolds in tissue engineering [14, 15] (Table 1).

5 Requirements of Tissue Scaffolds Vis-à-Vis Properties of Textile Materials

Porous, biocompatible and/or biodegradable structure is characteristically represent scaffold that is further seeded by cells and cultured in incubating atmosphere for the growth of extracellular matrix (ECM) resembling the native tissue.

An ideal scaffold should have the following characteristics:

1. A comprehensive grid of interconnecting pores for smooth migration and multiplication of cell throughout the structure with complete attachment;

Table 2 Requirements of scaffold materials

Biological requirements	
Biocompatibility	Scaffold must be non-toxic and allow cell attachment, proliferation and differentiation
Biodegradability	Scaffold material must degrade into non-toxic biomaterials
Controlled degradation rate	The degradation rate of scaffold must be adjustable in order to match the rate of tissue regeneration
Porosity	Appropriate porosity, micro and macro structure of the pores and shape to allow tissue in-growth and vascularisation. Scaffolds must be designed to maximise porosity while maintaining mechanical properties
Mechanical and physical requirements	
Strength and stiffness	Sufficient strength and stiffness to withstand stresses in the host tissue environment. Mechanical properties of a scaffold must initially match with the properties of the target tissue to provide structural stability to the injured site
Surface finish	Adequate surface finish guaranteeing that a good biomechanical coupling is achieved between the scaffold and the tissue
Sterilisation	Easily sterilised either by exposure to high temperatures or by immersing in a sterilisation agent remaining unaffected by either of these processes. The sterilisation process must not alter the material's chemical composition, as it may affect bioactivity, biocompatibility or degradation properties

2. Pathways for easy transport of essential cell nutrients, oxygen throughout the structure and removal of waste formed;
3. Biocompatibility with a high affinity for cells to attach and proliferate;
4. Exact desired structural dimensions that are matching with requirements and
5. Rate of degradation in line with and desired mechanical strength.

The specifications required of the scaffold materials in terms of its properties is summarized in Table 2.

Properties required in a temporary tissue scaffold are concurrent with the properties of textile materials. Large surface area to volume ratio which is an essential property of an textile fibres provides supports cell attachment in the textile structures used as scaffolds. This large surface area also provide rapid diffusion of nutrient responsible for cell growth and survival. Cotton and silk fibres known as king of fibres and queen of fibres respectively, are both biodegradable in nature. These textile fibres have huge scope of application in the scaffold and tissue engineering. Thus textile can contribute significantly in the field of the scaffold and tissue engineering [29].

6 Conventional Textiles Used for Preparation of Scaffolds

6.1 *Silk Based Scaffolds*

Silk fibres are known to the textile industry as the naturally elegant, lustrous and also strong fibres which have everlasting demand in the textile industry. Hence silk is also known to the industry as "Queen of fibres". The silk consist of two proteins sericin which is the waxy outer covering the inner fibroin protein. Since the last two decades the silk protein has attracted many bio-material scientists. The fibroin protein is explored to a larger extent than the sericin though the use of sericin as biomaterial for tissue engineering is also an area of interest among the researchers [30].

6.1.1 Characteristics of Silk

As a fibre, silk has many distinct and useful properties which sustained the use of silk in textile industry. The use of silk as a bio-material attracted the researchers due to some of these various properties. The important characteristics of silk fibroin that has made it popular among the biomaterial scientists can be summarised as follows:

- The most favourable property of the silk protein is its ability to get in aqueous mediums for the film formation or other forms of material and further processing it into insoluble form without much difficulty. Processing in water base is favourable for loading of sensitive drugs and biosensors. Thus, material based on silk can be prepared using water as solvent under the normal atmospheric condition using neutral pH and without applying any shear force.
- Conformational transition of α-helix and random coil to highly stable β-sheets can be done through water vapour annealing, mechanical stretching and ultrasonic treatments can be done in silk. This enables the prepared silk forms with good resistance to dissolution, thermal and enzymatic degradation. Genetically alterable composition and sequence to moderate specific features, such as molecular weight, crystallinity and solubility can also be performed on silk to increase the resistance.
- Combination of physical properties of silk that rival many high performance fibers. This is mainly due to the balance between the modulus, breaking strength and elongation.
- Considerably slower rates of degradation in vitro and in vivo as compared to the synthetic biomaterials eminent today are useful in biodegradable scaffolds in which slow tissue in-growth is desirable. The bioresorption ability of the silk is also very high without any harmful acidic bi-products as produced by polylactides or polyglycolides [31].
- Enhanced environmental stability of silk fibers in comparison to globular proteins due to the extensive hydrogen bonding and crystallinity are also favourable factors for the use of silk as tissue scaffold.

6.1.2 Various Forms of Silk Protein and Their Applicability in Tissue Scaffolds

Indigenous Fibres

The raw silk obtained in the form of fibres from the silk worm is degummed and can be used in tissue engineering in various forms such as rope, cable, textured yarn and braided fabric. A non-woven structure can be formed using silk cocoon by dissolving them partially to use as a cell supporting template. In this type of structure arrangement of filaments is maintained to impart porosity in nonwoven structure, which is an important requirement of tissue engineering scaffold. A knitted silk structure to reinforce 3-D porous tissue engineering scaffolds can also be used in ligament which is a load bearing tissue.

Regenerated Silk

Concentrated solution of various salts like LiBr, $CaCL_2$/ethanol/water, LiSCN or ionic liquids have been used as solvent to dissolve silk to prepare different form of regenerated silk protein available for engineering tissue for different applications.

Films

Films from silk fibrion can be prepared by casting the silk solution obtained by dissolving the protein. To improve β-sheet crystallinity, different techniques such as controlled drying, water annealing, and alcohol immersion are used and thus to increase the water stability of the regenerated films. Surface properties required for tissue engineering applications can be obtained by using lithography and advanced printing systems.

Electro-Spun and Wet-Spun Fibers

The regenerated silk solution is used for spinning the nano fibres mats using electro-spinning. These mats are utilised for cell seeding or as 3-D constructs in grafting of blood vessels and nerve guides. The regenerated silk fibres having micrometer scale can be produced with higher production rate than nano-fibers by employing spinning using solvent spinning or micro-fluidic solution spinning. Regeneration of fibers enables tuning of fiber morphology and subsequently application focussed properties. Bio-molecules can also be incorporated during spinning [29].

Hydrogels

The sol–gel transitions of aqueous silk fibroin solution in the presence of acids, dehydrating agents, ions, sonication or lyophilisation are used to form silk hydrogel. The major limitation of poor mechanical properties of these hydrogels is tackled by blending it with various other biopolymers like PVA, polyacrylamide, gelatin, chitosan, and collagen [32].

3-D Porous Scaffolds

Porous 3-D sponges are the ideal and most sought-after structures for tissue engineering scaffolds as they closely mimic the in vivo physiological micro-environment. Silk scaffolds are prepared using various techniques like freeze drying, porogen leaching and solid free form fabrication techniques. Freeze drying technique is studied widely for further improving the characteristics of the scaffolds, solvent casting/particle leaching or gas-foaming methods are also researched.

Silk Particles

Silk protein is also used in the form of particles which are obtained subjecting the regenerated silk solution to various drying techniques like freeze drying and grinding, spray drying, jet breaking, self-assembly and freeze-thawing. The silk particles can be obtained directly from silk fibres by milling technique without using any hazardous chemical.

But it has limitation of control over size of the particles. The silk particles are used in drug delivery systems. Milled particles are particularly used for reinforcing scaffolds to improve mechanical properties and cellular outcomes.

6.1.3 Application of Silk Fibroin

The various techniques used for construction of tissue scaffolds using silk as a starting material is described in the above section. Application of these scaffolds prepared using fibroin along with other biomaterials for the various tissue scaffolds is summarized in Table 3.

6.2 Wool Based Scaffolds

Wool is a well-established animal fibre in the textile industry. The protective covering or sheepskins or of other hairy mammals such as goats and camels are the sources of wool fibres. An ancient man in due course cultured to make yarn and fabric from

Table 3 Method of preparation of silk fibroin scaffolds and its application for various tissue regeneration [32–35]

Composition (SF—silk fibroin)	Method	Tissue/organ
SF	Lyophilisation	Neural
SF	Lyophilized, redissolved in Hexa-Fluoro-Iso-propanol (HFIP), dried	Bone
SF	Chemical cross linking/lyophilisation	Bone
SF	Lyophilisation	Cartilage
SF	Dried 24 h	Cartilage
SF	Lyophilisation 24 h (microporous silk sponges)	Ligament and tendon
SF fibres	Braided on polyvinyl chloride rod	Vascular
SF/collagen	Physical cross linking (thin films)	Hepatic tissue
SF/collagen	Chemical cross linking (genipin powder as cross linker)	Bone
SF/hyaluronic acid	Physical cross linking (silk conduits)	Neural
SF/chitosan	Lyophilisation (sheet structure)	Neural
SF/chitosan	Chemical cross linking (genipin powder as cross linker)	Cartilage
SF/alginate	Lyophilisation (stripe-type porous morphology)	Skin
SF/hyaluronic acid	Chemical cross linking (genipin powder as cross linker) (patched geometry)	Cardiac
SF/collagen/heparin	Lyophilisation (stripe-type porous morphology)	Skin
SF/cyclodextrin	Lyophilisation (channel structure)	No application yet
SF/PEO	Electrospun silk	Vascular
SF/PEO	Dried to form thin Fibroin membranes	No application yet
SF/PEO	Electrospun silk	Bone
SF/P(LLA-CL)	Electrospun silk (conduit)	Neural
SF/PLGA	Electrospun silk	Ligament and tendon
SF/cyclic olefin copolymer	Dried	Ocular
SF/PNIPAM	Chemical cross linking (BIS)	No application yet

their fibre covering such as sheepskins. Wool fibres is composed of the animal protein such as keratin with molecular weight ranging from 45 to 60 KDa of the micro-fibrils from the cortical cells to 6–28 kDa of the protein from the matrix [36].

The application of keratin in the field of tissue engineering has been late boosted due to its limited water solubility and the number of methods available for its extraction and processing. It did not gain the momentum as tissue scaffold as that of other biomaterials. However, in last ten years, due to the development in the extraction and application methods, keratin or modified keratin has emerged as one of most valuable biomaterial or scaffold in the field of cell cultivation and tissue engineering [10].

Keratin occupies about 50% by weight of the cortical cells (outer layer of wool). Compared to other proteins, keratin is abundantly available with less extraction cost. The inherent properties of wool keratin such as highly conserved superstructure, intrinsic ability to self-assemble and polymerize into porous fibrous scaffolds are extremely efficient for the reproducible architecture, dimensionality and porosity of keratin-based materials.

In comparison with other bio-materials, keratin biomaterials possess many distinct advantages, including a unique chemistry yielded by their high sulphur content, excellent biocompatibility, propensity for self-assembly, and intrinsic cellular recognition. The wool keratin as bio-material is researched in various forms and described in the following sections for its application as tissue scaffolds [37].

6.2.1 Keratin Films

The structural and biological properties of keratin protein film extracted from wool have been explored for last few years. The major limitations are flexibility and strength of the films. Addition of natural and synthetic polymers to keratin blended systems like chitosan, silk fibroin, poly-(hydroxyl butylate co-hydroxy valerate), polyethylene oxide, polyvinyl alcohol, gelatin are researched for the enhancing the properties of keratin films. Different techniques other than solvent casting such as compression moulding and substrate coating are employed for preparation of keratin and keratin blended films.

6.2.2 Keratin Sponges

The extracted keratin is lyophilized to form porous sponges which can be applied as support for tissue growth. The wool keratin which contains Arginine-glycine-aspartate protein sequence which are favourable for cell adhesion and promote healing thus increase in cell compatibility, attachment and proliferation of cells. The keratin protein is also used by modifying the protein with various chemical treatments. The research is been carried out to functionalise the sponges using iodoacetic acid, 2-bromoethylamine, and iodoacetamide to produce carboxyl-, amino-, and amido-sponges, respectively.

6.2.3 Keratin Fibres by Wet and Electro-spinning

The traditional fibre wet-spinning technique used for manufacturing synthetic fibres has been studied for creating single fibre strand for its application as biomaterials. The low mechanical strength keratin protein is a hindrance for spinning which is overcome by blending with synthetic and natural polymers. Formation of keratin nano-fibres using the electro-spinning technique is also explored by the researchers. The poor mechanical properties of keratin protein have been improved by blending with synthetic or natural polymers. It also improves the processability of keratin for fibre formation. The important feature as noted by the researchers was that the blending of the keratin protein with silk fibroin not only improved the processability of keratin but also increased the macromolecular interaction hence induced the formation of network structures leading to finer nano-fibres with smaller diameters.

6.2.4 Keratin Hydrogel

The self-assembling property of keratin is been used for preparing the hydrogel of keratin protein through regenerated keratin. The keratin is extracted in suitable solvent and subjected to dialysis with distilled water in controlled environment. Keratins have the inclination to polymerize in an aqueous environment to form hydrogels which is induced by addition of metal ion reagent. Also, alone disulfide and hydrogen bonds can formed a hydrogel keratin, without any cross-linking agent.

6.2.5 Application of Keratin Scaffolds

The overall research directed in the use of wool keratin as tissue engineering scaffold can be summarized in the following table.

Composition	Method	Cells/tissue/organ
Keratin/glycerol	Freeze drying	Fibroblasts
Keratin/chitosan/gelatin	Freeze drying	Soft tissue
Keratin	Freeze drying	PC12 cells, HOS cells and murine embryonic fibroblast
Keratin/poly-(L-lactic acid)	Solvent casting	Human bone marrow mesenchymal stem cells
Keratin/chitosan	Lyophilization	Fibroblast cells
Keratin/multiwalled carbon nanotube/Poly(lactic-co-glycolic acid)	Electrospinning	Guided bone regeneration
Keratin/poly-L-lactide	Electrospinning	Osteoblasts

(continued)

(continued)

Composition	Method	Cells/tissue/organ
Keratin/polyhydroxy butyrate cohydroxy valerate	Electrospinning	Fibroblasts
Keratin	Film formation	Ocular surface reconstruction
Keratin/poly(L-lactide co-glycolide)	Solvent casting/salt leaching	Cartilage tissue
Keratin/calcium phosphate	Lyophilization	Osteoblast

6.3 Cellulose Based Scaffolds

Cellulose in textiles is a populist term for referring to the polymer structure of most of the natural fibres like cotton, jute, hemp, flax and regenerated fibres like viscose rayon. These fibres contain up to 60–80% cellulose as a main component. Cellulose from these sources is extracted in micro and nano form and employed for regeneration of tissues. Utilization of wood pulp as raw material found to generate very high load on the ecosystem though method is well established and has proven industrial scale up application. To achieve better results wood cellulose and cotton cellulose hydrolysis promoted on industrial scale with assistance of mineral acids. The focus has been shifted on renewable sources as raw material for production of microcrystalline cellulose due to environmental concerns. Hemicelluloses and lignin are other two important components where cellulose enclosed within mentioned components, in the form of micro fibrils. Food, cosmetics and medical product manufacturing industries utilizes Micro/Nano cellulose on a larger scale compared to that of other industries [38, 39]. In these industries Micro/nano cellulose used more frequently as a suspension stabilizer, water retainer and as a flow characteristics controller in system where final product as a medical tablet and having functionalities such as reinforcing agent.

In medical field nano-cellulose has been explored to various directions. The use of nanocellulose extracted from wood based raw materials is highly prevalent in blood vessel growth, reconstruction of nerve and dental region, skins replacements for burnings and wounds, drugs releasing system, scaffolds for tissue engineering; stent covering and bone reconstruction [40].

Cellulose as a biomaterial has idealistic properties for tissue engineering applications. The cellulose has tunable chemical, physical as well as mechanical properties. The distinct characteristic of cellulose is that, since it is the most abundantly available polymer on earth, easily produced, inert by nature, thus these materials can be low cost platform for tissue engineering. The extracted cellulose also fulfils the important criteria of bio-compatibility, bioactivity, and biomechanical characteristics. Biodegradability is still a debatable issue with regard to cellulosed based medical scaffolds since cellulose is not degradable in human and thus cannot form temporary scaffolds.

Viscose cellulose sponge as a scaffold for cartilage tissue engineering was demonstrated by Pulkkinen et al. [41]. Sponges alone and with its combination with recombinant human type II collagen cross linked inside the material were found to be biocompatible and successfully showed potential cell proliferation could be achievable. Functionalisation of cellulose is also taken up by many researchers for improving the applicability of cellulose in tissue engineering. The primary hydroxyl groups present on the cellulose is the distinguishing factor which makes the functionalisation possible [42].

Oxidation of the hydroxyl group to a carboxylic acid [43] could achieved assisted with TEMPO (2,2,6,6-tetramethylpiperidine-1-oxyl radical). Through grafting of glycidyl trimethylammonium chloride to the surface to introduce a positive charge [44], thus cationization achieved. Hydrolysis assisted with sulphuric acid leads to formation of sulphate half esters on derivatization a range of cellulose esters and ethers are produced [45, 46]. Modifications of cellulose materials have in depth have been reviewed and its major applications for purification of water, rheology modification and drug delivery have been targeted [47, 48]. Recently its potential use in tissue engineering has been a novel attempt.

Cellulose has a crucial advantage that it has been processed into a wide spectrum of material. One of special ability of cellulose nano-crystals is that it could be used to make hydro-gels and could be easily casted as nano fibre from electro spinning technique or regenerated films such that it could be casted as porous 3D structures. There is a its potential application in field of specific tissue culture application by virtue of its different mechanical and physical properties [49, 50].

Cellulose materials to be used as potential tissue scaffolds can be divided into three main categories as modifications onto cellulose when targeted as per recent literature

6.3.1 Physical Modifications—Composites and Blends

Blending cellulose powder or dispersion or solution which another material, often a polymer or inorganic component composite scaffolds can be prepared which results in endurance of additional functionalities to it, such as instance introducing a charge, altering topography, or varying the mechanical properties which resulted in a creation of a family of cellulose based composite material [51–53].

6.3.2 Biochemical Modifications—Grafting of Bio Molecules onto the Surface

Though cellulose based tissues scaffolds have beneficial functional properties but its hydrophilic nature and low non-specific protein binding affinity on account of which mammalian cell lines do not readily absorb onto cellulose surfaces. These severe limitations could be successfully overcome by introducing bio molecules, viz growth factors, matrix ligands, functionalised onto the scaffold surface; FBS either contained in the cell growth media for of initial cell adherence [54].

6.3.3 Chemical Modifications—Introducing New Functional Groups

By virtue of presence of tri-primary alcoholic group in an anhydroglucose unit make cellulose to allow any kind of functionalization as mostly exposed at surfaces i.e. nanocrystals nanofibrils, or sheet, thus enabling novel physic-chemical properties of fabricated scaffolds.

On subjecting oxidation process, nano-cellulose fibrils allows easier water dispersion thus its processing ability enhancement takes place proficiently. The oxidized cellulose along with hydroxyapatite and gelatin in the form of hydrogel has potential application in bone tissue engineering [55]. Acidified sodium periodate assisted oxidation lead to formation of dialdehyde cellulose moieties (DAC) and on blending it with collagen scaffold formation takes place by crosslinking and leading to 3D porous sponge having functionalities such as dielectric behaviour suggesting that fabricated scaffold material ideal for neural tissue engineering and has potential application in regeneration of nervous system [56].

6.4 Polyester Based Scaffolds

As a textile professional, polyester directs the thought process towards the strong, supple fabric which has covered almost 50% of the market of synthetic fibres. However the degradability of this material has raised high alarms for the eco-systems. Polyester is made by condensation polymerisation of ethylene glycol and terepthalic acid. This aromatic polyester fibre does not have biodegrading ability. The fabric sample laid in a composting environment for 3 months does not lose its structure whereas cotton shows about 50% weight loss [57]. The application of this aromatic polyester (containing aromatic rings) is thus limited in bio-medical field. These are used in the form of membranes and meshes (review 2015). Due to its high strength, it has the application in the anterior cruciate ligament as permanent implant. The Lars ligament (Dijon, France) and Leeds-Keiow or Poly-Tape are the two commercial products that are clinically studied by researchers throughout the globe. The health authorities of Canada, Europe and several other countries have approved the Lars ligament which is a second generation and non-absorbable synthetic ligament device. It has limited to USA for certain rage of application [58]. A product specifically designed for ACL reconstruction with stiffness of 200 N/mm that is similar to natural ACL has been developed by collaborative research ventured by University of Leeds and the Keio University [59]. Other than these applications, the aromatic polyester fibre used in textiles has very limited applications in field of tissue engineering since it does not fulfil one of the primary needs of degrading in vivo.

The material scientists took the polyester terminology to a different horizon. The aliphatic acids were introduced for esterification to increase the degradability of the polyesters and thus a broad range of aliphatic polyesters were introduced and quickly absorbed into the medical devices. The most common degradable polymers poly(glycolic acid) (PGA), poly(L-lactic acid) (PLLA) and poly(D-lactic acid) (PDLA) are results of esterification of aliphatic glycolic acid or lactic acid. It has found that most of biobased or biodegradable polyesters are having aliphatic nature. Polylactic acid (PLA), polyglycolic acid (PGA), poly-ε-caprolactone (PCL), polyhydroxybutyrate (PHB), and poly(3-hydroxy valerate) are among some bio based polyesters gained commercial potential on their potential research performance. The most extensively studied biodegradable thermoplastic polyesters. Among these are PHB and PLA have been widely used.

High hydrophilic nature of aliphatic polyester predominantly seen when exposed to moderate to high water uptake on exposure to moist environment and it has been well characterized that these chemical species are having low melting point, glass transition temperature and possess poor hydrolytic stability leading to its poor mechanical properties and stabilities and this is reason why these moieties are blended with more stable polymers or some time biodegradable polymers are copolymerized with aromatic building blocks (aromatic anhydride and acids), i.e. adipic acid, terephthalic acid, and 1,4-butane diol are monomeric entities to synthesize poly(butylene adipate-co-terephthalate) (PBAT). It is well known that both are biodegradable, biocompatible but having relatively high melting point (160–180 °C). It has been observed that its applications have narrow range due to their brittleness and narrow processing window. Thus, in literature those have been blended with different polymeric system have been reported.

The property of biodegradability which is the motivating factor for the directed efforts towards aliphatic polyesters has many facets. The disposal of the degraded products is related to the degrading time. Degradation times have been observed to be in the range of several months to several years depending upon polymeric systems and conditions. Another crucial property is the tensile strength which is related to molecular volume, higher packing density, higher is the strength of the polyester. Molecular weight (MW) of any polymeric system has a prime importance. On alteration of MW it has observed that polyester possess varying mechanical functionalities, i.e. On monitoring MW tensile strength of PLA could found in range of 1–150 MPa. Tacticity also play a major role. Optical activity exhibited by most of the aliphatic polyesters on virtue of an asymmetrical carbon atom in its repetitive unit, i.e., of alteration of L- and D-units leading to obtain isotactic L-PLA or D-PLA and syndiotactic DL-PLA. These optical polymeric isomers have found to possess different mechanical properties. It has been reported that DL-PLA is having lesser tensile strength and Young's modulus of order of half to one third than that of L-PLA. To improve processing and end use application aliphatic polyesters are mostly blended with other resins. To achieve low cost and increase biodegradability polyesters could be blended with different starches. Unidirectional and biodegradable composite materials have been casted by employing biodegradable polyesters as the matrix resins. For reinforcement natural fibres such as kenaf, elephant grass, flax, hemp, jute and bamboo predominantly used.

6.4.1 Degradation Mechanism

The prime modes of degradation of these aliphatic polyesters are non-enzymatic hydrolysis. And further its degradation found to be catalyzed by its degradation products. It has been observed that monomer structure has limited role as far as degradation of polymer is concern, while molecular weight, crystallinity, fibre structure and substituting groups have prime role in degradation of polymer. Very high biomaterial fraternity is observed in case of PGA, PLLA and PDLA as their monomers and degradation products found to be physiological metabolites. Restrictions in the permitted amount have been result of acidic character of these degradation products.

7 Application of Textiles in Specific Organs Engineering

7.1 Textile Scaffolds as Tissue Adhesives

'Adhesion' is defined by the physicist as the force of attraction in molecules in the area of contact between unlike bodies that acts to hold them together and 'bio-adhesion' is simply the adhesive phenomena where at least one of the adherent is biological. The materials are attached to each other by interfacial forces for an extended period of time [1, 2].

Surgical suture (Commonly called Stitches) is a medical device or thread used to hold body tissues together after an injury or surgery. Application generally involves using a needle with an attached length of thread [60]. But surgical suturing itself inflicts trauma to tissues essentially when multiple passes are necessary. The post-operative integrity of the sutures may also be problematic, loosening or breaking required timely removal. It also requires prolong operative time and technical skill for effective suture placement.

Tissue adhesives serve as suitable alternatives to distressing treatment of suturing or stapling. It is more patient friendly, quick and simple technique eliminating the requirement of local anaesthesia. Overall the trauma associated with suturing can be eliminated [61].

7.1.1 Properties Required for Tissue Adhesive Materials

The properties which are a mandatory requirement for tissue adhesives are firm adhesion, tensile strength, biocompatibility, permeable to fluids and metabolites to prevent necrosis. Non-inflammation, high healing rate, lower post-operative wound infection and control bleeding are also crucial for designing an effective tissue adhesive. Crucial factors of adhesion and cohesion accompanied by intimate contact with wound site is favoured as a tissue adhesive [62].

Natural Polymers

The polymeric materials that may consist various substances but prominently feature proteins and polysaccharides are called "Natural bio-adhesives". The functionality of a bio-adhesive can be enhanced by introducing actives such as steroids, anti-inflammatory agents, pH sensitive peptides and small proteins such as insulin and local treatments to alleviate pain [62].

(A) Protein based polymers: collagen, albumin, gelatin
(B) Polysaccharides: alginates, cyclodextrines, agarose, hyaluronic acid, starch, chitosan, dextran cellulose

Synthetic Polymers

(A) Polyesters: Polylactic acid, polycaprolactone, poly-doxanones, polyglycolic acid, polyhydroxyl butyrate
(B) Polyanhydride: Polysebacic acid, polyadipic acid, polyester phthalic acid and various copolymers
(C) Polyamides: Poly-imino carbonates, poly amino acids.
(D) Phosphorous based polymers: Polyphosphates, polyphosphazenes, polyphosphonates [62].

Replacement of suture by tissue adhesive is a new approach which is highly recommended. Research with epoxy resins, acrylics and polyurethanes, to design an adhesive that would allow prompt and strong bonding have not proved useful because of inadequate bonding strength and poor biocompatibility. However, biological tissue adhesives should be used with caution in since they inhibit new tissue formation, cause a foreign body reaction, and may impede fracture healing [63].

7.2 *Textile Scaffolds for Skin Recovery and Replacements*

The purpose of skin tissue engineering is to develop the skin in vitro or in vivo preserving its all constituents' spatial and functions along with its cosmetic properties. Burning is a serious public health issue globally. As per World health organization reports 195,000 deaths occur each year because of fire alone with more deaths from electrical burns, scalds, and other forms of burns. The deaths related to fire ranks among top 15 causes among children and young adults aged 5–29 years. Millions are left with lifelong disabilities and disfigurements resulting stigma and rejection. Thus the magnitude of accidents causing burn is very high thus a very high requirement of artificial skins [64].

The most important features of scaffold for skin regeneration are mechanical stability and elasticity. The use of fibrous structures enables the promotion of angiogenesis (formation of new blood vessels) by mechanical stimulation in vivo [28]. Researchers have employed knitted fabrics structures by warp knitting of PLGA as a reinforcement. This resulted in increased mechanical stability of scaffold and inhibited wound contraction, effectively promoted cell infiltration, neotissue formation and blood vessel ingrowth in animal study model with mechanical strength up to 75% of normal skin. Stark et al. reinforced collagen hydrogels by hyaluronic acid fibers and seeded with skin fibroblasts to form a dermis scaffold. This scaffold showed improved architecture and stability, which provided the basis for skin regeneration and homeostasis [65].

7.3 Textile Scaffolds as Wound Dressings

Break or cut of any tissue can be defined as wound. Skin wound management is one of the earliest medical activities of humans. With advancement in biological and material science more technologies have emerged in past centuries decade for treating various types of wounds [66]. The market size ($28.7 billion in 2013) of wound management products forces the boost new development in this field. An ideal wound dressing material should be (i) oxygen permeable (ii) maintaining moist environment; (iii) absorbing exuding liquids; (iv) biocompatible and non-allergenic; (v) inhibit the growth of micro-organisms; (vi) able to provide proper stimulation and growth factors during different stages of wound healing; (vii) have suitable mechanical properties to maintain conformal contact between the dressing and the wound preventing any potential discomfort. Because of oxygen permeable microstructure which has ability to absorb exudates, textile materials have been widely used as wound care material. The textile materials can be scaled up for the use in wounds with various sizes. The textile system has the ability to alter its mechanical properties as per the generated construct is another advantage allows fabrication of dressing to maintain their contact with wound alongwith flexibility and elasticity. Traditionally cotton, polyester, silk, PGA and polyurethanes are the materials used in textile materials in the form of gauze and bandages in medical field. The conventional wound dressing materials show permeability to air and exudates providing support to the wound during healing. Hemostatic properties can be imparted in textiles textile based wound dressing materials by incorporation of suitable reagents or applying mechanical force to physically close the wound. These conventional textiles material has few limitations such as they are fail to maintain moisture and prevent bacterial growth in wound area. To counter this problem fibres or fabrics can be treated with anti-microbial agents [67]. Bio-textile has been used in dressing for wound management since years which is a promising substrate in the field of advanced wound healing applications. Bio-textile in combinations with advanced biomaterials, drug delivery and fibre based electronic has a promising future for developing smart textiles which can monitor the condition of wound and respond accordingly.

Electro spinning has become one of the most popular processes to produce medical textiles in the form of wound dressings [68]. This is a simple and effective method to produce nano-scale fibrous mats with controlled pore size and structure, from both natural and synthetic origin polymers. This technique has gain much attention because of its versatility, reproducibility, volume-to-surface ratio and submicron range. Recently, functionalizing these electrospun wound dressings with active compounds that accelerate wound healing and tissue regeneration has become the major goal [69]. In moist healing concept, alginate fiber becomes one of the most important fibers in the wound dressing. The incorporation of biological agents into the fiber used for nonwoven wound dressings provide a means for directly introducing such agents to the wound without a separate application and with no additional discomfort to the patient. Many researchers have explored the wound healing ability of the alginate fiber with different modification [70].

7.4 Textile Scaffolds for Bone Recovery and Regeneration

Bone constitutes the framework of human body because it is among the hard tissues. It is one of tissues which have one the most transplantation. The numbers of grafting surgeries related to bone are increasing exponentially. The site of the bone tissue, the size of the defect or damage, mechanical stresses and soft tissue that cover the damaged site are the factors which need to be consider while designing scaffold for bone tissue. The microscopic analysis of bone represents that bone is constituted of a dense shell of cortical bone to support and protection. The interior of bone is made up of porous cancellous bone which optimises weight transfer and minimise the friction at the articulating joints. The mineral phase present in the bone is responsible for the stiffness of the bone. This arrangement of the ordered organisation makes bone enables for the superior mechanical properties to those of its single components. Thus researchers need extensive study to develop a strategy for recovery and regeneration of the bone tissue mimicking this ordered organisation.

Biomaterials for bone regeneration should be biocompatible, biodegradable substances that support cell attachment, spread, proliferate (osteoconductive), and control cell differentiation into osteogenic lineages. Various materials has been used for the preparation of scaffolds recovery and regeneration [71].

Miscellaneous materials which resemble the host environment for regeneration of the damaged bone tissue are being researched and developed. The materials primarily focussed as biomaterials for bone tissue engineering include (i) bio-ceramic materials which show high compression ability, osteo-conductivity, containing bone integrating chemicals constitution of hydroxyapatite and calcium phosphate; (ii) natural polymers like collagens, fibrin, elastin, alginate, hyaluronic acid are used in clinical trials due to the their intrinsic bio-compatibility and negligible adverse immunological reaction; (iii) synthetic polymers and their modification based on monomer constituent, ratio of the co-polymers and functionalisation of side chains are also used for bone tissue engineering and (iv) hydrogels which constitute hydrated polymer chains giving the advantage of cell delivery and growth factors are also used. The biomaterials support the adhesion of cells throughout the structure and all the essential nutrients constructing the ECM matrix [72].

The bone being a highly complex and profoundly designed structure the fabrication of the scaffold skeleton is by advanced techniques like 3D printing, electrospinning or solid free formation (SFF) techniques. Thermally induced phase separation (TIPS), microsphere sintering, solvent casting/particle leaching and scaffold coating are also studied for the manufacture of scaffolds. For a textile professional point of view, the techniques of embroidering and non-woven have gained a momentum in the field of bone tissue engineering showing encouraging results [72].

The technique of embroidery makes the user available many variables so as to construct a bone like structure. These variables consist of stitching length and assembly, density of the stitching as per the model design and thread size empowers the designer to change the scaffold design giving desired porosity, surface area, and mechanical properties. 3D structure can be created by stacking, inter-weaving one ply on another. Strength of the designed scaffold can also be controlled through assembly and stitch mode. Interconnecting porous structure without compromising any of other parameters can be obtained with ease using embroidery technology [73].

Non-woven technology widely used in the manufacturing of medical scaffold and regenerative medicine field, thus the importance of nonwoven technology for bone tissue engineering is very significant. Nonwoven fabrics are highly porous in nature and their low density fibrous structure makes their overall density less than 0.40 g/cm^3. The internal porous structure of nonwoven can be altered in terms of interconnections of pores and pore size distribution during manufacturing. The most of the fibres in the nonwoven structures are arranged in the direction of either in length or width, few very processes are available which can arrange the orientation of fibres through the thickness of nonwoven such as air laid, carded webs, vertically lapped using needle punching and hydro-entanglement techniques. The orientation of fibres altered during manufacturing of nonwoven decides the properties of nonwoven fabric like direction of mechanical properties and transportation of fluid. Different types of nonwoven structures can be produced for the scaffold applications using various manufacturing routes.

There is extensive research has been conducted throughout years in the world in the field of applications of scaffold in bone tissue engineering. However integration of all properties and functions in a single biomaterial system is a challenge for the researchers.

7.5 Textile Scaffolds for Vascular Tissue Recovery and Regeneration

The blood vessels or the vascular system form an indispensible part of the cardiovascular system that is the crucial part of the human body responsible for the transport of the nutrients and oxygen and waste materials throughout the body. A diseased blood vessel can lead to the death of the person. The vascular diseases are mostly treated by using an autologous graft (blood vessel from patient's body). However, about 20–30% of patients do not have adequately sized vein or artery to perform the surgery. The tissue engineered vascular grafts (TEVG) have lower incidences of infectious complications than other grafts. It shows higher withstanding duration towards degradation. The TEVGs also maintain adequate suture strength and is suturable. The vascular scaffolds can also be prepared by the traditional textile methods, including wet, dry, or melt spinning, nonwoven bonding, weaving, knitting or braiding. Most of the times, weaving techniques are used when high mechanical properties are required [74].

The properties for an ideal TEVG is mimicking the extra-cellular matrix environment, mechanical strength to sustain the high vascular fluid movements, cell attachment and optimum pore size for growth of cells. The important characteristic for use of biomaterial as TEVG is resistance to thrombosis, inflammation, and neo-intimal proliferation. TEVG play a very crucial role of bridging, guiding cell-mediated remodeling to reproduce the structure and organization. Elasticity of the vascular grafts is a significant factors while consideration [74].

8 Future Challenges and Scope of Growth for Textiles in Medical Scaffold

The application of textile materials in tissue engineering scaffold is very important because of its amicable properties of the substitution of human tissue and organs. Properties of textile vary as per the structure of fibre or fabric. Where high elasticity is required knitted fabric can be used. This kind of elastic property is required for the implants used for the reinforcement of an organ subjected to the dynamic stress like vascular graft. Nonwoven textile structures resemble the connective tissue in the body along with mazy structure and very good water absorption which are desirable properties for the materials used in scaffolds and tissue engineering. Scaffold structures are prepared from various natural biomaterial or material originated from the nature by using various techniques. Many new innovative biomaterials have been prepared for the application as tissue engineering scaffolds by various scientists through recent years. These materials can be used as textile scaffold. 3D printing is one of the newer techniques which is used in the manufacturing of the scaffolds from the various recently synthesised novel biomaterials. The structure of textile scaffold is the most important factor which depends on the properties required in scaffold.

Thus various textile structures can be synergistically used to meet the requirement of the host tissue.

The current global market for tissue engineering and regeneration is accounted to be around $24.7 billion in 2018. The market is very promising and predictions state that the expected growth rate (CAGR) is predicted up to 34.8%with the market size of $109.9 billion by 2023. This chapter give in detail idea about the advancement of the textile industry in this field and from the market size it can be stated affirmatively that the textile technology will have the highest share in tissue engineering scaffold. The variation required in relation to the specific requirement of the damaged tissue makes the challenge of this field exciting and the research arena to highest magnitude of multi-disciplinary activity.

Though the medicine in our age has advanced to the stage of recovering lost crucial functions of body, the challenge of the medicine is like the vast sky and the understanding all the requirement related to specific process, design and application is yet not complete in itself. Imbibing textile technology is giving advantages in terms of regulatory measures, toxicity and safety profiles of these materials which are in all sense human friendly thus aiding in the development. The performance profiles such as degradation, interaction behaviour of cell materials are areas which need vigorous research. Performance characteristics are pushing research and development in the design of new materials to meet specific performance criteria in tissue engineering [75, 78].

References

1. Goldade VA (2017) Antimicrobial fibers for textile clothing and medicine: current state. Impact factor. Int Sci J Theor Appl Sci
2. Grand View Research, Medical textiles market size, share & trends analysis report by fabric (non-woven, knitted, woven), by Application And Segment Forecasts 2019–2025 (2019), Report ID: 978-1-68038-830-5, Grand View Research, Inc, Apr 2019
3. Mohammed AJ (2018) Comment A call to action: improving women's, children's, and adolescents' health in the Muslim world. Lancet 6736(18):1–2
4. Gupta BS (1998) Medical textile structures: an overview, medical plastics and biomaterials magazine
5. Chinta SK, Veena KV (2013) Impact of textiles in medical field. Int J Latest Trends Eng Technol 2:142–145
6. Gajanand E, Soni LK, Dixit VK (2014) Biodegradable polymers: a smart strategy for today's crucial needs. Crititcal Rev Pharm Sci 3:1
7. Madiwale P, Rachana S, Adivarekar R (July–Aug 2014) Tissue engineering scaffolds—an introduction (Chap. 1). J Textile Assoc 136–137
8. Loh QL, Choong C (2013) Three-dimensional scaffolds for tissue engineering applications: role of porosity and pore size. Tissue Eng Part B Rev 19(6):485–502
9. Porta GD, Reverchon E, Maffulli N (2017) Biomaterials and supercritical fluid technologies: which perspectives to fabricate artificial extracellular matrix? Curr Pharm Design 23(27):3759–3771
10. Madiwale P, Rachana S, Adivarekar R (2014) Tissue engineering scaffold - an introduction (Chap. 1). J Text Assoc 136–137

11. Domres B, Kistler D, Rutczynska J (2007) Intermingled skin grafting: a valid transplantation method at low cost. Ann Burns Fire Disasters 20(3):149–154
12. Pasqualini U, Pasqualini ME (2009) Treatise of implant dentistry: Italian tribute to modern implantology
13. Madiwale P, Rachana S, Adivarekar R (Nov–Dec 2014) Bio-polymers for tissue engineering scaffold (Chap. 3). J Textile Assoc 314–319
14. Madiwale P, Rachana S, Adivarekar R (Nov–Dec 2014) Role of textiles in tissue engineering scaffolds (Chap. 2). J Textile Assoc 203–205
15. Madiwale P, Rachana S, Adivarekar R, (Jan–Feb 2015) Techniques for manufacturing of tissue engineering scaffolds (Chap. 4). J Text Assoc 1:362–367
16. Serrano-Aroca Á, Vera-Donoso C, Moreno-Manzano V (2018) Bioengineering approaches for bladder regeneration. Int J Mol Sci 19(6):1796
17. Liu X, Ma PX (2004) Polymeric scaffolds for bone tissue engineering. Ann Biomed Eng 32(3):477–486
18. Zhou H, Tang Y, Wang Z, Zhang P, Zhu Q (2018) Cotton-like micro- and nanoscale poly(lactic acid) nonwoven fibers fabricated by centrifugal melt-spinning for tissue engineering. RSC Adv 8(10):5166–5179
19. Chen G et al (2003) The use of a novel PLGA fiber/collagen composite web as a scaffold for engineering of articular cartilage tissue with adjustable thickness. J Biomed Mater Res 67A(4):1170–1180
20. Zhang L, Morsi Y, Wang Y, Li Y, Ramakrishna S (2013) Review scaffold design and stem cells for tooth regeneration. Jpn Dent Sci Rev 49(1):14–26
21. Wu S, Duan B, Qin X, Butcher JT (2017) Living nano-micro fibrous woven fabric/hydrogel composite scaffolds for heart valve engineering. Acta Biomater 51:89–100
22. Kluin J et al (2017) In situ heart valve tissue engineering using a bioresorbable elastomeric implant—from material design to 12 months follow-up in sheep. Biomaterials 125:101–117
23. Wu S, Wang Y, Streubel PN, Duan B (2017) Living nanofiber yarn-based woven biotextiles for tendon tissue engineering using cell tri-culture and mechanical stimulation. Acta Biomater 62:102–115
24. Alshomer F, Chaves C, Kalaskar DM (2018) Advances in tendon and ligament tissue engineering: materials perspective. J Mater 2018:1–17
25. Jammalamadaka U, Tappa K (2018) Recent advances in biomaterials for 3D printing and tissue engineering. J Funct Biomater 9(1):22
26. Lewis PL, Shah RN (2016) 3D printing for liver tissue engineering: current approaches and future challenges. Curr Transplant Rep 3(1):100–108
27. Subramanian A, Krishnan U, Sethuraman S (2009) Development of biomaterial scaffold for nerve tissue engineering: biomaterial mediated neural regeneration. J Biomed Sci 16(1):108
28. Madiwale P, Rachana S, Adivarekar R (2015) Textile scaffolds for skin recovery and replacement (Chap. 8). J Text Assoc 76(3)
29. Rouwkema J, Koopman BFJM, Blitterswijk CAV, Dhert WJA, Malda J (2009) Supply of nutrients to cells in engineered tissues. Biotechnol Genet Eng Rev 26(1)163–178
30. Madiwale P, Rachana S, Adivarekar R (March–Apr 2015) Silk based scaffolds (Chap. 5). J Text Assoc 2:437–440
31. Nisal A, Sayyad R, Dhavale P, Khude B, Deshpande R, Mapare V, Shukla S, Venugopalan P (2018) Silk fibroin micro-particle scaffolds with superior compression modulus and slow bioresorption for effective bone regeneration. Sci Rep 8(1)
32. Kundu B, Rajkhowa R, Kundu SC, Wang X (2013) Silk fibroin biomaterials for tissue regenerations. Adv Drug Deliv Rev 65(4):457–470
33. Naskar D, Barua RR, Ghosh AK, Kundu SC (2014) Introduction to silk biomaterials. Woodhead Publishing Limited
34. Vepari C, Kaplan DL (2007) Silk as a biomaterial. Progr Polym Sci 32(8–9):991–1007
35. Lawrence BD (2014) Processing of Bombyx mori silk for biomedical applications. Woodhead Publishing Limited

36. Bragulla HH, Homberger DG (2009) Structure and functions of keratin proteins in simple, stratified, keratinized and cornified epithelia. J Anat 214(4):516–559
37. Madiwale P, Rachana S, Adivarekar R (2015) Wool based scaffolds (Chap. 6). J Text Assoc 76(1)30–33
38. Singh GP, Madiwale PV, Adivarekar RV (2018) Preparation and characterization of microcrystalline cellulose (MCC) from renewable source. Curr Appl Polym Sci 1(2):152–158
39. Singh GP, Madiwale PV, Adivarekar RV (2017) Investigation of properties of cellulosic fibres extracted from Saccharum munja grass and its application potential. Int J Fiber Text Res 7(1):30–37
40. Madiwale P, Rachana S, Adivarekar (2016) Textile scaffolds in dentistry (Chap. 11). J Text Assoc 385–389
41. Pulkkinen H et al (2006) Cellulose sponge as a scaffold for cartilage tissue engineering. 1 Jan. 2006:S29 – S35. Print
42. Novotna K, Havelka P, Sopuch T et al (2013) Cellulose-based materials as scaffolds for tissue engineering. Cellulose 20:2263–2278
43. Isogai H, Saito A, Fukuzumi T (2011) TEMPO-oxidized cellulose nanofibers. Nanoscale 3(1):71–75
44. Courtenay JC et al (2018) Unravelling cationic cellulose nanofibril hydrogel structure: NMR spectroscopy and small angle neutron scattering analyses; Soft Matter 14:255–263
45. Capron I, Cathala B (2013) Surfactant-free high internal phase emulsions stabilized by cellulose nanocrystals, Biomacromolecules 14(2):291–296
46. Braun B, Dorgan JR (2009) Single-step method for the isolation and surface functionalization of cellulosic nanowhiskers, Biomacromolecules 10(2):334–341
47. Eyley S, Thielemans W (2014) Surface modification of cellulose nanocrystals. Nanoscale 6:7764–7779
48. Crawford RJ, Edler KJ, Lindhoud S, Scott L, Unali G (2012) Formation of shear thinning gels from partially oxidised cellulose nanofibrils. Green Chem 2012, 14:300–303
49. Torres FG, Commeaux S, Troncoso OP (2012) Biocompatibility of bacterial cellulose based biomaterials. J Funct Biomater 3(4):864–878
50. Torres-rendon JG et al (2015) Bioactive gyroid scaffolds formed by sacrificial templating of nanocellulose and nanochitin hydrogels as instructive platforms for biomimetic tissue engineering. Adv Materi 27(19):2989–2995
51. Han D, Gouma P (2006) Electrospun bioscaffolds that mimic the topology of extracellular matrix. Nanomedicine 2(1):37–41
52. Chau M et al (2016) Composite hydrogels with tunable anisotropic morphologies and mechanical properties. Chem Materi 2016 28(10):3406–3415
53. Kumar A, Rao KM, Han SS (2017) Tissue engineering applications. Chem Eng J
54. Courtenay J, Sharma J, Scott J (2018) Recent advances in modified cellulose for tissue culture applications. Molecules 23(3):654
55. Park M, Lee D, Shin S, Hyun J (2015) Colloids and surfaces B: biointerfaces effect of negatively charged cellulose nanofibers on the dispersion of hydroxyapatite nanoparticles for scaffolds in bone tissue engineering. Colloids Surf B 130:222–228
56. Bodin A, Ahrenstedt L, Fink H, Brumer H, Risberg B, Gatenholm P (2007) Modification of nanocellulose with a xyloglucan—RGD conjugate enhances adhesion and proliferation of endothelial cells: implications for tissue engineering. Biomacromolecules 8(12), 3697–3704
57. Li L, Frey M, Browning KJ, C. Incorporated, and U. States (2010) Biodegradability study on cotton and polyester fabrics. J Eng Fibers Fabr 5(4)
58. Chen J, Xu J, Wang A (2009) Scaffolds for tendon and ligament repair: review of the efficacy of commercial products. Exp Rev Med Dev 6(1):61–73
59. Matsumoto H, Fujikawa K (1982) Cruciate series for the for the knee anterior ligament reconstruction, vol L, pp 161–166
60. Padsalgikar A (2017) Plastics in medical devices for cardiovascular applications. Plastic Design Library

61. Biranje S, Madiwale P, Shukla R, Adivarekar R (2015) Textile scaffolds as tissue adhesives (Chap. 7). J Text Assoc 76(2)
62. Prajapati TK, Patel KS (2017) Natural and synthetic polymers used in bioadhesive delivery system, Int.J. for Innovative Research in Multidisciplinary Field, 3:9, 76–86
63. Lee EJ et al (2017) Progress in materials science application of materials as medical devices with localized drug delivery capabilities for enhanced wound repair. Prog Mater Sci 89:392–410
64. Tyson AF et al (2013) Survival after burn in a sub-Saharan burn unit: challenges and opportunities. Burns 39(8):1619–1625
65. Pei B, Wang W, Fan Y, Wang X, Watari F, Li X (2017) Fiber-reinforced scaffolds in soft tissue engineering. Regener Biomater 4:257–268
66. Biranje S.S., Madiwale PV, Patankar KC, Chhabra R, Dandekar-Jain P, Adivarekar RV (2019) Hemostasis and anti-necrotic activity of wound-healing dressing containing chitosan nanoparticles. Int J Biol Macromol 121
67. Sutar T, Madiwale P, Shukla R, Adivarekar RV (2015) Textile scaffolds as wound healing bandages (Chap. 9). J Text Assoc 76(4)
68. Biranje S, Madiwale P, Adivarekar RV (2018) Porous electrospun casein/PVA nanofibrous mat for its potential application as wound dressing material. J Porous Mater
69. Santosh B, Madiwale P, Adivarekar RV (2017) Preparation and characterization of chitosan/PVA polymeric film for its potential application as wound dressing material. Indian J Sci Res 14(2):250–256
70. Rajendran S, Anand SC (2006) Contribution of textiles to medical and healthcare products and developing innovative medical devices. Indian J Fibre Text Res 31(1):215–229
71. Yang XB et al (2003) Novel osteoinductive biomimetic scaffolds stimulate human osteoprogenitor activity—implications. Conn Tissue Res 44:312–317
72. Madiwale P, Rachana S, Adivarekar R (2016) Textile scaffolds for bone recovery and replacement (Chap. 10). J Text Assoc 318–321
73. Breier AC (2015) Embroidery technology for hard-tissue scaffolds. In: Biomedical txtiles for othopaedic and surgical applications (pp 23–43)
74. Madiwale P, Rachana S, Adivarekar R (2016) Textile scaffolds for vascular tissue recovery (Chap. 12). J Text Assoc 1:47–51
75. Madiwale P, Shukla R, Adivarekar R (2016) Challenges and further scope for textiles in medical scaffolds (Chap. 13). 123–126
76. Chandrasekharan M, Yong J, Shyan M (2005) Recent advances and current developments in tissue scaffolding. Bio-med Mater Eng 15:159–177
77. Pei B, Wang W, Fan Y, Wang X, Watari F, Li X (2017) Fiber-reinforced scaffolds in soft tissue engineering. Regener Biomater 4(4):257–268
78. Aibibu D, Hild M, Wöltje M, Cherif C (2016) Textile cell-free scaffolds for in situ tissue engineering applications, J Mater Sci Mater Med. 27(3):63

Fabrication of Superhydrophobic Textiles

Munir Ashraf and Shagufta Riaz

Abstract Surface functionalization of natural and synthetic polymeric textile has become the most appealing and contemporary research area to develop multifunctional textiles according to the global demand. Textiles surfaces have been modified for properties like antibacterial, UV protection, antistatic, wrinkle resistant, superhydrophobic, hydrophilic, self-cleaning etc. by application of different inorganic and organic compounds. In this regards rigorous research has been conducted on finding best technique and their application method to mimic the nature so that superhydrophobic textile could be developed according to the demands of the da. In this chapter the fundamentals of hydrophobicity/wettability has been described to give comprehensive knowledge and the applied organic and inorganic material have been described briefly that have been used for fabrication of superhydrophobic textile.

Keywords Nanoroughness · Superhydrophobicity · Lotus effect · Nanostructures · Self-cleaning

1 Introduction

Generally on the basis of wettability surfaces have been categorized as hydrophilic and hydrophobic. Hydrophilic surfaces are water loving in nature, when water droplet falls on, it spreads and absorbed by the surface very quickly instead of rolling off onto the surface. The property of wettability or repellency is mainly dependent upon the surface contact angle. If the striking water droplets make an angle less than 90° then the water droplets would spread over the surface. But, in case of surface angle greater than 90°, the water droplets would not spread and some could rolled off onto the surface showing the hydrophobic characteristics of the material. However, if the angle would be 150° or more than the water would not be absorbed by the surface showing superhydrophobic nature of material. On super hydrophobic surfaces the water droplet completely roll off.

M. Ashraf · S. Riaz (✉)
Functional Textiles Research Group, National Textile University, 37610 Faisalabad, Pakistan
e-mail: shaguftariaz84@gmail.com

M. Shahid and R. Adivarekar (eds.), *Advances in Functional Finishing of Textiles*, Textile Science and Clothing Technology, https://doi.org/10.1007/978-981-15-3669-4_8

Superhydrophobic textiles have grabbed the greater attention of researchers and textile engineers to be used in different fields of life [1–3] New gates were opened for research on superhydrophobicity also known as physical self-cleaning after the proposal of lotus leaves model unveiling that a particular surface topography and low surface energy of material is required for a surface to be superhydrophobic. Thus, micro and nano hierarchical nanoroughness is vital for physical self-cleaning surfaces [4].

Thus recent trends to mimic the nature, have attracted the researchers to develop superhydrophobic textiles with water contact angle greater than 150° and contact angle hysteresis less than 10° using various techniques [5–7]. The loosely bound dirt, dust, and soil could be attached with the rolling droplet thus causing self-cleaning of material. Along with self-cleaning character such superhydrophobic surface are also stain resistant due to less contact of stain with the surface and very high water contact angle. Therefore, such surfaces could be used in areas [8–11] for water-oil separation, anti-fogging, anticorrosion, and oil repellency etc.

Surfaces could be made hydrophobic by just lowering the surface energy of substrates. But, superhydrophobic surfaces need development of hierarchical roughness along with lowering of surface energy [12]. In recent year, development of different manipulation techniques played an important part to fabricate micro and nano surface roughness to create artificial superhydrophobic materials. Surface morphology can be changed by different methods such as lithography, plasma treatment, sol-gel technique, layer by layer deposition, electrospinning, phase separation, template synthesis and many more [4, 13, 14].

In this chapter, the hydrophobic fundamentals, switchable super hydrophobicity and hydrophilicity have been described. Natural material based, organic and inorganic material based superhydrophobic textiles have been discussed in order to give the overview of advancements in hydrophobicity and material used until now. There is still a room for discovering the materials and application methods to make textile superhydrophobic.

2 Water Repellency

The property due to which the water does not absorbed by the natural or synthetic material when come in contact is called water repellency and it could be easily rolled off or bounced back from the solid surface. The repellant character would be determined by how much surface area of fabric is covered and what the angle of impact is. Apart from synthetic materials, the natural materials do not have this property and there is need to coat water repellent agents on the exterior of fabric or sometimes infused in the fibers (Fig. 1).

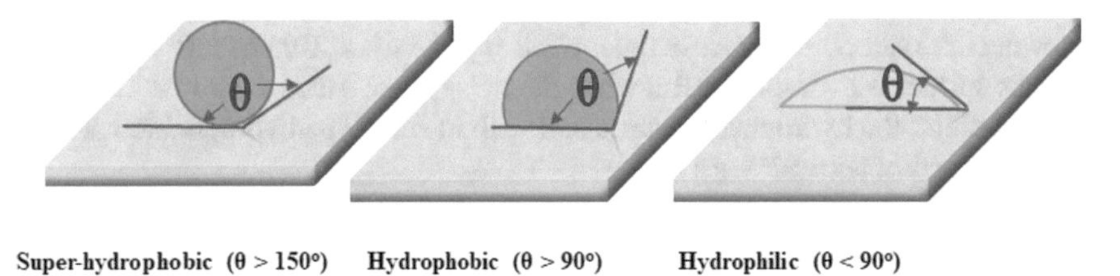

Fig. 1 Wettability of textile substrate as a measure water contact angle. "Reprinted with permission from Ref. [82], Copyright (2015) Elsevier Ltd"

2.1 *Fundamentals of Hydrophobicity*

Different theories were given by different researchers explaining the fundamentals of surface wettability or repellency. Young's modulus explains the surface wettability for flat surfaces by showing the thermodynamic equilibrium of surface free energy with the water droplet at the smooth surface. The other two models of Wenzel and Cassie-Baxter, on the other hand explain the surface wettability of rough surface. The Young's model was considered as the most basic theoretical model describing the surface wettability and the Young's equation was predicted in 1805 [15]. The Young's equation shows a relationship of static water contact angle with interfacial tension between a vapor phase and solid and interfacial tension between liquid and vapor phase for a water droplet that exists at the smooth, homogenous and flat surface as shown in Eq. (1) [15]

$$\gamma^{la}\cos\theta_z + \gamma^{sl} = \gamma^{sa} \tag{1}$$

or alternately; $\text{Cos}\theta_z = \gamma^{sa} - \gamma^{sl} / \gamma^{la}$

Here θ_z is the static contact angle at the solid surface and γ^{sa} is the interfacial tension of solid surface and air, γ^{sl} is interfacial tension of solid and liquid while γ^{la} is interfacial tension between liquid and air as shown in Fig. 2. It means the water contact angle is dependent upon types of surface tensions: interfacial tension of liquid-air, solid-air and liquid-solid. When $\gamma^{sa} - \gamma^{sl} = \gamma^{la}$ the angle θ_z would be 0°

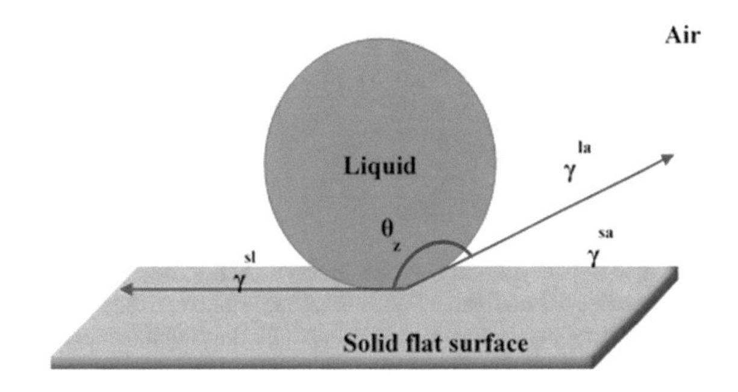

Fig. 2 Schematic of Young's model for static contact angle at infinitely smooth and flat surface in relation to interfacial surface tensions i.e. liquid-solid, solid-air and liquid-air

showing the hydrophilic nature of the surface. If $\gamma^{sa} - \gamma^{sl} > 0$ then the $0° < \theta_z < 90°$ still the surface is hydrophilic. But when the $\gamma^{sa} - \gamma^{sl} < 0$ then the $90° < \theta_z < 180°$ would exhibit the hydrophobic nature and with ideally superhydrophobic nature of the solid surface with $\gamma^{sa} - \gamma^{sl} = -\gamma^{la}$.

This ideal young's model could be perfect only for infinitely smooth and flat surfaces where the influence of surface structure of hydrophobicity is totally neglected and the chemical composition of the surface is only consideration along with some environmental effect. There were some limitations of this model because a textiles treated even with C9 perfluorocarbon to attain lowest possible surface free energy (5–10 dyne/cm) exhibited a maximum contact angle of 120° [16] but, there were some plant leaves that exhibited water contact angle greater than 160° thus, nanoroughness of the surface was also another consideration along with lower surface free energy that should be taken into account [17, 18] Because the water contact angle calculated through Young's equation was different from the actual contact angle of the surface due to the presence of surface roughness therefore, surface roughness was considered for the actual contact angle of the same superhydrophobic surface. When a water droplet falls on a rough surface, it can have two states i.e. composite and noncomposite contact surface as shown in Fig. 3.

Because the Young's model is applicable to flat surface but, every surface could not be infinitely smooth, therefore, roughness factor was included then in 1936 by another researcher Wenzel [19], who tried to overcome the Young's model limitations. He proposed the theory in which it was mentioned that the actual area of the surface increases due to the presence of nano-roughness due to which the total

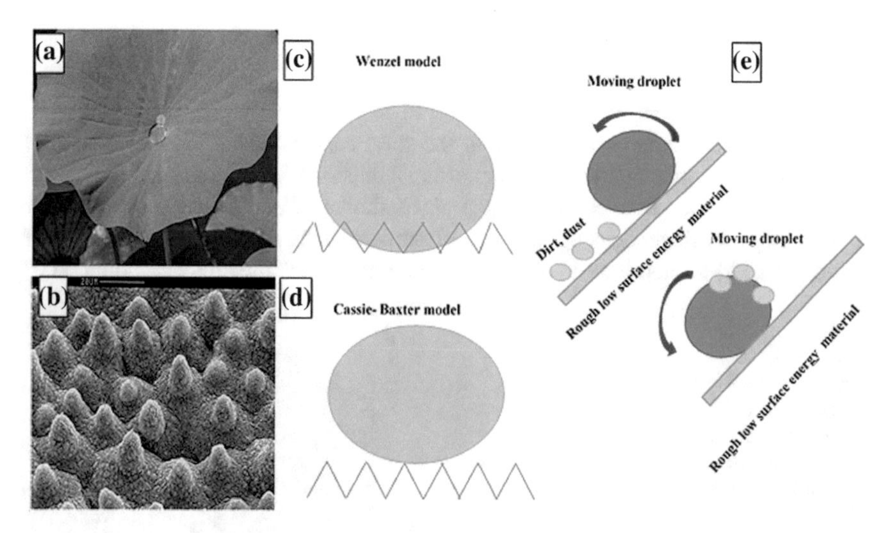

Fig. 3 **a** Superhydrophobic lotus leaf **b** SEM micrograph showing hierarchical nanoroughness and wax needles of lotus leaf. **c** Wet contact angle between the rough surface and water droplet (Wenzel's model). **d** Non-wet contact angle between the rough surface and water drop let (Cassie-Baxter's model). **e** Rolling droplet on material mimicking lotus illustrating physical self-cleaning

solid-liquid are could be higher than the apparent surface contact area and as a result the wettability/repellency of the surface could increase. He proposed that the rough surface can be filled by the water droplet on the solid roughened surface (Fig. 3c) and the surface free energy of droplet could be given by Eq. (2)

$$\delta E = r\left(\gamma^{sl} - \gamma^{sa}\right)dx + \gamma^{la}\cos\theta_w \cdot dx \tag{2}$$

In Eq. (2) the δE represents the change in surface energy of the wetting system and dx represents the small moving distance of the water droplet. In a balanced system, the relationship between the intrinsic water contact angle θ_z and apparent water contact angle θ_w is shown by Eq. (3) because of the smallest the surface free energy (δE = 0)

$$\text{Cos}\theta_w = r\left(\gamma^{sa} - \gamma^{sl}\right)/\gamma^{sl} \tag{3}$$

Here in Eq. (3), θ_w is the angle on solid surface known as Wenzel's apparent water contact angle and r is the roughness factor of the surface i.e. the ratio of actual surface area to the apparent surface area. It was obvious from the Wenzel's model that functions and the effects of soil–air interfacial surface tension and solid–liquid interfacial surface tension on surface free energy were now different and changed by surface roughness directly, leading to the different contact angle on the roughened solid surface than on infinitely flat and smooth solid surface. If there is a possibility of filling the grooves by water droplet, then a non-composite state could be formed between water droplet and the surface grove showing that the actual contact area of solid to liquid is higher than the apparent geometric area. The roughness factor "r" is always greater than 1 (r > 1), if the contact angle is greater than 90° i.e. θ_z > 90° then surface roughness is responsible for the hydrophobicity of material but in case of contact angle lesser that 90° i.e. $\theta_z < 90°$, then the surface roughness will be responsible for wettability of the surface. From Eq. (3) it is obvious that the $\cos\theta_w$ of a rough surface could be greater than 1 i.e. $\cos\theta_w > 1$, or the chemical composition of the surface could be different. In such cases the Wenzel's model is not applicable because in some cases the intrinsic contact angle is even more than 180° i.e. $\theta_w > 180°$ thus the Wenzel's model does not obey the real phenomenon under such conditions.

Thus to address these issues and the limitations of Wenzel's model instead of non-composite contact, composite contact model was proposed in 1944 by Cassie-and Baxter [20] in which it was suggested that the water droplet could not fully fill the surface grooves because the surface roughness is nano-scaled and the water droplet is larger than these grooves. The liquid-solid contact is basically composed of solid-air and solid-liquid contact surface due to presence of air in the grooves under the water drop as shown in Fig. 3d.

The variation in surface free energy of the system is shown by Eq. (4) when three phase contact line of water droplet moves a small distance

$$\delta E = f_1 (\gamma^{sl} - \gamma^{sa})dx + f_2\gamma^{la} dx + \gamma^{la} \cos \theta_c.dx \quad (4)$$

Here, $f_1$is the ratio of solid–liquid interface contact area to the actual surface area under droplet and $f_2$is the ratios of solid–air interface contact area to the actual surface area under water droplet within unit area.

Under balanced conditions of system the $\delta E = 0$. Rendering to Young's equation, Cassie-Baxter contact angle and apparent contact angle are associated as shown by Eq. (5).

$$\cos\theta_c = f_1 \cos \theta_z - f_2 \quad (5)$$

In theory, $f_1 + f_2 = 1$,Eq. (5) could be changes as

$$\cos\theta_c = f_1(\cos\theta_z + 1) - 1 \quad (6)$$

the ratio of air contact area to the rough surface would be infinite when $f_1$tends to be "0", in that case the Cassie-Baxter contact angle will be the highest i.e. 180° and the water droplet would be ideally spherical at the nanoroughened surface, making it ideally superhydrophobic. Roughness factor is still a consideration here, so it can be introduced in Cassie–Baxter equation. But the equation could be simplified and given as Eq. (7), due to presence of secondary or multilevel structures on the rough surface and the curvature present in real on liquid–air interface [21, 22].

$$\cos\theta_c = r\varphi\cos\theta_{z+}\varphi - 1 \quad (7)$$

In Eq. (7), φ is the solid-liquid interface projection area, if $f_1 = 1$,then Cassie-Baxter's equation transforms into the Wenzel's equation. Thus, all the surface that are superhydrophobic must have hierarchical nanoruoghness. Hence, Eq. (7) is near to natural model. A relationship between contact area ratio of solid–liquid and contact angle hysteresis was presented by Nosonovsky and Bhushan that validated that superhydrophobic surfaces must have composite interface [23].

Along with surface roughness and surface energy, surface morphology is another important factor to be described for the wettability of surface. So, the surface wetting state cannot be defined properly if the surface morphology is not fully known i.e. array pillar structure will have different property than the parallel groove structure. It means surface roughness and surface morphology both have high impact on the water contact angle. The inherent water contact angle of infinitely smooth and flat surface could not exceed 120°, so, Wenzel's model could not be applicable under such conditions. The Cassie-Baxter model could increase the intrinsic contact angle over secondary or multilevel structure. Yet, the Cassie-Baxter's model is not applicable to all regions, when a material in inherently hydrophilic then water droplet can immerse the nano grooves and the composite model can be converted to noncomposite model under high ambient pressure.

As the lotus leaf has self-cleaning property but, the self-cleaning does not happen if the water droplet slides on the rough surface of lotus leaf but arises only when the water droplet rolls off as shown in Fig. 3e. During the rolling off the dirt clings to the dropping water drops making the surface clean. The water droplet could be rolled off at the surface of leaf if the nano-scaled structures has enough softness to avoid the droplet's puncturing, sufficient strength to hold the water, and enough association to impulse the drop's rolling [17]. The air trapped in grooves and wax needles at nano-structures further reduce the chances of water droplet to stick on it, increasing the water contact angle making the surface superhydrophobic according to the real time conditions of Cassie-Baxter model.

2.2 Superhydrophibic Textiles

The textiles with water contact angle of more than 150° and contact angle hysteresis lower than 10° have been developed commercially for different indoor and outdoor activities. In past the main emphasis was on lowering the surface energy by coating some repellent agent. But, now the researchers have been mimicking the nature and creation of hierarchical nanoroughness at microfibrous surface is mainly emphasized. Thus, different inorganic metal and metal oxide nanoparticles have been coated along with surface energy lowering chemicals to make superhydrophobic textiles. The hydrophilic and hydrophobic textile has been shown in Fig. 4.

2.2.1 Organic Compound Coated Superhydrophobic Textiles

Hydrophobic textiles have been developed by applying various water-repellent/oil-repellent agents that change the surface free energy of the substrate and making it water resistant. The lowering of surface energy is basic mechanism and actual approach behind textile superhydrophobicity. Depending upon the fibers, yarns and

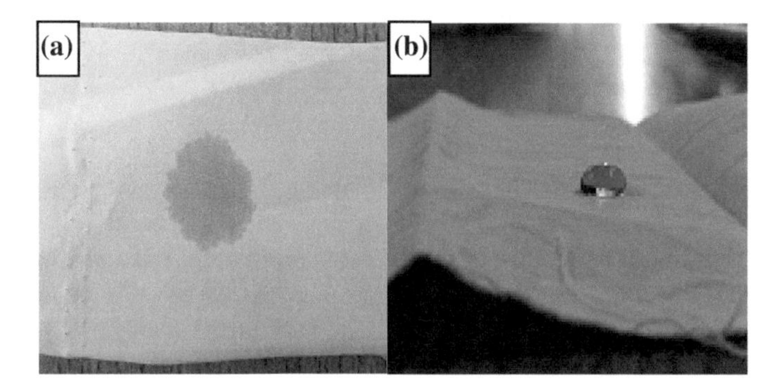

Fig. 4 **a** Hydrophilic cotton fabric. **b** Hydrophobic cotton fabric

weave structure the textile surface texture could be controlled. The roughness can be adjusted at submicron level by maneuvering the fiber numbers and composition, yarn diameter, and weaving mechanism. Thus, due to this micro roughness, hydrophobic characteristics could be granted by application of organic repellents. Until now different water and oil repellents have been introduced like paraffin waxes, stearic acid-melamine, silicon and fluorocarbon based repelling agents. Surface energy can be lowered by different ways like the mechanical incorporation of repellents into fibrous structure or onto fabric surface filling the fiber and yarn interstices. The paraffin repellents are integrated in this way. Then chemical reaction of repellent with textile is another approach to impart superhydrophobicity such as stearic acid. Silicon and fluorocarbon based repelling agents have been applied in the form of thin film on fiber surface [24].

These repelling agents could impart hydrophobicity by lowering the surface free energy of textiles depending upon their chemical composition. A study was conducted in which polyester fabric was coated with siloxane by conventional pad-dry and cure method. The polyester fabric with 2 μm yarn diameter upon coating with repellent agent showed superhydrophobicity with WCA 170° and contact angle hysteresis of below 5°. This research was done to validate the claim that microfiberous textile surface without having nanoroughness can exhibit superhydrophobic properties, because the air trapped inside microfibers was large enough that water droplets could not be absorbed and rolled off easily at the surface [25].

Paraffin Repellents

The paraffin waxes; known as earliest repellent agents, have been used to make textile hydrophobic and various researches have been conducted in this regards [26, 27]. These repellents are usually emulsions that contain aluminium, barium, or zirconium salts of fatty acid normally stearic acid. Because these finishes are making polar-nonpolar junction with textile therefore their adhesion could be increased. In this finishing mixture of repellent the paraffinic part is attracted towards the hydrophobic region and the polar end is attracted towards the fiber surface. These emulsions have been applied on textile by padding or exhaustion techniques. Though this finishing make the textile repellent in cost effective way but it increases the problem of inflammability and also less durable.

Stearic Acid–Melamine Repellents

Another primitive class of repellents that were more durable to laundering without having much effect on hand feel are mainly prepared by reaction of stearic acid with formaldehyde and melamine. In these repellent finishes, inherent hydrophobic nature of stearic acid provides the water repellency to the substrate finished with these repellants, and the durability of them is due to the N-methylol groups that reacts with cellulose for permanent binding [28]. Though these repellent agents give durable hydrophobicity to the textile but, there is issue with their application that they release formaldehyde and cause loss of mechanical strength such as decrease of tear strength, abrasion resistance along with change in shade percent of dyed fabrics. Due to the health hazards or carcinogenic nature of released of formaldehyde their application is limited now [24, 29]. The minimum concentration of about 0.1 ppm [30]. Stearic acid-melamine could be used at spacious and airy workplaces to be effectively applied on textiles by exhaustion or most commonly padding to make it hydrophobic. But still by proper monitoring and taking precautionary measures during application worker's health could be saved. Some time they are used in combination with other repellents like fluorochemicals in order to reduce the consumption of these chemicals and to improve the overall performance of textile.

Silicone Water Repellents

The superb water repellency could be achieved by treating textiles with hydrophobic silicone compositions. On silicon molecules long chain hydrocarbon serve the function of hydrophobicity and the silicon could enter the surface of textile the hydrophobic groups could be delivered deeply inside or through the substrate. Also the edge of stability of the silicon make them the best choice though to make textile durable hydrophobic but breathable with soft hand feel. Silicons or siloxanes and fluorocarbon based repelling agents are the most widely used. Silanes have –O–Si–O– backbones and the attached alkyl groups (can be with different chain lengths) mainly provide water repellency. They are responsible to lower the surface energy of textile to about 20 dyne/cm imparting hydrophobic nature to substrate.

The polydimethylsiloxanes are mostly used to provide the hydrophobic layer at the fabric surface as they form chemical bonding (usually hydrogen bonding) with the cellulose even at low concentration about 0.5–1% o.w.f. they could serve the purpose of hydrophobicity. The reaction of siloxanes with cellulosic textile require temperature of 120–250 °C for complete attachment with the textile and time required could be from several minutes to several hours to obtain satisfactory hydrophobicity.

Fluorocarbon-Based Repellents

Fluorocarbon based repellents are mostly copolymers that are mainly synthesized by perfluoroalkyl groups incorporated into urethane or acrylic monomers. The perfluoroalkyl groups were originally synthesized by electrochemical fluorination but, now a days they have been prepared by telomerisation. These monomers are then polymerized to from hydrophobic finishes. These repelling agents are providing the fibers with lowest surface free energy, thus, making fabric oil, water and soil repellent. Fluorine containing esters when come in contact with the fibers they arrange themselves vertically on textile surface, lowering the surface energy depending upon the distribution and orientation of perfluoroalkyl groups. The finish when applied

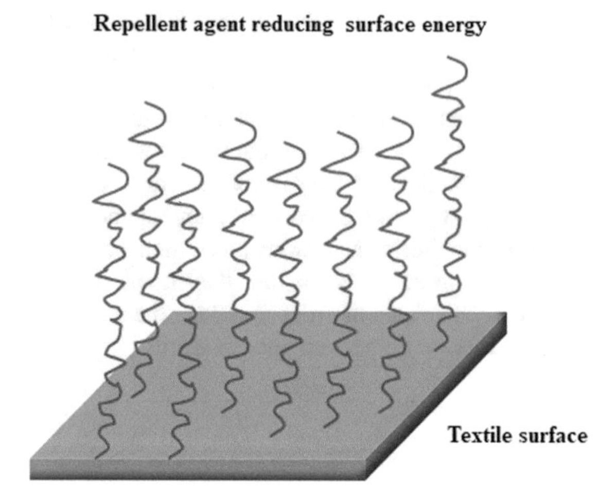

Fig. 5 Schematic showing textile substrate treated with fluorocarbons to lower the surface energy to make it hydrophobic

finally at the fabric surface should be in condensed CF3 outer form to provide maximum repellency. Because they are the providing the lowest surface energy and along with water they are oil and soil resistant therefore they have been most widely used repelling agents [31, 32] (Fig. 5).

The water repellent finishes based on C8 fluorocarbon were most commonly used in past. However it was hazardous to human body in decomposition of C8 fluorocarbon thus now potentially considered as carcinogenic most specifically perfluorooctanoic acid or pentadecafluorooctanoic acid and perfluorooctane sulphonate or heptadecafluoro-1-octanesulfonic acid. Restriction have been made on the use of such fluorocarbon based water repellent finishes [32, 33]. To improve the solubility and the emulsification commonly ethylene; a smaller spacer group, can be modified and also the side chain length of perfluorinated side chain should be kept smaller i.e. 8–10 carbon.

2.2.2 Inorganic Compounds Coated Superhydrophobic Textiles

In recent years, to attain the properties by mimicking nature has been the main concern of textile researchers. They have been imparting unusual attributes in textiles by getting inspiration and adapting the nature of different insects and plants like rose petal, lotus effect, cicada wings, and gecko feet that have micro and nanoscaled surface structure with low surface energy making themself-cleaning and superhydrophobic [17, 34–37]. In this regard different low energy chemicals could be used to fabricate hydrophobic textiles. But, surface characterization using inorganic compounds is of particular importance that at nano-scale can be achieved due to new developments in surface analysis techniques. Nanomaterials are gaining more attention because of unique physical, biological and chemical properties of matter that at molecular and atomic level changes drastically. Properties shown by nanomaterials are different

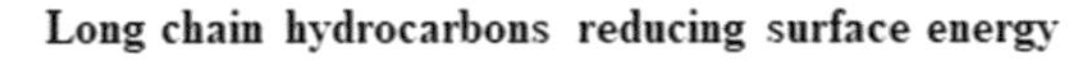

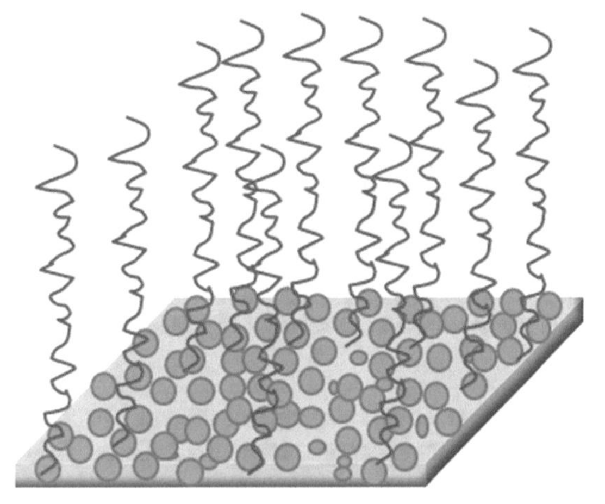

Fig. 6 Schematic diagram showing textile with nanoroughness and lowering of surface energy with long chain hydrocarbons

from bulk material [38, 39], due to larger surface area and pronounced quantum size effects they possess dynamically changed optical, electrical and magnetic properties. With ultra-small size they are more reactive and thus have been used for many textile and many other applications [40]. Different NPs have been applied for textile functionalization to impart different attributes. Along with some repellent agent to lower surface energy NPs have been used for creation of mico- and nanoroughness to make textile superhdrophobic [41–45] the schematic illustration of nanoroughness and lower surface energy for superhydrophobicity is given in Fig. 6.

Nanoparticles application onto textile is a two steps process: firstly their synthesis and in second step their homogenous and stable dispersion is coated onto textile by conventional coating processes. These coating include: (1) padding, (2) Sol-gel, (3) washing, (4) transfer printing and (5) spraying. Pad-dry-cure and sol-gel are the most widely used and simple methods by which the nanoparticle dispersion is applied on textile by simply spin or dipping to form a nano coating.

Physical self-cleaning is the one of most favorable attribute attained by making surface hydrophobic with water contact angle greater than 90°. Superhydrophobic textiles gained special consideration in scientific and industrial populations due to various applications in these fields. The SEM micrograph showed (Fig. 7) the untreated textiles (Fig. 7a) and the treated textile (Fig. 7b) that have been fabricated by modified NPs that are hydrophobized by treating some chemical to lower the surface energy of textile treated with inorganic NPs. the measure contact angle (Fig. 7c) showing the superhydrophobic nature.

Different nanostructures could be deposited on the textile to generate nanoroughness like SiO_2, ZnO, TiO_2 and CNT [20, 21, 24, 25]. The Silica nanoparticles have found wide application in superhydrophobic and self-cleaning textile due to its inert

Fig. 7 Textile substrate **a** untreated. **b** Treated with inorganic metal oxide nanoparticles and repellent agent. **c** Water contact angle of treated textile showing superhydrophobic nature

nature and water contact angle of around 170° could be achieved by their application. The inorganic nanoparticles like ZnO and TiO_2 have been practiced to generate nanoroughness on the textiles and subsequently treating with repellent agent make them superhydrophobic. The main advantage of using ZnO nanostructures is there ease of fabrication with different surface morphologies in cost effective way. Such nanostructures could be synthesized by various methods such as hydrothermal, electrochemical deposition, and thermal oxidation and many others. They can also be multifunctional because of their photocatalytic nature they can impart biological, UV protection and chemical self-cleaning [46–48]. To make superhydrophobic textiles, microwave assisted hydrothermal approach was used by which ZnO nanowires were grown on cotton fabric and then the fabric was functionalized with repellent agent i.e. stearic acid and WCA of 150° was obtained that confirmed the superhydrophobic nature of textiles [49].

Silane coupling agents are well known for their repellent nature therefore, Ashraf et al. [3] done a study in which ZnO nanorods were grown on seeded polyester fabric and then the modified fabric was subsequently treated with octadecyletrimethoxysilane by chemical vapor deposition to obtain a WCA of 158° and sliding angle 1° showing the lotus leaf effect of polyester fabric. Same kind of work was done by Park et al. [50] but with different repellent. ZnO nanorods were grown on nylon fabric and then treatment with n-dodecyltrimethoxysilane made nylon fabric superhydrophibc. Instead of ZnO nanoparticles mostly nanorods and nanowires are grown on textile for superhydrophobic character. Nanorods array was synthesized on cotton substrate using wet chemical route, then n-dodecyltrimethoxysilane was used for textile post-treatment. The Superhydrophobic textile was made with a WCA of 161° and roll-off angle of 9°.

TiO_2nanoparticles have been used as multifunctional agent for different types of textile because of band gap of 3.2–3.35 eV these nanoparticles are highly photocatalytic. TiO_2 NPs are used to create roughness on micrometer-scale fibers to make textile water repellent. Textile finished with nanocomposite of TiO_2 and polytetrafluoroethylene exhibited awesome self-cleaning and wettability change of structure. Co-deposition method was adopted onto structured substrate and multifunctional properties were obtained with water contact angle higher than 150° [51]. Mostly the textile modification by use of TiO_2 nanoparticles is a two-step process in which TiO_2 nanoparticles have been coated on cotton micrometer-scale fibers to create the nanoroughness and after this modification in next step post-treatment with some repellent like dodecafluoroheptyl-propyl-trimethoxysilane is done to develop excellent water repellent textiles with a WCA of 160° [52] along with other properties such as UV protection and dye degradation. Inspired by the hierarchical alignment of lotus leaf, different morphological structures of TiO_2 have also been applied onto textile to investigate the structural effect on its properties. In this regard a study was done by Huang et al. [53] in which marry-gold like TiO_2 nano-flowers were synthesized and deposited uniformly on cotton fabric by hydrothermal process, via facile strategy. Then the treated cotton fabric was coated with mixed methanolic solution of 1H,1H,2H,2H-per-flourodecyltriethoxysilane (F17) to prepare a robust superhydrophobic fabric with a water contact more than 160° and sliding angle less than 10°. Furthermore, the modified fabric demonstrated outstanding UV shielding, self-cleaning and water-oil separation.

Though the photocatalysts have been applied for multifunctional properties on textiles, but, the effect of these functional materials on the properties of textile is controversial. They generate highly reactive species know as reactive oxygen species (ROS; $OH^{\bullet}$, $O^{\bullet -2}$) by absorbing electromagnetic radiation in UV region and these ROS are responsible for organic decay. As cellulose is also organic thus these ROS could also have detrimental effect on their contact with the cellulosic fibers [51, 54]. Thus the mechanical properties such as tear strength, tensile strength, abrasion resistance could be lost when the cellulosic textile is treated with photocatalysts like TiO_2 and ZnO etc. due to crystalline changes of cellulosic structure [55].

Due to the negative effects on dyed fabrics and the loss in mechanical properties mostly the SiO_2 nanoparticles have been used for manufacturing of superhydrophobic textile due to inert nature of SiO_2 nanoparticles. A researcher Stöber developed the Sol-gel process for nanoparticle's synthesis that has been used now by many researchers to generate nanoroughness at the substrate with few modifications. TEOS: tetraethyl orthosilicate and some organic solvents or distilled water are the most commonly used precursors for fabrication of SiO_2 nanoparticles by sol-gel process. After synthesis these nanoparticles have been applied by conventional pad-dry-cure method onto textile for surface morphological change by creation of nanoroughness. Then post treatment by repellents made textile hydrophobic with water contact angle of greater than 130° [56]. In early experimentation, the water contact angle was lower than 150° mostly and the textiles were not superhydrophobic. Therefore the rigorous researches were carried on to mimic the lotus leaf and to attain the angle more than 150°. Several chemicals were tried and it was known after intensive research that fluoroalkylsiloxane could be best repellents providing superhydrophobic nature to textiles. Thus Hao et al. [57] reported application of SiO_2 nanoparticles for nanoroughness and treatment with fluoroalkylsiloxane that lowered the surface energy enough to increase WCA 138°–156.5°. Thus, an approach was adopted that was near to lotus leaf effect.

The extent of hydrophobicity could be dependent upon the chain length of repellent silane coupling agents. To confirm this and to find the best hydrophobe a research was done in which silanes with different chain lengths were used to lower the surface energy along with silica nanoparticles. Cross linker was added to enhance the durability of nanoparticles. From the study it was confirmed that n-dodecyltrimethoxysilane was the best hydrophobe due to longest chain length among seven repelling agents such as n-octadecyltrimethoxysilane, n-dodecyltrimethoxysilane, n-hexadecyltrimethoxysilane, n-butyltrimethoxysilane, n-ethyltrimethoxysilane, n-octyltrimethoxysilane, and methyltrimethoxysilane [44].

As the nanoparticle do not have the affinity for textiles, therefore, some binders have been used [58–60] to increase their adhesion with the substrate mentioned previously in this chapter. Due to the application of binders the comfort and properties like tear strength have been compromised [61]. Therefore, there was a need to find a way out to improve the durability without affecting the inherent characteristics of the textile. The researchers tried to functionalize the nanoparticles with silane coupling agents to directly bind them with the textile without using any crosslinker. Riaz et al. [1] investigated the comfort properties by developing highly durable superhydrophobic textiles by application of 3-(Trimethoxysilyl) propyl-N,N,N-dimethyloctadecyl ammonium chloride and 3-Glycidoxypropyl)trimethoxysilane modified SiO_2 nanoparticle at cotton fabric. It was found that water contact angle of nearly 150° was obtained without compromising the comfort properties and functionality up to 20 industrial washing cycles was obtained.

To make the surface very near to lotus effect binary hierarchical roughness in more than one layer was also tried to be created at the textile to attain maximum WCA. In this regard, different sizes of SiO_2 nanoparticles were used with two silanes. First layer of 3-aminopropyl triethoxysilane functionalized SiO_2 nanoparticles was applied onto textile and then 2nd layer of 3-glycidoxypropyltrimethoxysilane functionalized SiO_2 nanoparticles was deposited onto the 1st layer. The deposition of different sized nanoparticles created the binary hierarchical roughness at the substrate. Due to direct relationship between surface morphology, surface roughness and water contact angle it was possible to tailor the water contact angle [62]. Xue et al. [63] also presented same type of study in which binary roughness was created on epoxy modified cotton fabric by applying amino and epoxy-functionalized silica nanoparticles on textile. Then the surface energy of treated textile was further lowered by post treating with stearic acid and the superhydrophobic textile with water contact angle nearly 170° was obtained.

Along with pad-dry-cure method, spraying have also been used for the coating of modified nanoparticles on the textile. SiO_2 nanoparticles dispersed in toluene, functionalized with trichlorododecylsilane that grafted dodecyltrichloro group on nanoparticle surface were sprayed on cotton after drying at room temperature the modified textile confirmed superhydrophobicity exhibiting water contact angle more than 160° also with contact angle hysteresis less than 10° [64].

For textile modification by attachment of nanomaterials various methods have been adopted and reported with merits and demerits of every technique.

The sol-gel method have always been most well-known and well applied to adhere nanoparticles with textile to make it superhydrophobic. But, the main issue with this technique is that the inherent properties of textile more or less have been distorted because of this application technique [65, 66]. Different methods have been adopted with minimum effect on intrinsic textile properties such as air permeability, bending rigidity, hand feel, thermal and moisture management, appearance etc. with greater durability and homogenous distribution of nanoparticles. Layer by Layer self-assembly deposition is one of them in which electrostatic force of attraction is basic reason to bind the nanoparticles at substrate. Zhao et al. [67] deposited silica NPs using polyelectrolyte by layer by layer electrostatic self-assembly and created nanoroughness at the surface. The post treatment of SiO_2 nanoparticles coated textile was done with fluoroalkylsilane and superhydrophobic textile was developed water contact angle higher than 150° and also the sliding angle of 10° was obtained. As the surface morphology is also an important factor for the superhydrophobicity, therefore, it could be controlled by changing the number of layers on textiles and the water contact angle of 170° could be achieved for textile modified with epichlorohydrin and SiO^2 nanoparticles deposited by layer by layer assembly. Half of these nanoparticles were 3-aminopropyltriethoxysiloxane and the remaining were 3-glycidoxypropyl trimethoxysilane modified [68]. Nevertheless, the results are good but the technique is limited due to higher production cost and the longer time for processing.

Though until now the silica nanoparticles have been most widely used inorganic material to fabricate superhydrophobic textiles due to its inertness. But the textiles treated with sol-gel synthesized SiO_2 are stiffer with less bending and comfort properties. Also decrease in mechanical properties like tear strength has been reported in literature, due to cross-linking. But, even then the functionality is higher than the demerits therefore, the practice to use these nanoparticles is still higher and the use will be continued until the fabrication of some new material came into existence.

2.2.3 Plant Based Organic Repellents

Because the mostly used water repelling agents are not biodegradable and sustainable, also, they are not easy to produce therefore, researchers have been trying to find out natural, sustainable and cost effective materials to make textiles superhydrophobic. Phytic acid containing six phosphate groups is naturally occurring component present in many plant tissues, seeds, and also in legumes, grains and cereal. Several metal ions Ce^{III}, Ag^{I}, Fe^{III}, Sn^{IV} and Zr^{IV} could have the ability to bind the phytic acid just like a crosslinker and form the insoluble complex aggregations. This insoluble metal complex aggregations were used to generate hierarchical nanoroughness at the surface of cotton fabric. To enhance the superhydrophobic character further treatment was done polydimethylsiloxane to make it superhydrophobic due to the its several hydrophobic alkyl chain [69]. Schematic presentation of the whole process is given in Fig. 8.

Researchers at Queensland University of Technology (QUT) have identified a biodegradable and sustainable water proof coating going to commercialize soon. They were analyzing possible applications for bagasse that is fibrous husk discarded after sugar making. They realized that the lignin that is gigantic molecule could be

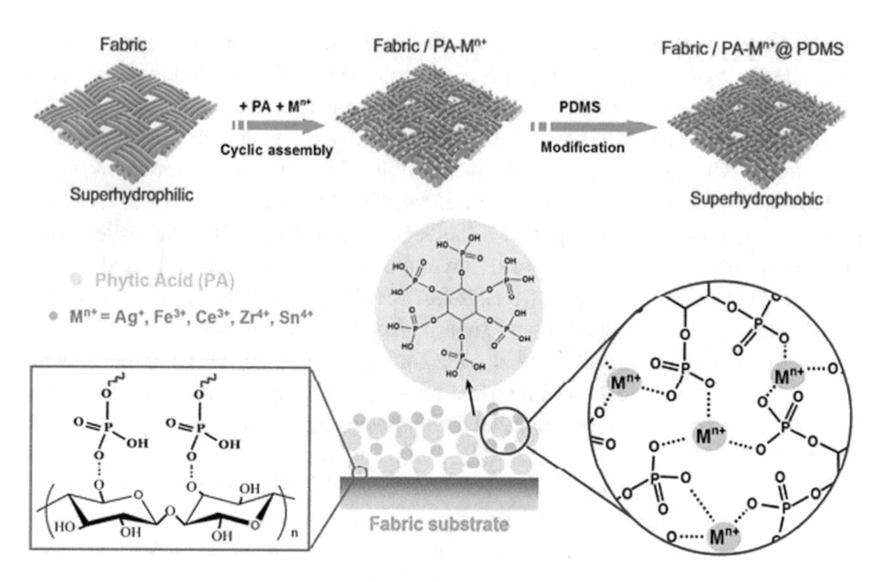

Fig. 8 Schematic showing the development of superhydrophobic cotton fabric coated with PA-Mn + @PDMS. "Reprinted with permission from Ref. [69], Copyright (2017) American Chemical Society"

used as water repellent coating that is found in most of the plants and few algae. If textile researchers work on it, the lignin formulation could very useful to make textile hydrophobic. Zein is obtained from maize that is water-insoluble protein, leguminis found in leguminous seeds like peas, is a casein-like protein and lectins is found in many plants are glycoproteins. Research could be done to use these materials to make them textile alone or along with other water repelling agents.

2.3 *Commercialized Superhydrophobic Textiles*

Until now the first commercialized textile was developed by *Nanotex®*. They by creating nanoroughness and utilizing low surface energy chemicals have developed commercial multifunctional textile with amazing moisture management, superhydrophobic and odor defusing features. They created nano-whiskers, that are hydrocarbons and their size was 1/1000 of usual cotton fiber that have been added to the fabric to make it water-repellent without affecting the mechanical properties of cotton. The pore size of developed nano-whiskers is usually below than the water droplets but greater than a water molecule thus water could not enter the whiskers and remain at the surface of textile. But, the moisture and air can pass through making it breathable hydrophobic, stain repellent, retaining the natural comfort, softness and breathability [70].

However, if a very high pressure was applied the liquid could pass through it. Another textile based Swiss Company Schoeller synthesized Nanosphere and coated them on textile to make it water repellent. The lotus effect was provided onto Nanosphere coated fabric that is involved impregnation of 3-D surface structure and gel forming additives that repel water and during rolling off all the dirt and dust particles are attached with the water droplet making the textile super hydrophobic and self-cleaning [71].

2.4 *Switchable Hydrophilicity and Hydrophobicity*

The surface with water contact angle less than 90° is hydrophilic, less than 5° is superhydrophilic, higher than 80° is hydrophilic and even higher than 150° is super-hydrophilic. The wettability or repellency of surface is dependent upon the surface roughness and surface chemistry. The switchable surface which combines the effect of both superhydrophilic and superhydrophobic have been recently reported that are smart, adaptable and stimuli responsive by change of surface chemistry. Such surfaces can be used in many applications such as drug delivery, oil-water separation, protein concentrators, and microfluidic pumps. There could be some triggering mechanism depending upon the surface chemistry that can reverse the surface's character under different conditions. Switchability of surface could be induced by any stimulus like light [72], pH [73], temperature [74, 75], electric potential [76, 77], solvent [78] and mechanical stress [79].

There have been thermos-responsive polymers used to approach the switching character of a material that undergo the phase transformation at certain temperature which is called lower critical solution temperature at which switching between hydrophobicity and hydrophilicity occurs. Tao Sun and his colleagues worked on thermos responsive PNIPAAm (poly N-isopropylacrylimide) thin film fabrication on roughened and smooth silicon surface with LCST of about 32–33 °C. The mechanism of switchability was adaptation from hydrogen bonding with solvent at temperature below LCST to internal hydrogen bonding above the LCST. Compared to flat surface the roughened surface of silicon with thin film of PNIPAAm gave the ability to surface to be switched between superhydrophilic at ~25 °C and superhydrophobic at ~40 °C [74].

Light sensitive switchable duperhydrophilic, superhydrophobic films have been synthesize by hydrothermal treatment of an aqueous titanium trichloride solution supersaturated with NaCl. These fabricated films on UV irradiation transformed from superhydrophobic to superhydrophilic. Because TiO_2 is a photocatalyst, on exposure to UV radiation the generated holes react with oxygen to form surface oxygen vacancies, thus water molecules coordinate kinetically making the surface superhydrophilic. As the surface is roughned by TiO_2 nanorods the water droplets fill the nanogrooves by replacing the trapped air. The water contact angle will be about below 5°. The surface will be in metastable state on water adsorption, which can be replaced by atmospheric oxygen [80] after the films are located in the dark.

Consequently, the surface will be transformed into its original state and the superhydrophilicity would be transformed into superhydrophobicity of the films again [81].

3 Conclusion

Superhydrophobic textiles inspired from lotus leaf have been developed and used in various fields of life. Organic and inorganic materials that have been used for hydrophobic textiles are discussed briefly in this chapter. Durable water repellent textiles containing organic repelling agents could not provide superhydrophobicity because they only lower the surface energy, but, nanoroughness and surface morphology should also to be considered for manufacturing of superhydrophobic textiles.

Also, the risk associated with organic repellents like stearic-acid melamine and fluorochemicals made their use limited in textile. Therefore, inorganic compounds in the form of metal oxide nanomaterial along with silicone based repelling agents are discussed in detail for development of superhydrophobic textile.

References

1. Riaz S, Ashraf M, Hussain T, Hussain MT (2019) Modification of silica nanoparticles to develop highly durable superhydrophobic and antibacterial cotton fabrics. Cellulose 26(8):5159–5175
2. Zhai L, Cebeci FC, Cohen RE, Rubner MF (2004) Stable superhydrophobic coatings from polyelectrolyte multilayers. Nano Lett 4(7):1349–1353
3. Ashraf M, Campagne C, Perwuelz A, Champagne P, Leriche A, Courtois C (2013) Development of superhydrophilic and superhydrophobic polyester fabric by growing Zinc Oxide nanorods. J Colloid Interface Sci 394(1):545–553
4. Feng X, Jiang L (2006) Design and creation of superwetting/antiwetting surfaces. Adv Mater 18(23):3063–3078
5. Xue CH, Li M, Guo XJ, Li X, An QF, Jia ST (2017) Fabrication of superhydrophobic textiles with high water pressure resistance. Surf Coatings Technol 310:134–142
6. Sohyun P, Jooyoun K, Chung Hee P (2015) Superhydrophobic textiles: review of theoretical definitions, fabrication and functional evaluation. J Eng Fabr Fibers 10(4):1–18
7. Su X, Li H, Lai X, Zhang L, Liao X, Wang J, Chen Z, He J, Zeng X (2018) Dual-functional superhydrophobic textiles with asymmetric roll-down/pinned states for water droplet transportation and oil-water separation. ACS Appl Mater Interfaces 10(4):4213–4221
8. Lei S, Shi Z, Ou J, Wang F, Xue M, Li W, Qiao G, Guan X, Zhang J (2017) Durable superhydrophobic cotton fabric for oil/water separation. Colloids Surfaces A Physicochem Eng Asp 533:249–254
9. Oh J-H, Ko T-J, Moon M-W, Park CH (2017) Nanostructured fabric with robust superhydrophobicity induced by a thermal hydrophobic ageing process †
10. Cortese B, Caschera D, Federici F, Ingo GM, Gigli G (2014) Superhydrophobic fabrics for oil-water separation through a diamond like carbon (DLC) coating. J Mater Chem A 2(19):6781–6789
11. Xiang T, Han Y, Guo Z, Wang R, Zheng S, Li S, Li C, Dai X (2018) Fabrication of inherent anticorrosion superhydrophobic surfaces on metals. ACS Sustain Chem Eng 6(4):5598–5606

12. Nosonovsky M (2007) Multiscale roughness and stability of superhydrophobic biomimetic interfaces. Langmuir 23(6):3157–3161
13. Pan C, Shen L, Shang S, Xing Y (2012) Preparation of superhydrophobic and UV blocking cotton fabric via sol-gel method and self-assembly. Appl Surf Sci 259:110–117
14. Ryu J, Kim K, Park JY, Hwang BG, Ko YC, Kim HJ, Han JS, Seo ER, Park YJ, Lee SJ (2017) Nearly perfect durable superhydrophobic surfaces fabricated by a simple one-step plasma treatment. Sci Rep 7(1):1981
15. Young T (1805) III. An essay on the cohesion of fluids. Philos. Trans. R. Soc. London 95:65–87
16. Zisman WA (1963) Influence of constitution on adhesion. Ind Eng Chem 55(10):18–38
17. Barthlott W, Neinhuis C (1997) Purity of the sacred lotus, or escape from contamination in biological surfaces. Planta 202(1):1–8
18. Bhushan B, Jung YC, Koch K (2009) Micro-, nano- and hierarchical structures for superhydrophobicity, self-cleaning and low adhesion. Philos Trans A Math Phys Eng. Sci. 367(1894):1631–1672
19. Wenzel RN (1936) Resistance of solid surfaces to wetting by water. Ind Eng Chem 28(8):988–994
20. Cassie ABD, Baxter S (1944) Wettability of porous surfaces. Trans Faraday Soc 40:546–551
21. Marmur A (2003) Wetting on hydrophobic rough surfaces: To be heterogeneous or not to be? Langmuir 19(20):8343–8348
22. Patankar NA (2004) Mimicking the lotus effect: Influence of double roughness structures and slender pillars. Langmuir 20(19):8209–8213
23. Michael N, Bhushan B (2007) Hierarchical roughness makes superhydrophobic states stable. Microelectron Eng 84(3):382–386
24. Schindler WD, Hauser PJ (2004) Chemical finishing of textiles: Woodhead publishing series in textiles, vol. null. CRC
25. Gao L, McCarthy TJ (2008) Teflon is hydrophilic. Comments on definitions of hydrophobic, shear versus tensile hydrophobicity, and wettability characterization. Langmuir 24(17):9183–9188
26. Seo K, Kim M, Kim DH (2014) Candle-based process for creating a stable superhydrophobic surface. Carbon N Y 68:583–596
27. Abo-Shosha MH, El-Hilw ZH, Aly AA, Amr A, Nagdy ASIE (2008) Paraffin wax emulsion as water repellent for cotton/polyester blended fabric. J Ind Text 37(4):315–325
28. Zero Discharge of Hazardous (2012) Durable water and soil repellent chemistry in the textile industry: A research report-P05 water repellency project
29. Agents classified by the IARC monographs, Volumes 1–124–IARC. [Online]. Available: https://monographs.iarc.fr/agents-classified-by-the-iarc/. Accessed on 10 Jul 2019
30. T. Cotton Foundation Memphis JW, Keeling JP, Bordovsky J, Everitt KF, Bronson RK, Boman, and Jr., Mullinix, BG (1997) J Cotton Sci 13(2), Cotton Foundation
31. Conder JM, de Voogt P, Mabury SA, Cousins IT, Buck RC, Franklin J, Berger U, Kannan K, Jensen AA, van Leeuwen SP (2011) Perfluoroalkyl and polyfluoroalkyl substances in the environment: Terminology, classification, and origins. Integr Environ Assess Manag 7(4):513–541
32. Baran JR (2002) Fluorinated surfactants and repellents: Second edition, revised and expanded surfactant science series. vol 97. By Erik Kissa (Consultant, Wilmington, DE). Marcel Dekker: New York. 2001. xiv + 616 pp. $195.00. ISBN 0-8247-0472-X., J Am Chem Soc 123(36): 8882
33. Commision European (2006) Directive 2006/122/ECOF the European parliament and of the council of 12 December 2006 amending for the 30th time council directive 76/769/EEC on the approximation of the laws, regulations and administrative provisions of the Member States relating to res. Off J Eur Union L 372:32–34
34. Wagner T, Neinhuis C, Barthlott W (1996) Wettability and contaminability of insect wings as a function of their surface sculptures. Acta Zool 77(3):213–225
35. Parker AR, Lawrence CR (2001) Water capture by a desert beetle. Nature 414(6859):33–34
36. Gao X, Jiang L (2004) Biophysics: Water-repellent legs of water striders. Nature 432(7013):36

37. Byun D, Hong J, Ko JH, Lee YJ, Park HC, Byun B-K, Lukes JR (2009) Wetting characteristics of insect wing surfaces. J Bionic Eng 6(1):63–70
38. Roduner E (2006) Size matters: Why nanomaterials are different. Chem Soc Rev 35(7):583–592
39. Roduner E (2006) Nanoscopic materials: Size-dependent phenomena. RSC Pub, Cambridge
40. Joshi M (2011) Nanotechnology: A new route to high performance textiles. pp. 272–293
41. Xue C-H, Jia S-T, Chen H-Z, Wang M (2016) Superhydrophobic cotton fabrics prepared by sol–gel coating of TiO2 and surface hydrophobization. Sci Technol Adv Mater 9(3):035001
42. Huang L, Lau SP, Yang HY, Leong ESP, Yu SF, Prawer S (2005) Stable superhydrophobic surface via carbon nanotubes coated with a ZnO thin film. J Phys Chem B 109(16):7746–7748
43. Wu L, Zhang J, Li B, Wang A (2013) Mimic nature, beyond nature: Facile synthesis of durable superhydrophobic textiles using organosilanes. J Mater Chem B 1(37):4756
44. Taylor P, Roe B, Kotek R, Zhang X (2012) Durable hydrophobic cotton surfaces prepared using silica nanoparticles and multifunctional silanes. J Text Inst 103(December):385–393
45. Berendjchi A, Khajavi R, Yazdanshenas ME (2011) Fabrication of superhydrophobic and antibacterial surface on cotton fabric by doped silica-based sols with nanoparticles of copper. Nanoscale Res Lett 6:594
46. A. P. Dr. Kumar BS (2015) Self-cleaning finish on cotton textile using sol-gel derived TiO2 nano finish\n. IOSR J Polym Text Eng 2(1): 01–05
47. Kathirvelu S, D'Souza L, Dhurai B (2008) A comparative study of multifunctional finishing of cotton and P/C blended fabrics treated with titanium dioxide/zinc oxide nanoparticles. Indian J Sci Technol 1(7):1–12
48. Bozzi A, Yuranova T, Kiwi J (2005) Self-cleaning of wool-polyamide and polyester textiles by TiO2-rutile modification under daylight irradiation at ambient temperature. J Photochem Photobiol A Chem 172(1):27–34
49. Ates ES, Unalan HE (2012) Zinc oxide nanowire enhanced multifunctional coatings for cotton fabrics. Thin Solid Films 520(14):4658–4661
50. Park Y, Park CH, Kim J (2014) A quantitative analysis on the surface roughness and the level of hydrophobicity for superhydrophobic ZnO nanorods grown textiles. Text Res J 84(16):1776–1788
51. Kamegawa T, Shimizu Y, Yamashita H (2012) Superhydrophobic surfaces with photocatalytic self-cleaning properties by nanocomposite coating of TiO2 and polytetrafluoroethylene. Adv Mater 24(27):3697–3700
52. Zhang Y, Li S, Huang F, Wang F, Duan W, Li J, Shen Y, Xie A (2012) Functionalization of cotton fabrics with rutile TiO2 nanoparticles: Applications for superhydrophobic, UV-shielding and self-cleaning properties. Russ J Phys Chem A 86(3):413–417
53. Huang JY, Li SH, Ge MZ, Wang LN, Xing TL, Chen GQ, Liu XF, Al-Deyab SS, Zhang KQ, Chen T, Lai YK (2015) Robust superhydrophobic TiO2 @fabrics for UV shielding, self-cleaning and oil–water separation. J Mater Chem A 3(6):2825–2832
54. Awungacha Lekelefac C, Busse N, Herrenbauer M, Czermak P (2015) Photocatalytic based degradation processes of lignin derivatives. Int J Photoenergy (2015): 1–18, Hindawi Publishing Corporation
55. Zhang H, Zhu LL, Sun RJ (2014) Structure and properties of cotton fibers modified with titanium sulfate and urea under hydrothermal conditions. J Eng Fiber Fabr 9(1):67–75
56. Bae GY, Min BG, Jeong YG, Lee SC, Jang JH, Koo GH (2009) Superhydrophobicity of cotton fabrics treated with silica nanoparticles and water-repellent agent. J Colloid Interface Sci 337(1):170–175
57. Hao LF, An QF, Xu W, Wang QJ (2010) Synthesis of fluoro-containing superhydrophobic cotton fabric with washing resistant property using nano-SiO2 sol-gel method. Adv Mater Res 121–122:23–26
58. Mihailović D, Šaponjić Z, Radoičić M, Radetić T, Jovančić P, Nedeljković J, Radetić M (2010) Functionalization of polyester fabrics with alginates and TiO2 nanoparticles. Carbohydr Polym 79(3):526–532

59. Nadanathangam V, Vigneshwaran N, Kumar S, Kathe AA, Varadarajan PV, Prasad V (2006) Functional finishing of cotton fabrics using zinc oxide-soluble starch nanocomposites at CIR-COT, Mumbai view project functional finishing of cotton fabrics using zinc oxide-soluble starch nanocomposites functional finishing of cotton fabrics using zinc oxide-soluble starch nanocomposites. Train Adv Microsc View Proj Train Adv Nanotechnol (17): 5087–5095
60. Vigneshwaran N, Nachane RP, Balasubramanya RH, Varadarajan PV (2006) A novel one-pot 'green' synthesis of stable silver nanoparticles using soluble starch. Carbohydr Res 341(12):2012–2018
61. Jalan V, Butola BS (2018) Influence of binder type on color characteristics of cotton fabric colored with a photochromic colorant. J Nat Fibers 15(2):229–238
62. Athauda TJ, Ozer RR (2012) Investigation of the effect of dual-size coatings on the hydrophobicity of cotton surface. Cellulose 19(3):1031–1040
63. Yu M, Gu G, Meng W-D, Qing F-L (2007) Superhydrophobic cotton fabric coating based on a complex layer of silica nanoparticles and perfluorooctylated quaternary ammonium silane coupling agent. Appl Surf Sci 253(7):3669–3673
64. Jeong SA, Kang TJ (2016) Superhydrophobic and transparent surfaces on cotton fabrics coated with silica nanoparticles for hierarchical roughness. Text Res J. pp. 1–9
65. Te Hsieh C, Wu FL, Yang SY (2008) Superhydrophobicity from composite nano/microstructures: Carbon fabrics coated with silica nanoparticles. Surf Coatings Technol 202(24):6103–6108
66. Zimmermann J, Reifler FA, Fortunato G, Gerhardt LC, Seeger S (2008) A simple, one-step approach to durable and robust superhydrophobic textiles. Adv Funct Mater 18(22):3662–3669
67. Zhao Y, Tang Y, Wang X, Lin T (2010) Superhydrophobic cotton fabric fabricated by electrostatic assembly of silica nanoparticles and its remarkable buoyancy. Appl Surf Sci 256(22):6736–6742
68. Xue C-H, Jia S-T, Zhang J, Tian L-Q (2009) Superhydrophobic surfaces on cotton textiles by complex coating of silica nanoparticles and hydrophobization. Thin Solid Films 517(16):4593–4598
69. Zhou C, Chen Z, Yang H, Hou K, Zeng X, Zheng Y, Cheng J (2017) Nature-inspired strategy toward superhydrophobic fabrics for versatile oil/water separation. ACS Appl Mater Interfaces 9(10):9184–9194
70. Nanotex–Stain, Moisture, Odor & Wrinkle Resistant Apparel Fabrics. [Online]. Available: https://www.nanotex.com/. Accessed on 07 Jul 2019
71. Schoeller Textil AG, Nanosphere–Technologies, Schoeller Textiles AG. *Schoeller Website*, 2019. [Online]. Available: https://www.schoeller-textiles.com/en/technologies/nanosphere. Accessed on 07 Jul 2019
72. Wang R, Hashimoto K, Fujishima A, Chikuni M, Kojima E, Kitamura A, Shimohigoshi M, Watanabe T (1997) Light-induced amphiphilic surfaces [4]. Nature 388(6641): 431–432. Nature Publishing Group, Jul-1997
73. Jiang Y, Wang Z, Yu X, Shi F, Xu H, Zhang X, Smet M, Dehaen W (2005) Self-assembled monolayers of dendron thiols for electrodeposition of gold nanostructures: Toward fabrication of superhydrophobic/superhydrophilic surfaces and pH-responsive surfaces. Langmuir 21(5):1986–1990
74. Sun T, Wang G, Feng L, Liu B, Ma Y, Jiang L, Zhu D (2004) Reversible switching between superhydrophilicity and superhydrophobicity. Angew Chemie-Int Ed 43(3):357–360
75. Xia F, Feng L, Wang S, Sun T, Song W, Jiang W, Jiang L (2006) Dual-responsive surfaces that switch between superhydrophilicity and superhydrophobicity. Adv Mater 18(4):432–436
76. Xu L, Chen W, Mulchandani A, Yan Y (2005) Reversible conversion of conducting polymer films from superhydrophobic to superhydrophilic. Angew Chemie-Int Ed 44(37):6009–6012
77. Russell TP (2002) Surface-responsive materials. Science 297(5583): 964–967. American Association for the Advancement of Science, 09-Aug-2002
78. Motornov M, Minko S, Eichhorn KJ, Nitschke M, Simon F, Stamm M (2003) Reversible tuning of wetting behavior of polymer surface with responsive polymer brushes. Langmuir 19(19):8077–8085

79. Zhang J, Lu X, Huang W, Han Y (2005) Reversible superhydrophobicity to superhydrophilicity transition by extending and unloading an elastic polyamide film. Macromol Rapid Commun 26(6):477–480
80. Wang R, Sakai N, Fujishima A, Watanabe T, Hashimoto K (1999) Studies of surface wettability conversion on TiO2 single-crystal surfaces. J Phys Chem B 103(12):2188–2194
81. Feng X, Zhai J, Jiang L (2005) The fabrication and switchable superhydrophobicity of TiO2 nanorod films. Angew Chemie-Int Ed 44(32):5115–5118
82. Zhang P, Lv FY (2015) A review of the recent advances in superhydrophobic surfaces and the emerging energy-related applications. Energy 82: 1068–1087, Pergamon, 15-Mar-2015
83. Liu B, Wang L, Gao Y, Tian T, Min J, Yao J, Xiang Z, Huang C, Hu C (2014) Synthesis and characterization of photoreactive silica nanoparticles for super-hydrophobic cotton fabrics application. Text Res J 85(8):795–803

Self-cleaning Finishes for Functional and Value Added Textile Materials

Subhankar Maity, Kunal Singha and Pintu Pandit

Abstract General consequences with textiles and clothing are becoming dirty after daily use and required frequent washing with detergent and water resulting wastage of time, money, and water. Recent trends is to develop self-cleaning textiles and clothing which can clean themselves without water and detergent. It is a natural tendency of textile surface to catch foreign particles as dirt as well as helps to grow bacteria. The broader characteristics of today's self-cleaning textiles are water and oil repellency and anti-bacterial efficacy. Nature has plenty of evidences of such self-cleaning effects, e.g. Lotus leaf. Learning from nature mimicking of lotus effect can be achieved in textile substrate to achieve super hydrophobic surface that cannot be wetted by liquid and liquid droplet rolls over the surface like a pearl and clean the surface as well. Such self-cleaning effect can be achieved by coating the textile substrates with some active agents as functional finish. Various nanoparticles like TiO_2, ZnO, Ag etc. can be applied on textile surface by various means to achieve such self-cleaning anti-microbial effects. This chapter review those nanotechnologies, materials, characteristics and limitations of such self-cleaning textiles in brief.

Keywords Self-cleaning · Nanotechnology · Photo-catalyst · Super-hydrophobic · Contact angle · Photo-catalytic · Lotus effect · Nanoparticle

1 Introduction

Self-cleaning textiles are the materials which can clean themselves without laundering. This property can be achieved in textiles by making their surface repellent to water and soil. Water and oil repellency is related to the surface energy and contact angle with the contact fluid. The contact angle is defined as the angle between the

S. Maity (✉)
Department of Textile Technology, Uttar Pradesh Textile Technology Institute, Kanpur 208001, India
e-mail: maity.textile@gmail.com

K. Singha · P. Pandit
Department of Textile Design, National Institute of Fashion Technology, Ministry of Textiles, Govt. of India, NIFT Campus, Mithapur Farms, Patna 800001, India

M. Shahid and R. Adivarekar (eds.), *Advances in Functional Finishing of Textiles*, Textile Science and Clothing Technology, https://doi.org/10.1007/978-981-15-3669-4_9

solid textile surface and droplet of contact-fluid. The wettability of the solid surface depends on the surface energy of the solid and the liquid. Depending on the difference in surface energy the contact angle is subtended in smooth surface. There are wettability theories proposed in literature for smooth and rough surfaces. For smooth surface, Young model (Eq. 1) was proposed which is based on the three interfacial energies per unit area which are in equilibrium at the droplet resting on solid surface.

$$\gamma_{sv} = \gamma_{sl} + \gamma_{lv}\cos\theta \tag{1}$$

where, θ is the contact angle as shown in Fig. 1, γ_{sv} and γ_{sl}, γ_{lv} are interfacial energies per unit area of the solid-vapor, solid-liquid, and liquid-vapor interfaces, respectively.

If the contact angle subtended by a fluid droplet on the solid surface is less than 90°, the solid surface is termed as a hydrophilic surface corresponding to the same fluid. When the contact angle is >90°, the surface is termed as a hydrophobic surface corresponding to the fluid. If the contact angle is approaching to zero the solid surface is becoming super hydrophilic and inversely, when the contact angle is >150° the surface is super hydrophobic. A super hydrophobic surface creates the water or oil repellency leading to self-cleaning effect. Various means of water and soil repellency finish of textiles have been one of the major research focus since few decades. This increasing interest towards development of self-cleaning textiles is due to their ability to reduce cost of cleaning and henceforth commercial success. There are various materials and techniques are available for preparation of water, oil and soil repellent textiles with self-cleaning effect. They are especially surface treatments to achieve the condition of limited wettability or repellency which leads to the concept of self-cleaning textiles. Super repellent surface or self-cleaning surface is already available in nature itself [1]. A common example of super hydrophobic self-cleaning effect in nature is lotus leaf, where water droplets can roll out of surface without wetting and contaminating. The surface of lotus leaf is thus have ability of self-cleaning by repelling water and dirt. We need to understand such natural phenomenon and apply the knowledge of chemistry and physics to achieve similar self-cleaning effect on textile surface. The lotus leaf consists of two levels of architecture viz. micro-scale bumps and nano-scale hair-like structures coupled with some waxy chemicals. Researchers are developing artificial self-cleaning textiles by the concept of this architecture of lotus leaf. There are irregular epicircular wax crystals present of the

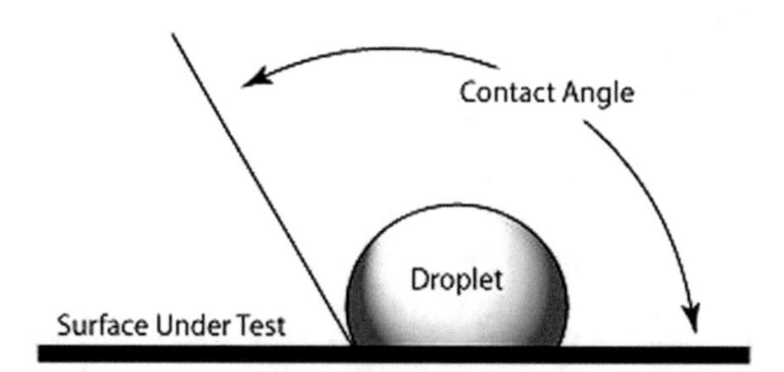

Fig. 1 Contact angle for surface wetting

surface of the lotus leaves making an uneven micro-texture. Air molecules are thus easily trapped in interstitial spaces of the rough surface resulting less adherence of water and dirt molecules by reducing adhesive forces. Water molecules appear like sphere over such surface and roll over easily and in the course remove the dirt particles away from the surface [2–9]. Self-cleaning effect can be achieved by textile materials either by preparing a hydrophobic surface or by some hydrophilic coatings. Both of these types of surface coatings can help to clean textile surface without laundering.

2 Principles of Preparation of Self-cleaning Textiles

There are two principal methods of preparing self-cleaning textiles. It can be prepared either by coating of textile surface with super hydrophobic materials (such as silicones, fluorocarbons etc.) or by coating with some functional hydrophilic materials by the route of nanotechnology.

A liquid droplet can wet a solid surface when the surface tension of a solid is higher than the surface tension of the liquid. Therefore, the surface tension of the solid need to be reduced than that of the liquid for achieving liquid repellency. Fluorocarbons are the carbon compounds which contain perfluorinated carbon chain possessing a very low surface tension of about 10 dyne/cm. During the application process, the fluorocarbons form a coat of thin layer around the textile surface. As a result, surface tension of the coated textiles becomes lower than that of water and water repellency effect is achieved. A water droplet then does not adhere to the textile surface and rolls out off. Silicones are actually organosilicon compounds which are highly explored for preparation of super hydrophobic textile surfaces. There are various approaches available in literature for preparation of super hydrophobic self-cleaning surfaces using silicones. PDMS (polydimethylsiloxane) is one of the popular silicone, can be used for surface modification of textiles by exciting CO_2 pulsed laser to introduce peroxide groups onto the PDMS surface to create a rough surface. These peroxide groups assist graft polymerization of 2-hydroxyethylmethacrylate (HEMA) onto the PDMS. By this method excellent hydrophobic surface achieved with water contact angle of about 175°. But these hydrophobic coating processes have drawback in terms of durability of the coat which is not satisfactory and in case of cotton material this is found to be very poor. Other demerit is hazardous effect of the fluorine compounds which reacts with biological issues and causes skin irritation [1, 3, 10–14]. Nanotechnology is relatively a new approach of achieving self-cleaning effect for textiles. This route is proved to be technically viable as well as economically successful. Various approaches of preparing self-cleaning textiles by this route are proposed in literature by various researchers. Most widely described approach is by the applications of photo catalyst like TiO_2. Other methods are using silver nanoparticles, carbon nanotubes, colloidal metal oxide, N halamine, microwaves irradiation, etc.

3 Photo-Catalytic Self-cleaning Effect

Unlike hydrophobic surfaces which are based on rolling over of water droplets to clean the adhering dirt from the surface, hydrophilic coatings of a photo-catalyst chemically break down the dirt/foreign molecules in the exposure of sunlight. This is the photocatalytic self-cleaning effect. Titanium dioxide (TiO_2) and zinc oxide (ZnO) nano particles are commonly coated over textile surface which are acting as photocatalysts. The thickness of the coating is typically in the range of 20 nm. When these nanoparticles are irradiated with ultra-violet rays of sunlight that has energy higher than their band gap then valance electrons are excited to jump into conduction band. These conductive electrons (e^-) form O^{2-} radical ions in presence of atmospheric oxygen. The O^{2-} radical ions are unstable and combined with contaminated dirt particles, pollutants, and micro-organisms which are generally organic compounds. The reaction is resulted in the formation of carbon dioxide (CO_2) and water (H_2O). TiO_2 or ZnO act as catalysts only and never used up in the reaction process and destroy dirt molecules, organic matters and micro-organisms from the textile surface providing self-cleaning effect in presence of sunlight. The mechanism is demonstrated in Fig. 2 [7, 15–23].

Self-cleaning anti-microbial cotton fabrics are prepared by impregnating the same in a dispersion solution of TiO_2 nanoparticles. This TiO_2 nanoparticle enriched fabrics can kill bacteria in sunlight [24]. The antibacterial functionality of the TiO_2-enriched cotton fabrics is attributed to the destruction of the bacteria cell wall and membrane by O^{2-} radical ions which are generated in course of photocatalytic reactions. In another study, it is reported that TiO_2 loaded cotton textiles eliminate the stains of wine, coffee, tea etc. by destroying chromophore(s) of the stains by the irradiation of ultra violet rays of sunlight [25]. Such self-cleaning functionality of the TiO_2 loaded textiles is attributed to the formation of highly oxidative intermediates generated at textile surface as shown in Fig. 3. The mechanism of dismissing of the stains and release of CO_2 due to light irradiation is shown in Eq. (2).

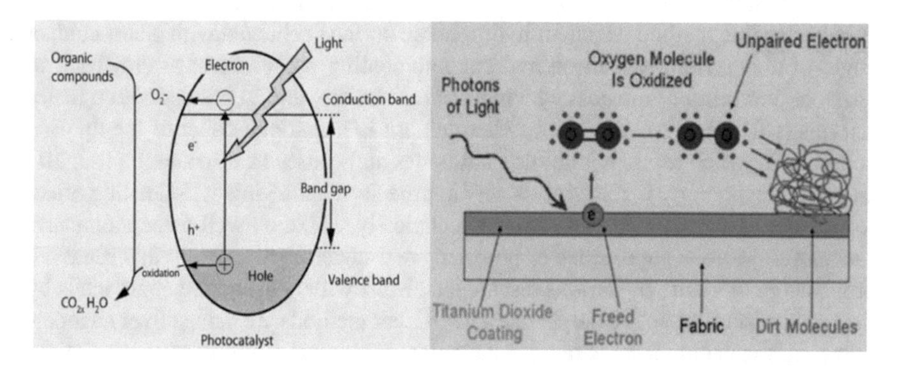

Fig. 2 Photocatalytic self-cleaning of titanium dioxide coated on textile surface [19]

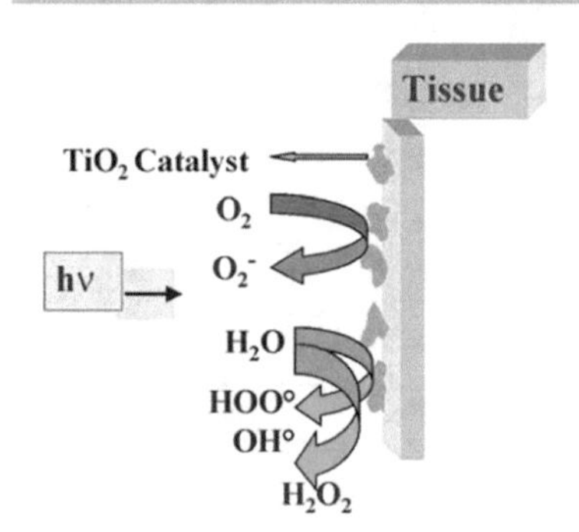

Fig. 3 Photocatalytic oxidative intermediates produced by TiO_2 on a cotton tissue in the presence of O_2 and H_2O vapor. Reprinted from Ref. [25], Copyright (2019) Elsevier

$$C_xH_yN_vS_w + hv + H_2O_w + O_2 \rightarrow CO_2 + H_2O + SO_p + NO_q \quad (2)$$

In other studies, self-cleaning functionality is bring about on cotton textiles by coating with TiO_2 film as well as loading of AgI particles which enables the fabric to be cleaned in visible light only [26]. This visible light irradiated photocatalytic effect of the AgI–Nano-TiO_2 coated cotton textiles is reported significantly better than simply TiO_2 treated cotton textiles. This reveals that AgI effectively assists the photocatalytic activity of TiO_2 and the effect sustains for several numbers of photodegradation cycles [26].

4 Self-cleaning Effect Using Microwaves

Nanoparticles have poor affinity to textile surface attributed to poor washing fastness. Microwave technology is developed to attach nanoparticles onto textile surface. The functionality of the nanoparticles is enhanced towards better self-cleaning functionality by attaching some chemicals those can repeal water, oil or bacteria. These duel nanoparticles-functional chemicals create protective layer over textile surface and kill bacteria, repel fluid and dirt. This technology is developed by scientists working in U.S. Air Force, and presently applied to prepare anti-microbial t-shirts and underwear which can be worn hygienically for weeks without washing as shown in Fig. 4 [19].

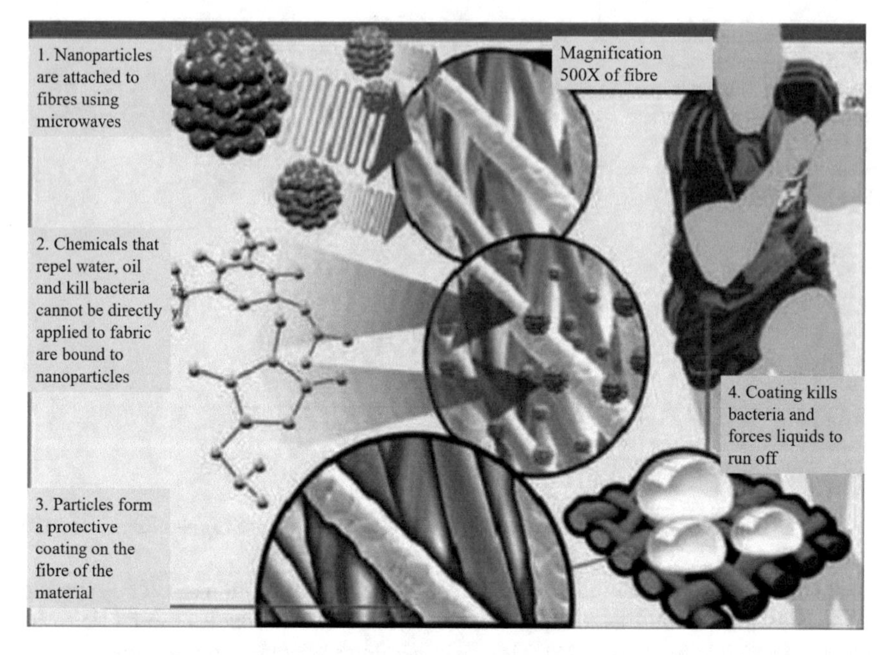

Fig. 4 Self-cleaning clothing fibers using microwaves [19]

5 Self-cleaning Effect Using Carbon Nanotubes

Mimicking of lotus leaf surface is a suitable method of preparing self-cleaning surface where nanosize rods are vertically arranged in regular pattern creating a rough micro-surface. Controlled assembling of carbon nanotubes over textile surface can serve the purpose and which is successfully achieved by researchers. By this method, carbon nanotubes are assembled on cotton surface and water contact angle greater than 150° is achieved. Such cotton fabrics are super hydrophobic in nature and carbon nanotube being electro-conductive the coated fabrics can exhibit various sensory functions [27]. The super hydrophobic functionality of micro-structured surface contained with carbon nanotubes is shown in Fig. 5 [28]. It is reported that the self-cleaning performance of the fabric does not deteriorate even after multiple use. In another study, it has been reported that fluorinated carbon nanotubes have better performance than that of ordinary carbon nanotubes. Fluorinated carbon nanotubes are vertically arranged on textile surface that exhibit better hydrophobicity and excellent self-cleaning performance [29].

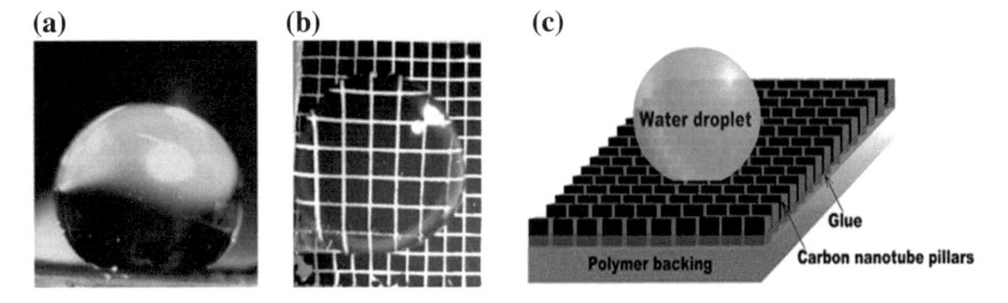

Fig. 5 Superhydrophobic performance of micro-scale carbon nanotube pillars of 250 μm in width and 100 μm in height. **a** A10μL water droplet sitting on the surface of vertically aligned carbon nanotubes. **b** Top view of water droplet sitting on the micro-scale carbon nanotube pillars. **c** Schematic diagram showing carbon nanotube pillars held at base by polymer adhesive and a water droplet sitting on top of pillars [28]. Reprinted/adapted from Ref. [28], Copyright (2019) American Chemical Society

6 Self-cleaning Effect Using Silver Nanoparticles

Silver (Ag) nanoparticles are loaded onto textile surface for multipurpose applications. Silver is an excellent antimicrobial agent. Being having higher specific area its nanoparticles is a highly active material. Coating textile materials with silver nanoparticles brings about self-cleaning effect. The silver particles destroy various organic compounds such as dirt, contaminants as well as micro-organisms resulting minimal washing of cloths. The Ag nanoparticles exhibit water repellency effect by creating nano-whiskers over textile surface which are made of hydrocarbons and have about 1/1000th of the size of a typical cotton fiber. The Ag nanoparticles create a fuzz effect on the textile fibre surface without deteriorating the tensile properties of the fibre as shown in Fig. 6. Ag nanoparticles are also used in conjunction with TiO_2 particles to coat over textile surface either in colloidal form or in particular

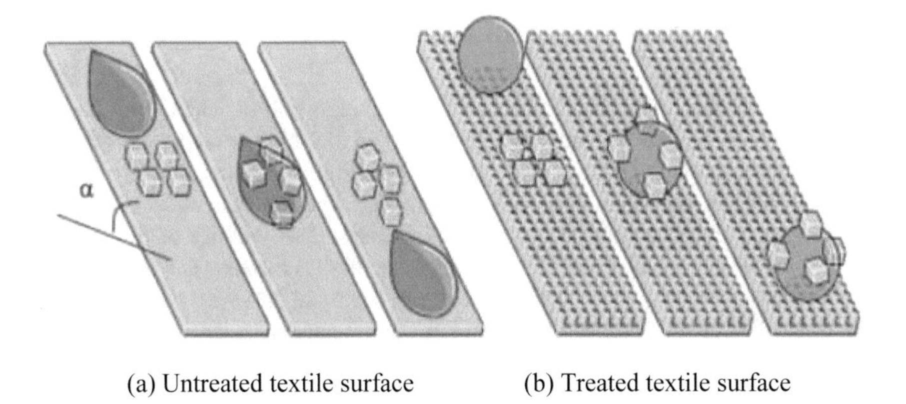

Fig. 6 Silver nano-particles used for self-cleaning textile

form for improved functionality. The fixation of the particles on textile surface can be improved by high temperature curing. The high temperature curing treatment on cotton or polyester fibres produces activated surface induced by oxygen containing diverse polar groups. These polar groups increase the synergy of Ag blending with TiO_2 on textile surface. High frequency plasma treatment in presence of oxygen and vacuum UV lead plasma treatments are also tried for increasing adhesion of Ag and TiO_2 on textile surface [30].

7 Self-cleaning Effect by Using Colloidal Metal Oxide

A colloidal solution of suitable metal oxide particles is to be prepared and textile fabric is to be dipped into it followed by a through heat treatment process to create certain roughness on fibre surface in nanometer scale. By this treatment fabric become water repellent with water contact angle above 150°. Synthetic textiles can be coated with TiO_2 by this colloid suspension method [31]. These TiO_2 coated fabrics are able to remove the stain of tea, coffee, wine etc. under visible light with time. The durability of the TiO_2 coating on textile surface achieved by this process is found to be satisfactory for multiple use. In another study, a blended colloidal solution is prepared by mixing TiO_2 powder in titanium isopropoxide (TTIP) colloid and wool/polyamide, polyester fabrics are processed through this colloid. The treated textiles are able to discolor wine and coffee stains under solar radiation [31]. The colloidal coating is found to be stable and excellent stain removal potential and it is also can perform under neon light. TiO_2–SiO_2 sol-gel preparation is proposed to be as a transparent photoactive coating that can apply to textiles at low temperature without damaging textile surface. This preparation is reported to be better photo-catalytic agent than that of TiO_2 alone [31].

8 Self-cleaning Effect by Using N Halamine

Chlorine is a well known disinfectant used for killing bacteria. The chlorine atom present is N-halamines are successfully explored for biocides applications. N-halamines are heterocyclic organic compounds containing at least one covalent bond between nitrogen atom and a halogen atom (N–X). In case of stable N-halamines the halogen is chlorine (N–Cl) in most of the cases. The stability of N–Cl bond depends on the chlorination reaction by which the bond is formed. The chlorination of amine, amide and imide groups are generally occurs in dilute hypochlorite solution. N-halamines are biocides which can kill a broad range of micro-organisms like bacteria, fungi, viruses etc. This anti-microbial effect is attributed to the capability of electrophilic substitution of chloride ion (–Cl) with hydrogen ion (–H) situated in N–Cl. This substitution reaction occurs in presence of water (H_2O) resulting to the transfer of Cl^+ ions that can bind to acceptor regions on microorganisms. As a result the enzymatic and metabolic processes of micro-organisms are hindered and they are dismissed. Once the N-halamines perform a reaction to kill bacteria, N–Cl bonds are converted to N–H bond which are inactive and does not have antimicrobial properties. Therefore, regeneration of the same is required by treatment with dilute hypochlorite solution. N-Halamines can be applied to broad range of textile substrates including cellulose, polyamide, and polyester to make them anti-microbial [32]. Though N-halamines contain chlorine it is not toxic since toxic chlorine gas is not generated during the process. The N-halamine-treated textiles can kill microorganisms almost instantly on contact, and therefore, they are found to be best suited for hygiene and medical applications like uniforms, bedding, towels, wipes etc.

9 Applications of Self-cleaning Textiles

Self-cleaning textiles and garments retain their original texture and feel after the chemical treatment. Textiles which can keep themselves clean can save a lot of water, detergent and energy by avoiding frequent washing. Using above mentioned technologies self-cleaning fabrics can prevent dirt, oil, and also act as a disinfectant. Moreover, the self-cleaning effect keeps the textiles long lasting and fresh looking than ordinary fabrics. Anti-dirt, anti-bacterial and self-cleaning clothes can be used in medical, sports, defense, and home textiles widely. Few commercial products are available in global market with this self-cleaning quality. One example is Mincor® TX TT which can be used for tailoring of outdoor textiles like tents, sunshades, flags umbrellas and sails. NanoTex® is another product suitable for apparels like men's dress materials aprons, gloves, shirts, etc. Nanosphere® is mainly used for preparation of men's shirts with self-cleaning effect. Nano-whisker surface is created on textile surface and water droplets along with dirt particles rest only on the peaks of the whisker, and as a result there is lower contact area with textile surface as shown in Figs. 7, 8 and 9. Due to lower contact area, surface adhesion is reduced significantly. Water droplet thus rolls out off the textile materials and dirt particles either repelled or can simply be rinsed off automatically. Such effect is called self-cleaning when textiles require no or very less washing. Washing conditions even will be very gentle at low temperature with minimum requirement of soap or detergent. Such products also possess excellent durability in terms of abrasion resistance and washing fastness. The aesthetic appeal, hand and breathability are not affected even after numerous washing cycles. Therefore, NanoSphere® is marketed as an ideal product for use outdoor apparel, sportswear, men's and women's wear, work wear, shoes cover and home furnishings [2].

Fig. 7 The NanoSphere® surface

Fig. 8 High level of water resistance

Fig. 9 Durable protective function

10 Limitations of Self-cleaning Fabric

Self-cleaning textiles are becoming practical and economical in the era of advancement of nanoscience and nanotechnology. Advancements in nanoscience and nanotechnology both practical and economical. Industries readily accept the technologies in commercial production of self-cleaning textiles which made the products readily available in market and this could really obsolete the washing machines, laundry detergents. However, there are few shortcomings limiting the efficiency and performance of self-cleaning. Wash fastness of the coated textiles is one of the limitations. All the agents are not suitable for all textile substrates. Another factor is the irradiation time in sunlight. Sunlight as a sole source of energy of photocatalytic self-cleaning required to be sufficient in terms of intensity and duration. A tea-shirt having tea stain required to be exposed in sunlight for a whole day for the removal of the stain. Therefore, it is a time consuming task. Sometimes, it is not problematic for military persons who wear the cloth and stand outside sun for prolong time during his duty and clean their clothes. The intensity of light also plays a big role because the excitation of electrons in the valence band of TiO_2 depends on it and unless the electrons hope to conduction band the cleaning process does not commence. The electrons in valance band must react with atmospheric oxygen which causes depletion of oxygen arising environmental concern.

11 Conclusion

There are various approaches reported in literature for preparation of self-cleaning textiles. Broadly they can be divided in two ways such as either by making surface super hydrophobic (coating with silicones, fluorocarbons etc.) making the surface repellent to water, oil soil etc. or by coating with some functional hydrophilic finishes by the route of nanotechnology. Nanotechnology has been found very promising in this regard because by application of fluorocarbons etc. on textile materials made them hazardous and cause skin irritation. Application of nanoparticles, carbon nanotubes, N-halamines etc. in various forms brings about the properties of self-cleaning textiles and researches are going on in this area for further improvement. At present, there are some limitations or demerits with existing self-cleaning textiles in practical and economical point of view. The finish cannot be applied to all kind of material and longer time is required for cleaning treatment in sufficient sunlight.

References

1. Forbes P (2008) Self-cleaning materials: lotus leaf-inspired nanotechnology. Scientific American Magazine, 30 July 2008, pp 1–5
2. Online available: www.nano-sphere.ch

3. Rao TN, Srinivasan G (2010) Anti-bacterial and self-cleaning textiles. Nanotech 1(1):1–4
4. Anonymous, The Lotus effect–self-cleaning for textiles. AATCC Rev 7(7):17–17 (2007)
5. Anonymous, Emulating nature: self-cleaning effects for textiles. Riv delle Tecnologie Tessili 21(6):59–62
6. Anonymous, BASF to showcase self-cleaning textiles. Indian Text J 117(8):90
7. Anonymous, Schoeller and Clariant partnership to exploit self-cleaning finish. Advances in textiles technology, pp 3–4 (2007)
8. Anonymous, Self-cleaning technical textiles. Future Mater Issue 2, pp 10–11 (2007)
9. Parthasarathi V (2008) Super functional finish for apparels. Text Mag 49(3):90–94
10. Du J, Zhang J, Tan P, Joo Lee H, Duncan S (2012) Self-cleaning super oleophobic surfaces. Fiber Society symposium, Spring, pp 1–2
11. Schuler P (2012) Creator of self-cleaning and antibacterial textile. Tut: Text l'Usage Tech 84:8–8
12. Habersack LM et al (2011) Oleophobic coating for self-cleaning, fluid-resistant textiles. Fiber Society symposium, pp 134–135
13. Xin JH (2005) Self-cleaning textile surface treatments using nanotechnology. Text Asia 36(11):60–61
14. Gulrajani ML (2006) Nano finishes. Indian J Fibre Text Res 31(1):187–201
15. Yuranova R et al (2006) Self-cleaning cotton textiles surfaces modified by photoactive SiO_2/TiO_2 coating. J Mol Catal A: Chem 244:160–167
16. Nakajima A, Koitzumi S, Watanabe T, Hashimoto K (2000) Pho-toinduced amphiphilic surface on polycrystalline anatase TiO_2 thin films. Langmuir 16:7048–7050
17. Kasanen J, Suvanto M, Pakkanen TT (2009) Self-cleaning, titanium dioxide based, multilayer coating fabricated on polymer and glass surfaces. J Appl Polym Sci 111:2597–2606
18. Liu N, Sun G, Zhu J (2011) Photo-induced self cleaning functions on 2-anthraquinone carboxylic treated cotton fabrics. J Mater Chem 21(39):15383. https://doi.org/10.1039/c1jm12805a
19. El-Khatib EM (2012) Antimicrobial and self-cleaning textiles using nanotechnology. Res J Text Apparel 16(3):156–174
20. Veronovski N et al (2006) Self-cleaning textiles based on nano TiO_2 coatings. TEKSTILEC 49(10):213–217
21. Qi K et al (2011) Photocatalytic self-cleaning textiles based on nanocrystalline titanium dioxide. Text Res J 81(1):101–110
22. Bozzi A, Yuranova T, Kiwi J (2005) Self-cleaning of wool-polyamide and polyester textiles by TiO_2 modification under daylight irradiation at ambient temperature. J Photochem Photobiol A 172(1):27–34. https://doi.org/10.1016/j.jphotochem.2004.11.010
23. Nörenberg R, Badura W, Sylvia VK (2007) Self-cleaning technical textiles. Tech Text Tech Textilen 50(2):123–124
24. Wu D et al (2009) Synthesis and characterization of self-cleaning cotton fabrics modified by TiO_2 through a facile approach. Surf Coat Technol 203:3728–3733
25. Meilert KT, Laub D, Kiwi J (2005) Photocatalytic self-cleaning of modified cotton textiles by TiO_2 clusters attached by chemical spacers. J Mol Catal A: Chem 237:101–108
26. Wu D, Long M (2011) Realizing visible-light-induced self-cleaning property of cotton through tools toward the development of molecular wires: a coating N-TiO_2 film and loading AgI particles. ACS Appl Mater Interfaces 3(12):4770–4774
27. Liu Y, Tang J, Wang R, Lu H, Li L, Kong Y, Qi K, Xin JH (2007) Artificial lotus leaf structures from assembling carbon nanotubes and their applications in hydrophobic textiles. J Mater Chem 17:1071–1078, https://doi.org/10.1039/b613914k
28. Sethi S et al (2008) Gecko-inspired carbon nanotube-based self-cleaning adhesives. Nano Lett 2
29. Gulrajani ML, Gupta D (2011) Emerging techniques for functional finishing of textiles. Indian J Fibre Text Res 36:388–397
30. Kiwi J, Pulgarin C (2010) Innovative self-cleaning and bactericide textiles. Catal Today 151(1/2):2–7. https://doi.org/10.1016/j.cattod.2010.01.032

31. Yuranova T et al (2007) Synthesis, activity and characterization of textiles showing self-cleaning activity under daylight irradiation. Catal Today 122:109–117
32. Fouda MMG (2010) Antibacterial modification of textiles using nanotechnology. A search for antibacterial agents 4:47–72. Online available: pubs.acs.org/doi/abs/https://doi.org/10.1021/ie101519u

Insights into Phosphorus-Containing Flame Retardants and Their Textile Applications

Mohd Yusuf

Abstract In recent years, flame retardants (FRs) focused on eco-friendliness, eco-viable and durable, are in great social demand and one of the most growing area of research interest on account of increased awareness towards environmental concerns. In this regard, strategies are considered onto FRs for textiles as well as other substrates with their applicability and selectivity. Phosphorus-based FRs provide a foundation for the directed design of nontoxic FRs mainly because of its versatility, for example, it can act in both the condensed and gas phase, as an additive or as a reactive component, in various oxidization states, and in synergy with numerous adjuvant elements. Various P-moieties make valuable contribution and combinations including elemental, inorganic salts and organophosphorus compounds. This chapter highlights general insight into phosphorus-based flame retardants for polymeric systems with future R&D opportunities.

Keywords Flame retardants · Textile finishing · Phosphorus compounds · OPFRs

1 Introduction

Flame-retardation ability to the polymeric substrates such as textiles (woven or non-woven), plastics, rubber and others, is vastly needed because they have been applied to a myriad of applications for general and engineering purposes. Specific chemical bonding was shown in several polymeric materials (*i.e.* cotton, linen, hemp; silk and wool; nylon; polypropylene; polyester etc.) acquired specificity towards implanting flame-retardancy. Textile materials have been applied worldwide in both civilian as well as military fields due to their inherent and excellent properties such as air-permeability, softness, comfortableness, hydrophilicity etc. In a general opinion, fire can be described as the combustion cycle which is illustrated by a fire triangle for which three components are necessary to be fire to occur; one is heated, secondly a combustible fuel and thirdly a combustive process. The most important factor in the combustive process is the air (oxygen). According to the fact that fire or flame is

M. Yusuf (✉)
Department of Chemistry, YMD College, MD University, Nuh, Haryana 122107, India
e-mail: yusuf1020@gmail.com

M. Shahid and R. Adivarekar (eds.), *Advances in Functional Finishing of Textiles*, Textile Science and Clothing Technology, https://doi.org/10.1007/978-981-15-3669-4_10

often initiated from the burning of the textile materials which subsequently results in burns and even loss of human life, causing serious damage to furniture, carpets, upholstery, buildings, properties, etc. [1, 2]. Thus, flame retardant (FR) finishing of textile substrates is extremely necessary for many applications for the prevention of fire and for protection of human life. Most of the polymeric materials such as cellulosics, wool, nylon, polyesters, polyurathanes possess higher flammability [3], and therefore, required high performance flame retardant finishes to overcome flammable aspects.

FRs provide fire resistance ability to the textiles through the heat absorbing, the covering effect, inhibition of chain reaction and gas dilution phase [4]. In general, there are many chemical treatments that are commonly employed to impart flame retardant finish for textile/polymeric materials. Main six categories of FRs are highly discussed and accepted, for example; halogenated, formaldehyde-based, P-based, N-based, Si-based and other mixed formulations [5]. However, the purpose of FR finishes is to reduce the amount of heat that is supplied to the polymer system to be below the level for flame stability [6]. Halogenated, phosphorus and formaldehydes based compounds such as Proban, THPC-TMM (Tetrakis(hydroxymethyl)phosphonium chloride-Trimethylolmelamine) and Pyrovatex CP, have been widely employed as the commonest FRs to impart durable fire-resistant ability to the cotton substrates [3–7].

In early 1990s serious environmental concerns have been noticed concerning halogenated FRs, especially brominated flame retardants (BFRs). It is found that under severe thermal stress or when they were burnt in accidental fires or uncontrolled combustion, BFRs could form halogen-based dioxins and furan derivatives [8, 9]. Furthermore, it is noteworthy that the environmental and health concerns limited not only of BFRs, but also of other types of flame retardants and have been studied extensively at a global scale. Several scientific meetings and conversations were organized in the late 90 s onto flame retardants: uses, risk assessments and safety globally until the transition to Registration, Evaluation, Authorisation and Restriction of Chemicals (REACH), a European Union Regulation Authority came existence in 2006. In 2008, REACH, [10] (REACH, 1907/2006/EC) entered "No data no market" slogan which requires the basic health and environmental data to be submitted for all chemicals before commercialization for their safety evaluation [10]. Halogenated FRs generate poisonous substances on fire and combustion [2] whereas formaldehyde-based FRs release formaldehyde which found carcinogenic [3]. They are found to have adverse health effects in animals and humans, including endocrine and thyroid disruption, immunotoxicity, reproductive toxicity, cancer, and adverse effects on fetal and child development and neurologic function [9–13].

As a result, health and environmental hazards associated with these FRs driven R&D for identifying and utilizing safer alternatives. Because of the social concerns onto eco-preservation using eco-friendly FRs, the new FRs needs to be halogen-free and formaldehyde-free. Phosphorus, nitrogen, and silicon-containing compounds are generally considered as environment-friendly FRs, because they do not generate harmful substances to human-ecosystem on burning with fire and their synergistic effects [6, 14]. P-based FRs are found very effective inert towards fire (most effective in high oxidation states) mainly because of its characteristics such as (i) low water solubility, (ii) low volatility, (iii) low dose requirement, (iv) less degradation to possibly hazardous substances, and (v) no toxic emissions [2, 6]. Mechanistically, phosphorous based flame retardants, during a fire form poly and meta-phosphoric acids which form an oxygen-barrier layer [15] and commonly used due to the environmental scrutiny halogenated and formaldehyde-free FRs.

The flame retardant mechanism described for phosphorus containing flame retardants includes both a condensed and a vapor phase mechanism depending on the type of phosphorus compound and the polymer. Specific applications for red phosphorus, organophosphates, chlorophosphates and bromophosphates are described. The use of triarylphosphates in PVC, modified polyphenylene oxide, and polycarbonate/ABS is described. The chlorophosphates are used in polyurethanes and the bromophosphates in engineering thermoplastics. Flammability and mechanical properties are given for specific polymers [16].

Phosphorous based flame resistant materials have long been used since the 1940s–1950s. P-based FRs exhibit excellent fire inertness ability and found effective both in the vapour and condensed phases. They vary in oxidation states (0 to +5) and can be classified into elemental, inorganic, and organic or organo-phosphorus [17] categories.

1.1 Elemental Phosphorus as FR

Phosphorus (P^0) has several allotropic forms [18] out of which white phosphorus (WP) and red phosphorus (RP) are most common. WP is a white, soft, waxy solid consists of tetrahedral P_4 molecules, in which each atom is bound to the other three atoms by a single bond. It ignites spontaneously and is very toxic and reactive in nature, and therefore cannot be used as FR [17].

Samples of WP always contain red phosphorus in a very little amount and accordingly appear yellow. On heating, WP can be converted into RP in the absence of air. It is harder, denser, more stable, less toxic, less reactive than WP and polymeric in structure with P_4 units [19]. Although, it ignites easily but possesses thermal stability up to 450 °C (approx.) and thus, the ability to be used as sufficient FR agent [18]. RP is observed as an efficient FR especially for oxygen-containing polymers that work in the vapor and condensed phases [19, 20]. Among the high performance flame retardants (HFFR) additives, RP is a type of powerful FR and has been significantly used for polymeric moieties other than textiles such as polyethylene [21], poly(ethylene terephthalate) [22], nylon [23, 24] etc.

Moreover, the combination of RP with other HFFR additives, metal hydroxide or intumescent FR can improve the overall fire retardancy of highly flammable substrates like polyolefins (PO) blends have been investigated and reported as effective [25]. However, the main disadvantages of RP are because of poor thermostability, the evolution of highly toxic phosphine (PH_3) during the reaction with moisture and the lack of compatibility with synthetic resins [26]. A novel technology was developed to prepare microencapsulated RP with suitable filling/supported agent to minimize the associated problems. Wu et al. [27] investigated microencapsulated RP as FR agent for synthetic polymers. In this study, they conclude that the microencapsulation of red phosphorus efficiently improved its water absorption, thermostability, ignition point, and decrease the amount of phosphine evolution with 5% amount [27]. A similar study was carried out by Liu and Wang [26]. In this study, a composite system of RP encapsulated by N-based FR was used for polyamide 6 (PA6) due to higher N-P synergistic effects. The action and mechanisms of the NFR-microencapsulated RPFR on PA6 were investigated in terms of limiting oxygen index (LOI) by using vertical burning experiment (UL94), thermogravimetric analysis (TGA), and scanning electron microscope (SEM) observations. It was concluded that the NFR-microencapsulated RPFR combination possessed desired flame retardancy because of effective char-formation of the condensed phase and it also showed satisfactory mechanical properties as the result of the good compatibility between flame retardant and PA6 resin.

1.2 Inorganic Phosphorus-Based FRs

Inorganic phosphorus-based FRs were developed and commonly used in the nineteenth century, mainly phosphates and polyphosphates. However, the great scientist Gay Lussac in 1821 used ammonium phosphate solution to impart flame retardancy of theater curtains [19, 20]. Ammonium phosphates (APs) possess fairly fire retarding ability and prevent afterglow. Monoammonium phosphate ($NH_4H_2PO_4$) and diammonium phosphate ($(NH_4)_2HPO_4$) or mixtures of these two phosphates have good water solubility and found very effective for many substrates as FR, for example, textiles, cellulosic fibers, wooden and paper products [19, 28]. With respect to susceptibility to bloom out of the material, matrix is a down-manner of APs. The low susceptibility of APs introduces ammonium polyphosphates (APPs) which have higher susceptibility. APPs are moderately soluble in water with several crystalline forms that differ in molecular weight ratio and particle size. APPs have been heated with a small amount of urea to enhance the solubility [19]. APPs are used as the principal ingredients in intumescent FR coatings because of their decomposition temperature (greater than 256 °C). The decomposition of APPs produces phosphorus acid that will interact with the carbon source to produce a carbonaceous char [19, 20]. APPs are cheaper, low toxic, quite thermally stable than their organic counterparts, and good thermal stability, and can be used for other non-textile materials such as plastics, rubber, paper, epoxy resins and wood [29].

1.3 Organic or Organophosphorus-Based FRs

Organophosphorus-based FRs gain increased attention preventing the risks of fire in recent times because of their significant efficacy and environmentally safer nature over halogenated and formaldehyde containing FRs [30]. Organophosphorus FRs have been used from last few decades and widely used for several various polymeric consumer products like plastics, textiles, polyurathanes (PU), polyamides (PA), polyethyleneterephthalate (PET), epoxy-materials etc. [31, 32]. Organophosphorus compounds containing P–C bonds are developed extensively as FR additives due to their excellent thermal and hydrolytic stability as well as ease of generation. In a mechanistic pathway it was observed that during the fire, phosphorus compounds break down to phosphorus acid which blocks polymer's oxygen functionality, and therefore lead to the char-formation in the high ratio [33].

Wendels et al. [34] has published an exhausted and comprehensive review on organophosphorus compounds for the recent developments in organophosphorus flame retardants (OPFRs) having P–C bonding with their synthesis pathways and applications [34].

Nowadays, various products, large in numbers are commercially available based on organophosphorous compound. On the basis of carbon unit/moiety OPFRs can be broadly categorized into five main classes (Table 1) [17, 32–36]:

(i) *organophosphates*,
(ii) *organophosphonates*,
(iii) *organophosphinites*,
(iv) *organoposphine oxide*, *and*
(v) *organophosphites*.

1.3.1 Organophosphate FRs

Organophosphate FRs with a wide range in their polarity, solubility and persistence have been produced and used as significant FRs in substitutes to the stringent regulation in the use of brominated FRs. These FRs are widely used as flame retardants in various consumer products such as textiles, electronics, industrial materials and furniture to prevent the high risk of fire [1, 32]. A large number of OPFRs have been fabricated with varied P–C assemblies. Some commonest OPFRs are shown in Table 1.

Aliphatic: P–$C_{aliphatic}$ bond containing organophosphorus compounds have been utilized as good FRs to polymeric materials such as Dimethyl phosphate (DMP), Diethyl phosphate (DEP), Trimethyl phosphate (TMP), Tripropyl phosphate (TPP), Tri-isopropyl phosphate (TIPP), Tri-*n*-butyl phosphate (TNBP) Tris(2-butoxyethyl) phosphate (TBOEP) etc.

Cl-aliphatic: Chlorine-based P–$C_{aliphatic}$ bond containing organophosphorus compounds have superior flame retardancy. Examples of this subclass include

Table 1 Various OPFRs classes

OPFRs classes	Chemical structures	
Organophosphates *Aliphatic*	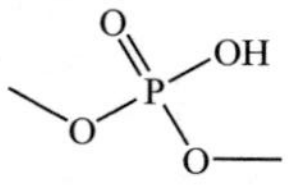 Dimethyl phosphate (DMP)	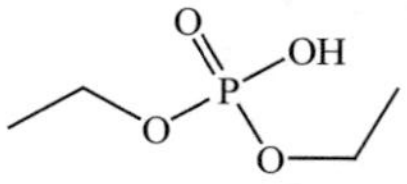 Diethyl phosphate (DEP)
	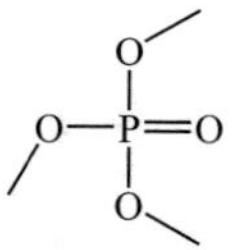 Trimethyl phosphate (TMP)	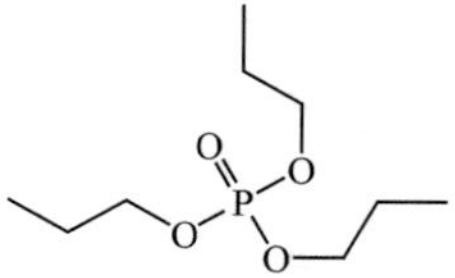 Tripropyl phosphate (TPP)
	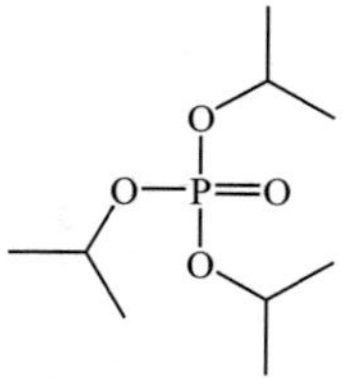 Tri-isopropyl phosphate (TIPP)	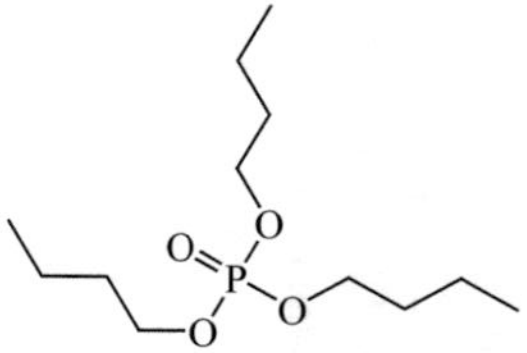 Tri-*n*-butyl phosphate (TNBP)

(continued)

Table 1 (continued)

OPFRs classes	Chemical structures	
	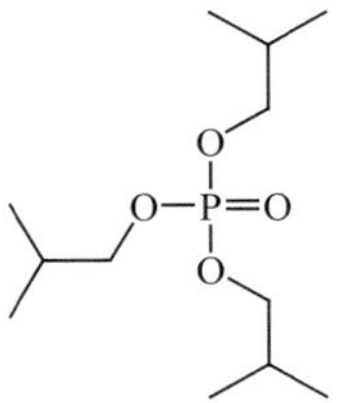 Triisobutyl phosphate (TIBP)	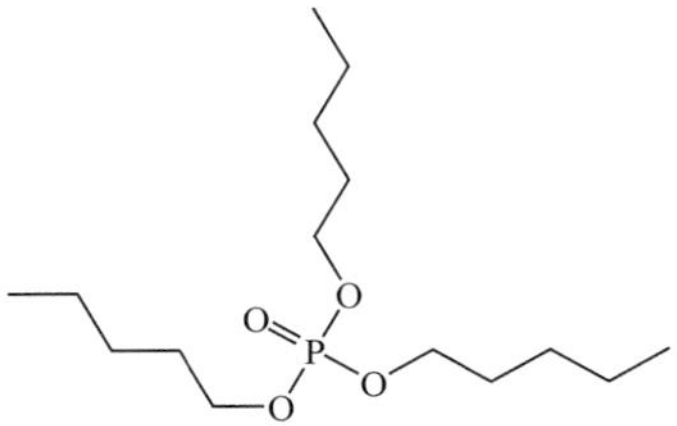 Tripentyl phosphate (TPeP)
	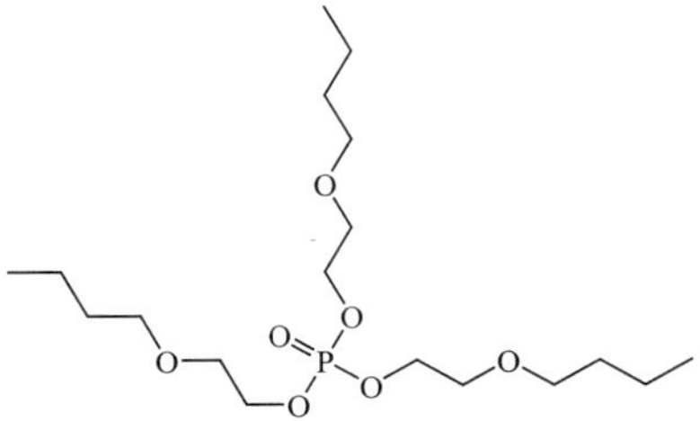 Tris(2-butoxyethyl) phosphate (TBOEP)	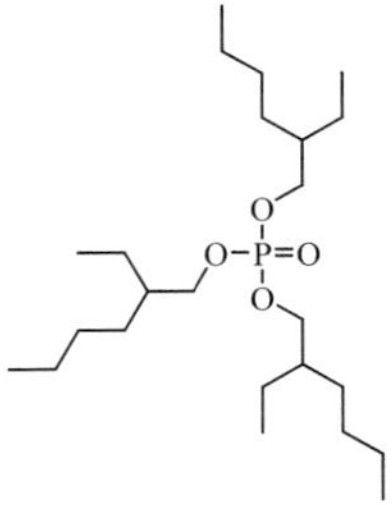 Tris(2-ethylhexyl) phosphate (TEHP)

(continued)

Table 1 (continued)

OPFRs classes	Chemical structures	
Cl-aliphatic	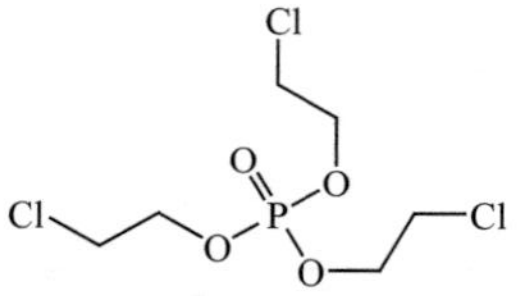 Tris(2-chloroethyl) phosphate (TCEP)	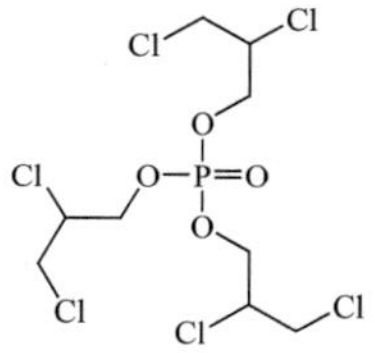 Tris(2,3-dichloropropyl) phosphate
	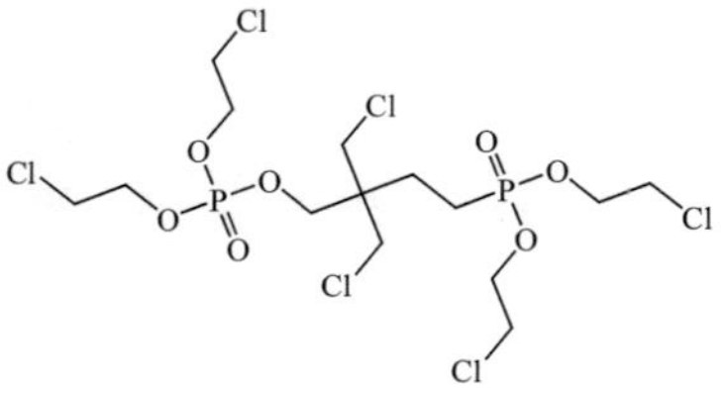 4-((bis(2-chloroethoxy))phosphoryl)-2,2-bis(chloromethyl)butyl bis(2-chloroethyl) phosphate (BCPBCBBCP)	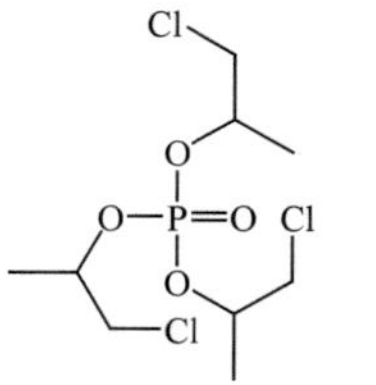 Tris(1-chloro-2-propyl) phosphate (TCPP)

(continued)

Table 1 (continued)

OPFRs classes	Chemical structures	
	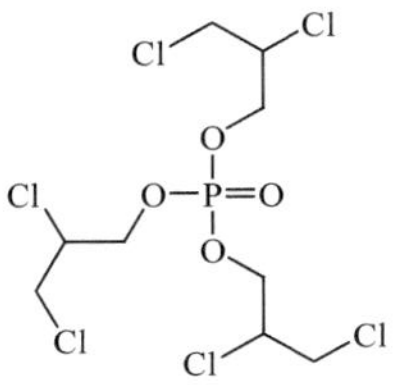 Tris(2,3-Dichloropropyl) phosphate (TDCPP)	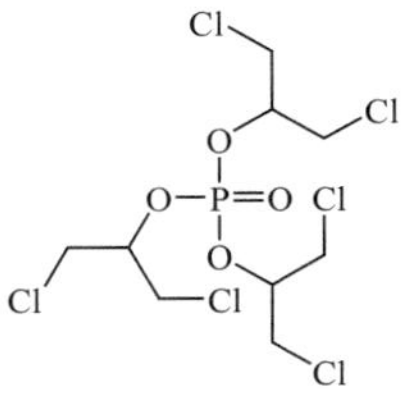 Tris(1,3-dichloro-2-propyl) phosphate (TDCPP)
	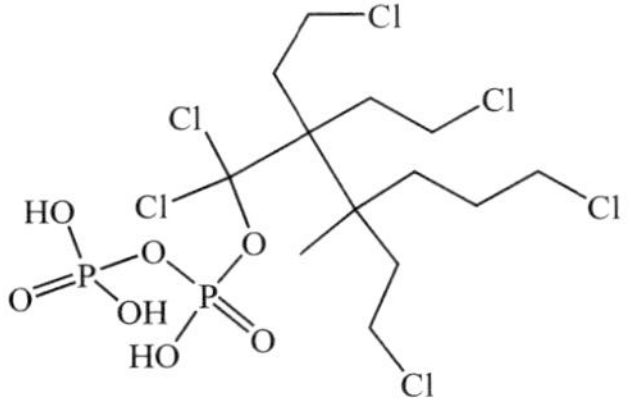 Tetrakis(2-Chloroethyl) dichlo-isopentyl diphosphate (TCDIDP)	

(continued)

Table 1 (continued)

OPFRs classes	Chemical structures	
Br-aliphatic	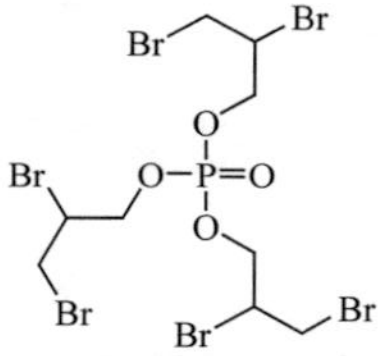 Br Br O Br O–P=O O Br Br Br Tris(2,3-dibromopropyl) phosphate (TDBPP)	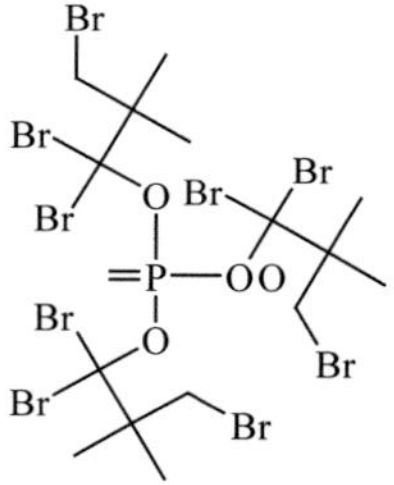 Br Br Br O Br Br =P—OO Br O Br Br Br Tris(tribromoneopentyl) phosphate (TTBPP)
Aromatic-aliphatic	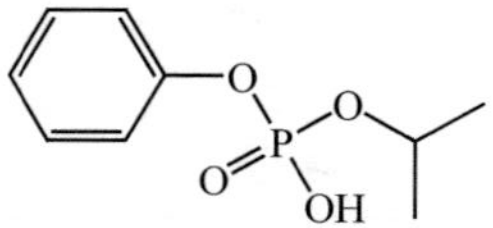 O O P O OH Phenylpropan-2-ylhydrogen phosphate (PPHP)	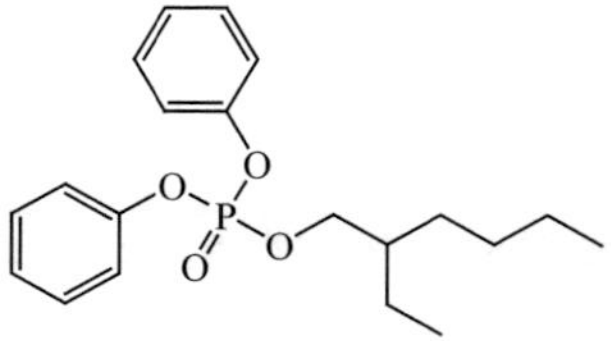 O O P O O 2-Ethylhexyldiphenyl phosphate (EHDPP)

(continued)

Table 1 (continued)

OPFRs classes	Chemical structures	
	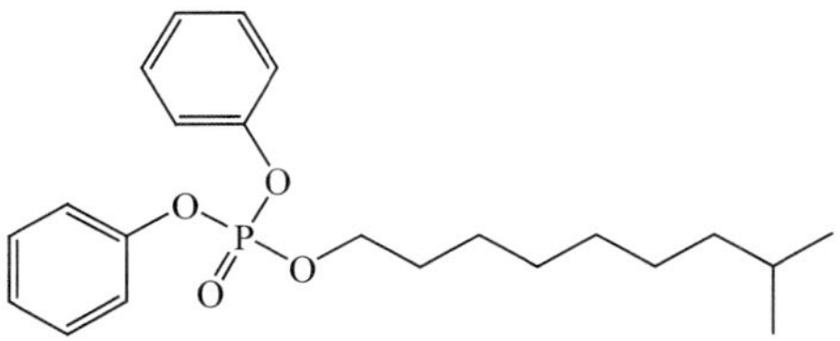 Isodecyldiphenyl phosphate (IDPP)	
Aromatic	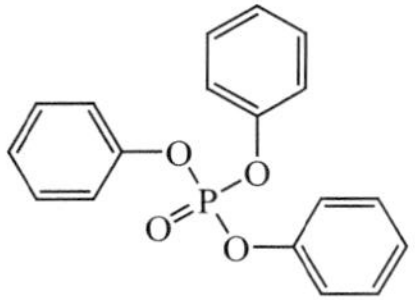 Triphenyl phosphate (TPP)	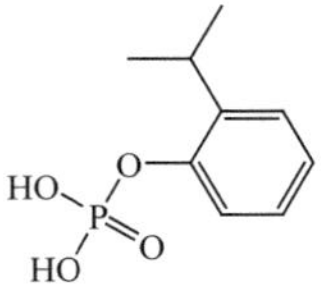 Isopropylphenyl phosphate (IPPP)
	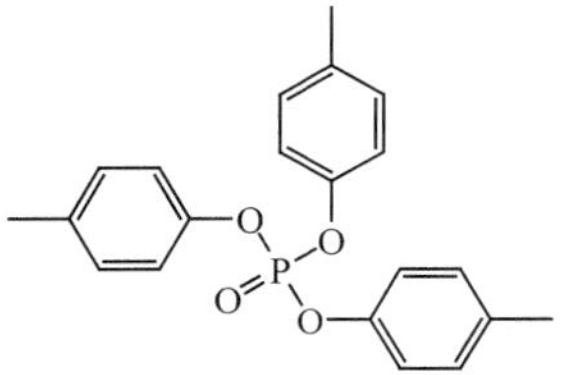 Tricresyl phosphate (TCP)	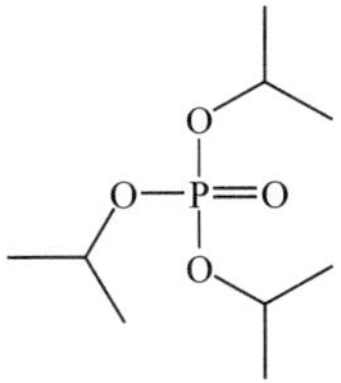 Triisopropyl phosphate (TIPP)

(continued)

Table 1 (continued)

OPFRs classes	Chemical structures	
	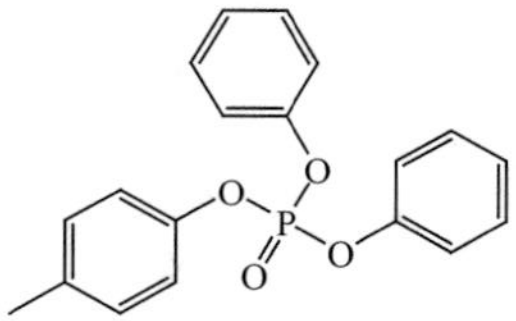 Cresyl diphenylphosphate (CDPP)	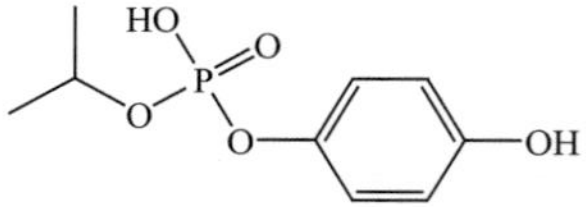 Phenol,4-(1-methylethyl) phosphate (PMEP)
	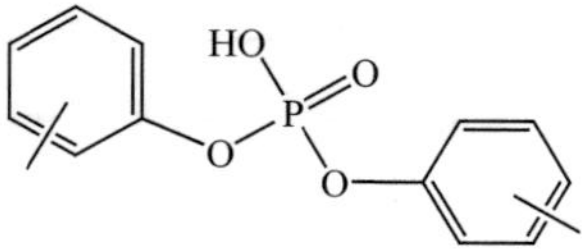 Alkylated diphenylphosphate (ADPP)	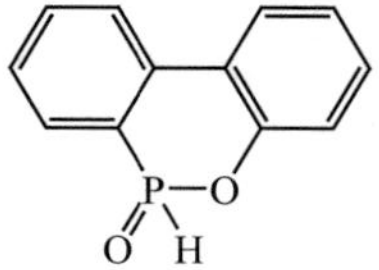 6H-Dibenz[c,e] [1, 2] oxaphosphorin,6-oxide (DOPO)
	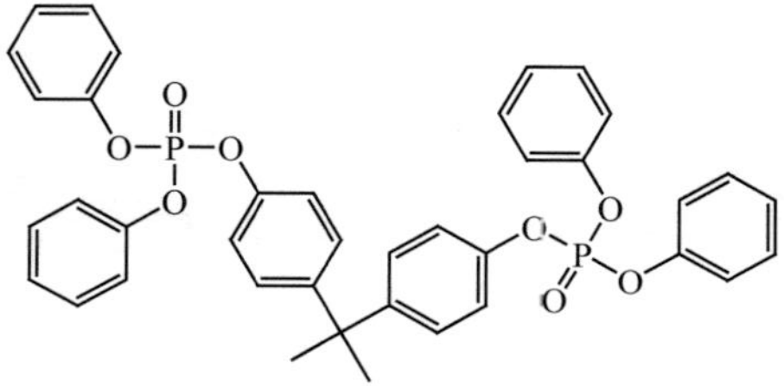 Bisphenol-A bis-(diphenylphosphate) (BPA-BDPP)	

(continued)

Table 1 (continued)

OPFRs classes	Chemical structures	
Organophosphonates *Aliphatic*	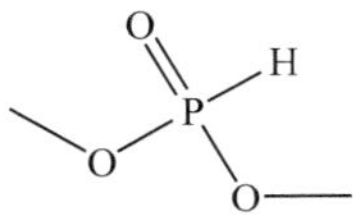 Dimethyl hydrogen phosphonate (DMHP)	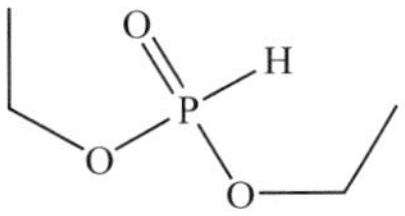 Diethyl hydrogen phosphonate (DEHP)
	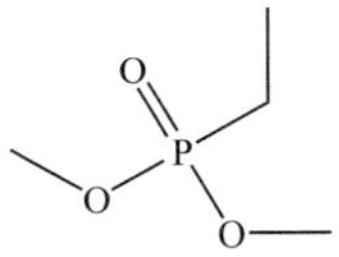 Dimethylethyl phosphonate (DMEP)	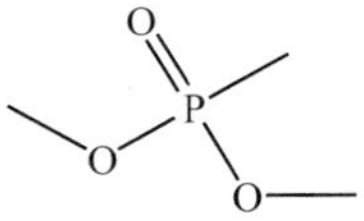 Dimethylmethyl phosphonate (DMMP)
	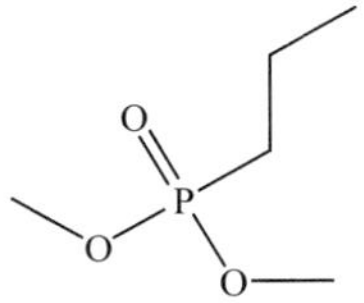 Dimethylpropyl phosphonate (DMPP)	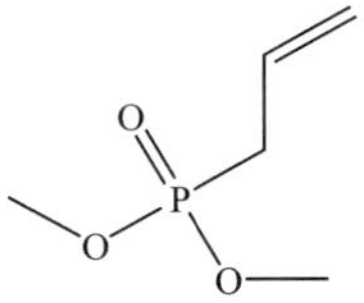 Dimethylallyl phosphonate (DMAP)

(continued)

Table 1 (continued)

OPFRs classes	Chemical structures	
	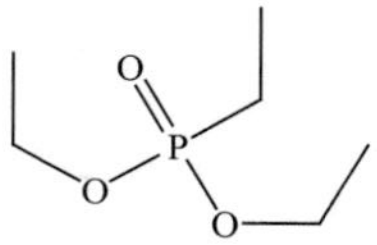 Diethylethyl phosphonate (DEEP)	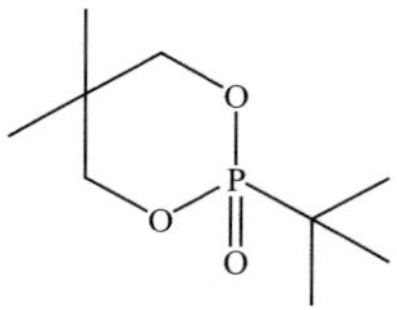 Cyclic *tert*-butyl phosphonate (CTBP)
Aromatic	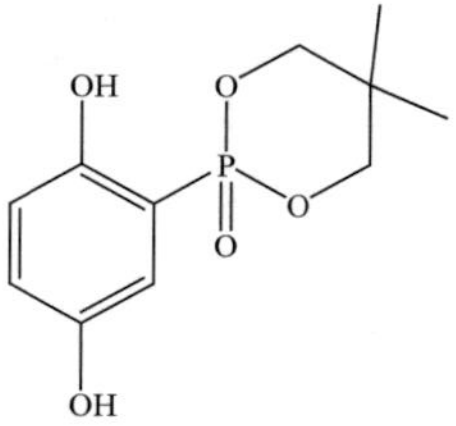 Ar-Phosphonate derivative 1	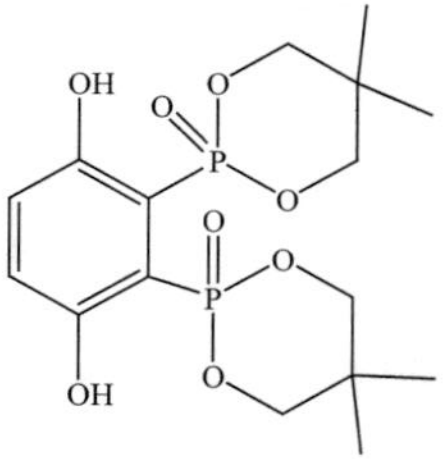 Ar-Phosphonate derivative 2

(continued)

Table 1 (continued)

OPFRs classes	Chemical structures
	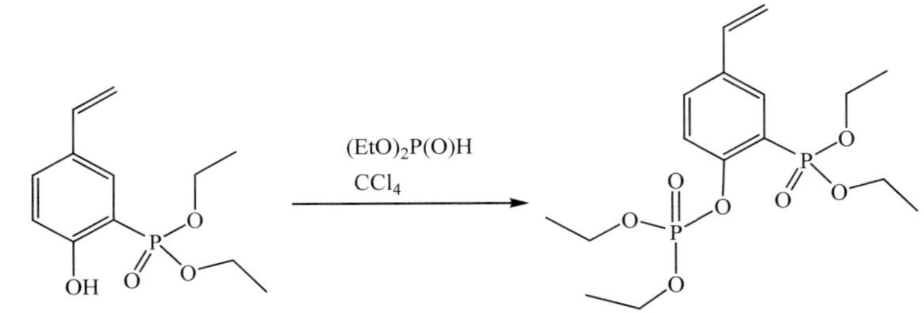 $(EtO)_2P(O)H$ CCl_4 Ar-Phosphonate derivative 3 → Ar-Phosphonate derivative 4
	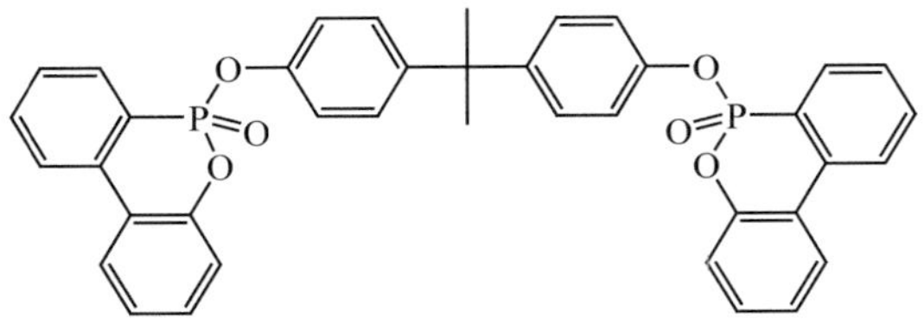 Ar-Phosphonate derivative 5 (Bisphosphonate)

(continued)

Table 1 (continued)

OPFRs classes	Chemical structures	
	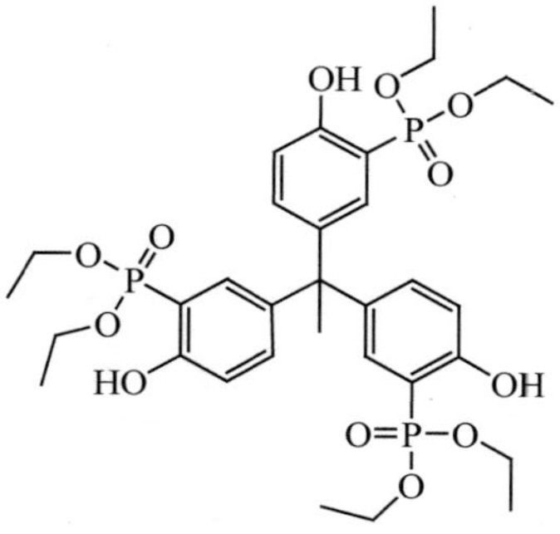 Ar-Phosphonate derivative 6 (Phenolic)	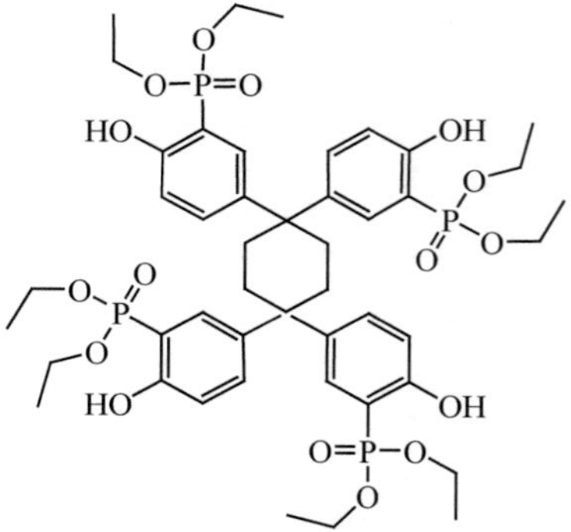 Ar-Phosphonate derivative 7 (Phenolic)
Organophosphinites	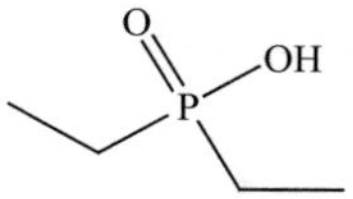 Diethylphosphinic acid (DEPA)	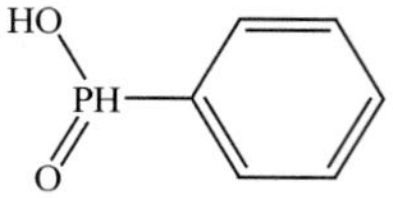 Phenylphosphinic acid (PPA)

(continued)

Table 1 (continued)

OPFRs classes	Chemical structures	
	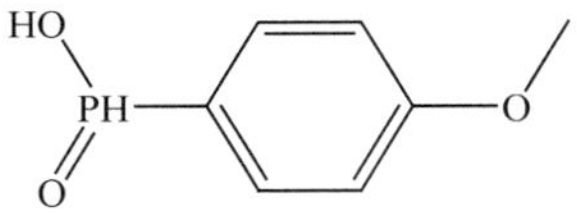 *p*-Methoxyphenylphosphinic acid (PMPPA)	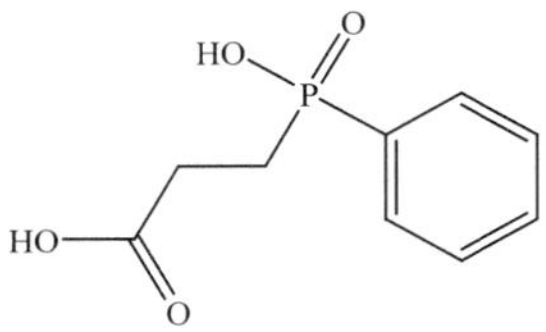 Carboxyethyl-phenylphosphinic acid (CEPPA)
	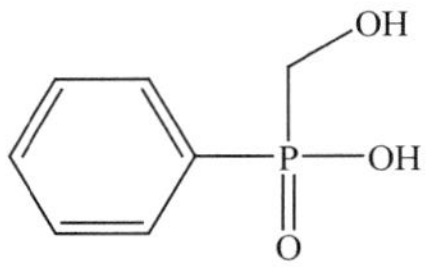 Hydroxymethylphenylphosphinic acid (HMPPA)	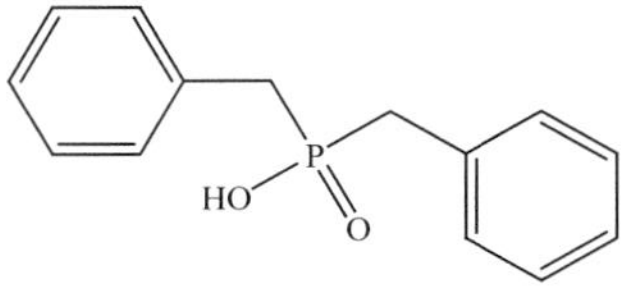 Dibenzylphosphinic acid (DBPA)
	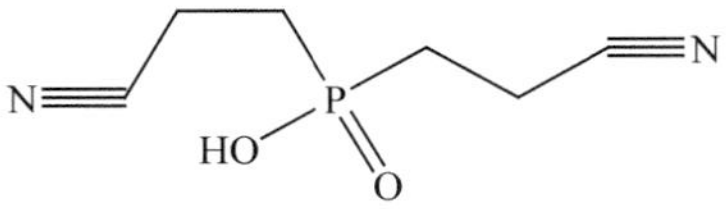 Bis(2-cyanoethyl)phosphinic acid (BCEPA)	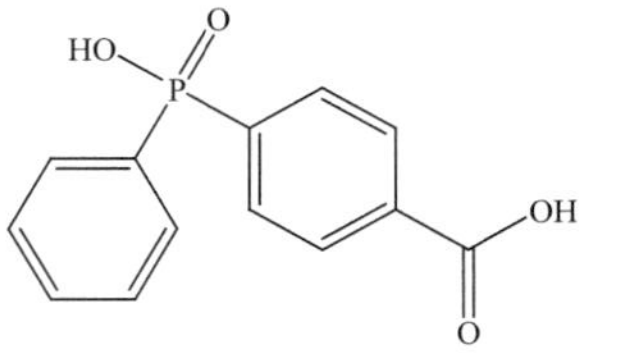 4-Carboxyphenylphenylphosphinic acid (CPPPA)

(continued)

Table 1 (continued)

OPFRs classes	Chemical structures	
Organoposphine oxide	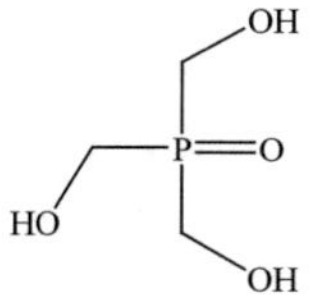 Tris(hydroxymethyl) phosphineoxide (THMPO)	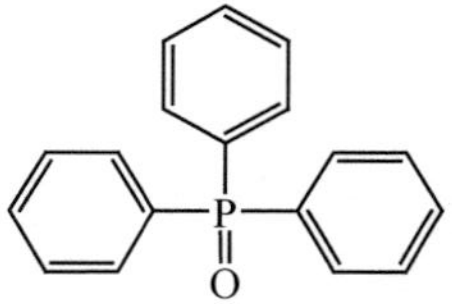 Triphenylphosphineoxide (TPPO)
	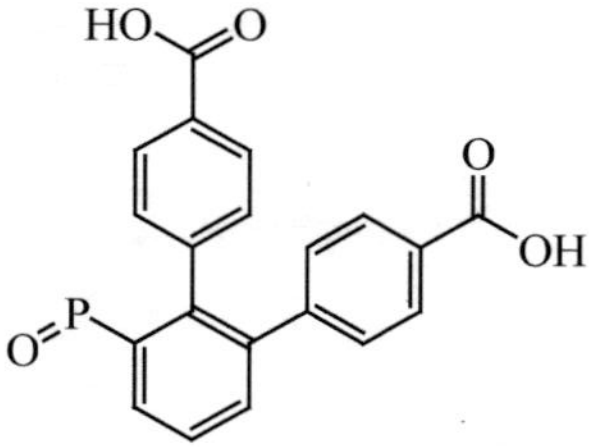 Bis(4-carboxyphenyl) phenylphosphine oxide (BCPPPO)	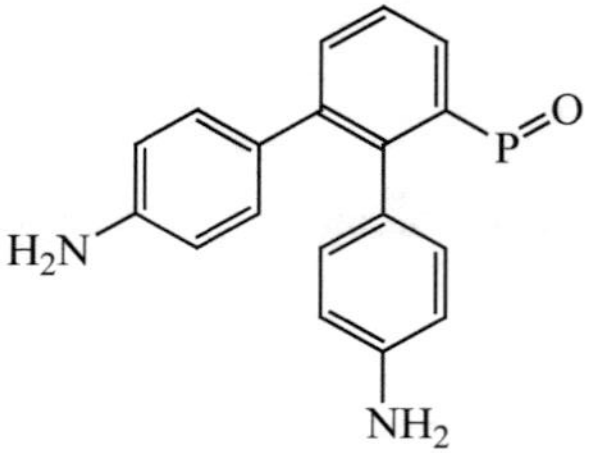 Bis(4-aminophenyl) phenylphosphine oxide (BAPPO)
Organophosphites	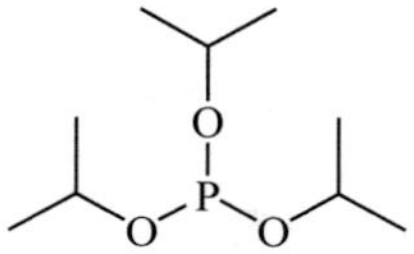 Triisopropyl Phosphite (TIPP)	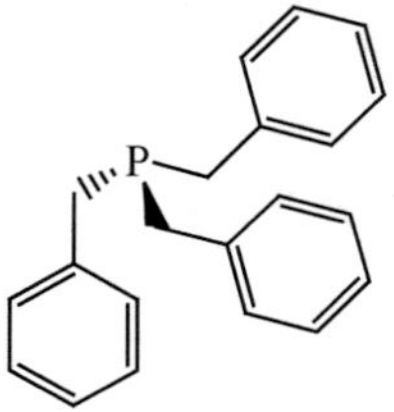 Tri-phenyl Phosphite (TPPi)

Tris(2,3-dibromopropyl) phosphate (TDBPP) and Tris(tribromoneopentyl) phosphate (TTBPP).

Br-aliphatic: Brromine-based P–$C_{aliphatic}$ bond containing FRs are found effective FRs, for example, Phenylpropan-2-ylhydrogen phosphate (PPHP), 2-Ethylhexyldiphenyl phosphate (EHDPP), Isodecyldiphenyl phosphate (IDPP) etc.

Aromatic: In the previous few decades aromatic phosphates possess strong flame-retardant properties [34]. Common examples are Triphenyl phosphate (TPP), Tricresyl phosphate (TCP), Triisopropyl phosphate (TIPP), Cresyl diphenylphosphate (CDPP), 6H-Dibenz[c,e] [1, 2] oxaphosphorin,6-oxide (DOPO), Bisphenol-A bis-(diphenylphosphate) (BPA-BDPP) etc.

1.3.2 Organophosphonate FRs

Organophosphonates were extensively developed and used as FRs. Many commercially accepted compounds based on organophosphonate specification have been generated and successfully employed as FRs and composed of two types:

Aliphatic: FRs with P–$C_{aliphatic}$ bond having phosphonate group are shown in Table 1 and commonly include Dimethyl hydrogen phosphonate (DMHP), Diethyl hydrogen phosphonate (DEHP), Dimethylethyl phosphonate (DMEP), Dimethylmethyl phosphonate (DMMP), Dimethylpropyl phosphonate (DMPP), Dimethylallyl phosphonate (DMAP) as well as a Cyclic tert-butyl phosphonate (TBP).

Aromatic: FRs with P–$C_{aromatic}$ bond having phosphonate group FRs were derived from the transformation of aliphatic moiety with aromatic units. A monosubstituted phosphorus compound (Ar-Phosphonate derivative 1, m.p. 184–187 °C) and a disubstituted phosphorus compound (Ar-Phosphonate derivative 2, m.p. 258–260 °C) were obtained as a mixture using modified separation and purification steps [37]. The phosphonate product diethyl-(2-hydroxy-5-vinylphenyl) phosphonate (Ar-Phosphonate derivative 4) was fabricated from diethyl-(4-vinylphenyl)phosphate (Ar-Phosphonate derivative 3) using ethoxy-P-acid [38]. Su et al. [39] patented the P-enabled esterification of phenol or polyhydroxybenzenes to develop several OPFRs. one example is compounds Ar-Phosphonate derivative 5 obtained from bisphenol A [39].

Another two phenolic white crystals/compounds (Ar-Phosphonate derivative 6, m.p. 169–171 °C and 7, m.p. 216–217 °C) were prepared through [1, 3]-sigmatropic rearrangement and were found to have high-performance against polymers [40].

1.3.3 Organophosphinites FRs

Recently, synthesis and applications of different phosphinic acid derivatives have been reported excel flame resistant abilities [41]. Examples of organophosphinites FRs are Diethylphosphinic acid (DEPA), Phenylphosphinic acid (PPA), *p*-Methoxyphenylphosphinic acid (PMPPA), Carboxyethyl-phenylphosphinic acid (CEPPA), Bis(2-cyanoethyl)phosphinic acid (BCEPA) etc.

1.3.4 Organoposphine Oxide FRs

Organoposphine oxide derivatives found to have limited FRs capabilities. Examples are Tris(hydroxymethyl) phosphineoxide (THMPO), Triphenylphosphineoxide (TPPO) and Bis(4-carboxyphenyl) phenylphosphine oxide (BCPPPO). In addition, Bis(4-aminophenyl) phenylphosphine oxide (BAPPO) was developed as a moderate water-soluble compound that found effectiveness towards Polyurathane-based materials with environmental susceptibility [42].

1.3.5 Organophosphites FRs

Like organoposphine oxide derivatives, organophosphites were also least responsive FRs because of their cholinergic neurotoxicity [43, 44]. Examples are Triisopropyl Phosphite (TIPP) and Tri-phenyl Phosphite (TPP).

2 Textile Applications of P-Based Flame Retardants

Organophosphorus compounds with P–C moieties have shown a wide range of design and development of exciting organophosphorus FRs due their high thermal and hydrolytic stability (P–C bond), ease of synthesis and suitability of processing even at high temperature. Extensive works have been published regarding the creation of P–C bond and their potential applications [1–6, 11–13, 22, 31–35]. These P–C containing organophosphorus compounds have shown several applications along with fire resistancy, for example, reagents, catalysts, pesticides, insecticides, herbicides, surfactants, lubricants and even more [1–4, 45].

Textiles and clothing are prepared from various fiber forming substrates either natural origin polymers such as cellulose and protein, or a wide variety of semi-synthetic and synthetic polymers such as cellulose acetate, polyesters, polyamides, polyolefins, polyacrylonitriles, polyaramids, polylactides, polyetherketones etc. All of these polymers are allocate a frequent limitation, combustible under normal environmental conditions and sometimes pose serious fire hazards in case of fire accidents.

In the present scenario, fire-caused deaths are a growing global problem. The Fire Administration Authority of US, has reported recently on the basis of 24 industrialized nations as the average rate of fire-related deaths and concluded that 10.7 per million populations every year have been observed [46, 47]. Additionally, The National Fire Protection Association claims that home structure involved fires are the main cause of fire-related death [47].

Textile materials provide an excellent source of fuel during the burning process, are found to be a rich source of inflammable or ideal fire carriers like hydrocarbons. The potential hazards and risks associated with textiles are described in depth by various researchers [1, 2, 7, 11]. In this prospect, textiles with lower flammability are still experiencing some changes like the improvement in effectiveness and the replacement of toxic chemical products with counterparts that have a low environmental impact and, more sustainable [4, 5]. Health and environmental concerns associated with halogenated as well as formaldehyde-based FRs driven R&D for identifying and utilizing safer alternatives. Because of the social concerns onto eco-preservation using eco-friendly FRs, the new FRs needs to be halogen-free and formaldehyde-free. Phosphorus, nitrogen, and silicon-containing compounds are generally considered as environment-friendly FRs, due to their safer-nature for human-ecosystem and synergistic effects [6, 14]. The effectiveness of P-based FRs towards fire mainly because of its characteristics, for example, low water solubility, low volatility, less dose requirement, less degradation to possibly hazardous substances, and no toxic emissions [2, 6, 31]. P-based FRs, play a key role possibly in combination with silicon- or nitrogen-containing structures, to the design of new and efficient FRs for textile substrates. Mechanistically, phosphorous based FRs, during a fire form poly and meta-phosphoric acids which form an oxygen-barrier layer [15].

Phosphorus based FRs have been found very reactive to inhibit fire and are used as thermosets for many substrates such as unsaturated epoxy resins, polyesters or polyurethanes. These type of substrates contain activated functional groups (i.e. halogens, alcohols, epoxy, amines etc.), which allow incorporation into the polymer matrix during the process [42, 43]. In case of cotton fibre, organic assembled phosphorus compounds (i.e. Pyrovatex CP and Pyrovatex CP New) can either with the cotton fabric to form cross-linked adducts/linked structures with the fibers [48]. In a study, a formaldehyde-free, inorganic-organic hybrid FR was developed and markedly found inferior FR performance compared with conventional formaldehyde-containing organic phosphorus FR. Lessan et al. [49] investigated the flame retardant behavior of sodium hypophosphite (SHP)—nano-TiO_2 hybrid on woven cotton fabric through pad-dry-cure process. As a result, decreasing the flammability with increasing the char formation of the treated fabrics was observed [49].

Despite the use of toxic and not environmentally-friendly chemicals, high-molecular-weight proteins even DNA derived from animal or microbial sources have been investigated as "green" FRs for cotton fabrics [42, 49]. Current trends are made towards high-molecular-weight FRs based on P-moiety combined with polymeric/complex textile substrates impart multifunctional structures will aid in reducing flammability without a loss of their valuable properties. A novel organic phosphorus-based flame retardant has reported the enhancement of flame retardancy of cotton fabrics through the high-molecular-weight grafting of cellulose-phosphonic acid by Gao et al. [50] as an alternative to halogen-formaldehyde-based FRs [50]. In this study, an ammonium salt of hexamethylenediamine-N,N,N',N'-tetra(methylphosphonic acid) (AHDTMPA), was fabricated using the reaction of urea with hexamethylenediamine-N,N,N',N'-tetra(methylphosphonic acid) (HDTMPA). Further, new P–O–C covalent bonds were formed by this ammoniated salt reacted with the O-6 hydroxyls of glucose residues of cellulose. The resulted hybrid FR containing both P and N-moieties exhibited excellent flame retardant performance for cotton fabrics (70 g/L hybrid FR with limiting oxygen index (LOI) value of 36.0%) which remained relatively stable after 50 laundering cycles (70 g/L hybrid FR with LOI value 28.0%).

Some particular proteins such as phosphorus and sulphur-rich proteins (i.e. caseins and hydrophobins) derived from animal or microbial sources have been under investigations as a novel as well as green flame retardants for cotton fabrics. Alongi et al. [51] investigated caseins and hydrophobins as a novel and green flame retardants for cotton fabrics [51]. As a consequence, P-based-polymer matrix was achieved with improved flame retardancy, indicated by the increased total burning time as well as by the decreased total burning rate. In this study, the change in the flammability features of the fabric, favouring the dehydration of cellulose to form char as opposed to the depolymerization with further production of combustible volatile species were observed. The familiar results have been observed by the same research group when whey proteins were employed that homogeneously deposited on cotton fabric to impart FR properties, using the layer-by-layer technique [52].

In addition, bio-derived phytic acid exhibits the great potential to improve the flame retardancy of textile materials, but with low washing durability. To overcome the poor durability, Cheng et al. [53] investigated a reactive, efficient P-containing flame retardant using phytic acid, pentaerythritol and 1,2,3,4-butanetetracarboxylic acid [53]. The wool fabric treated with HPPHBTCA 0.14 mol/L HPPHBTCA had self-extinguishing performance even after 20 washing cycles during the vertical burning test, presenting good FR ability and resistance to washing with slight negligible effect on the whiteness, tensile strength and handle of wool fabric.

Therefore, phosphorus-containing compounds offer a novel route to prepare so-called green, eco-friendly and durable flame retardants for textiles or textile-based materials that inhibit or resist the spread of fire.

3 Conclusion and Future Outlook

Flame retardants have been added to the polymer systems to prevent the risk of fire and overall the use of FRs has substantially decreased the number of fires and fire fatalities in our social wardrobe. With increased awareness towards environmental concerns about FRs selectivity, phosphorus-based FRs provide a foundation for the directed design of nontoxic FRs mainly because of its versatility, for example, it can act in both the condensed and gas phase, as an additive or as a reactive component, in various oxidization states, and in synergy with numerous adjuvant elements.

Various P-moieties make a valuable contribution and various combinations including elemental, inorganic salts and organophosphorus compounds. Nowadays, various products, large in numbers are commercially available based on organophosphorus compound. Although the use of a new generation of chemicals known as

Organophosphorus Flame Retardants (OPFRs) is a worthy goal for controlling household fires, on one hand, the other hand, it is also important to control or prevent their toxic effects. Based on carbon unit/moiety OPFRs can be broadly categorized into five main classes; *organophosphates, organophosphonates, organophosphinites, organoposphine oxide,* and *organophosphites.* A vast variety of P–C bond containing efficient FRs are being developed; however, further R&D works are needed in terms of their economical, renewability and green synthetic pathways, environmental impacts, long term durability, acute and chronic toxicity etc. without a loss of valuable properties at laboratory as well as their possible larger exploration.

References

1. Horrocks AR (2011) Flame retardant challenges for textiles and fibres: new chemistry versus innovatory solutions. Polym Degrad Stab 96:377–392
2. Horrocks AR (1986) Flame-retardant finishing of textiles. Rev Prog Color Relat Top 16(1):62. https://doi.org/10.1111/j.1478-4408.1986.tb03745.x
3. Liu W, Chen L, Wang YZ (2012) A novel phosphorus-containing flame retardant for the formaldehyde-free treatment of cotton fabrics. Polym Degrad Stabi 97(12):2487–2491
4. Gaan S, Salimova V, Rupper P, Ritter A, Schmid H (2011) Flame retardant functional textiles. In: Pan N, Sun G (eds) Functional textiles for improved performance, protection and health. Woodhead Publishing, Cambridge, UK, pp 98–130
5. Bocchini S, Camino G (2010) Chapter 4 halogen-containing flame retardants. In: Wilkie CA, Morgan AB (eds) Fire retardancy of polymeric materials, 2 edn. CRC Press, Florida, USA
6. Lu SY, Hamerton I (2002) Recent developments in the chemistry of halogen-free flame retardant polymers. Prog Polym Sci 27(8):1661–1712
7. Yusuf M (2018) A review on flame retardant textile finishing: current and future trends. Curr Smart Mater 3(2):99–108
8. Söderström G, Marklund S (2002) PBCDD and PBCDF from incineration of waste-containing brominated flame retardants. Environ Sci Technol 36(9):1959–1964
9. de Wit CA (2002) An overview of brominated flame retardants in the environment. Chemosphere 46(5):583–624
10. https://ec.europa.eu/environment/chemicals/reach/reach_en.htm. Retrieved on 20 Sept 2019
11. Shaw S (2010) Halogenated flame retardants: do the fire safety benefits justify the risks? Rev Environ Health 25(4):261–306
12. Ülker OC, Ulker O (2019) Toxicity of formaldehyde, polybrominated diphenyl ethers (PBDEs) and phthalates in engineered wood products (EWPs) from the perspective of the green approach to materials: a review. BioResour 14(3):7465–7493
13. Castellano A, Colleoni C, Iacono G, Mezzi A, Plutino MR, Malucelli G, Rosace G (2019) Synthesis and characterization of a phosphorous/nitrogen based sol-gel coating as a novel halogen-and formaldehyde-free flame retardant finishing for cotton fabric. Polym Degrad Stab 162:148–159
14. Wang Y, Su Q, Wang H, Zhao X, Liang S (2019) Molded environment-friendly flame-retardant foaming material with high strength based on corn starch modified by crosslinking and grafting. J Appl Polym Sci 136(11):47193
15. Hull TR, Law RJ, Bergman Å (2014) Environmental drivers for replacement of halogenated flame retardants. In: Papaspyrides CD, Kiliaris P (eds) Polymer green flame retardants. Elsevier, Oxford, UK, pp 119–179
16. Green J (1992) A review of phosphorus-containing flame retardants. J Fire Sci 10(6):470–487
17. Morgan AB, Gilman JW (2013) An overview of flame retardancy of polymeric materials: application, technology, and future directions. Fire Mat 37(4):259–279

18. Holleman A, Wiberg N (1985) "XV 2.1.3". Lehrbuch der Anorganischen Chemie (33rd edn). de Gruyter (ed.). ISBN 3-11-012641-9
19. Weil ED, Levchik SV (2017) Phosphorus flame retardants. Kirk-Othmer Encycl Chem Technol 2:1–34
20. Granzow A (1978) Flame retardation by phosphorus compounds. Acc Chem Res 11:177–183
21. Peters EN (1979) Flame-retardant thermoplastics. I. Polyethylene–red phosphorus. J Appl Polym Sci 24(6):1457–1464
22. Yeh JT, Hsieh SH, Cheng YC, Yang MJ, Chen KN (1998) Combustion and smoke emission properties of poly (ethylene terephthalate) filled with phosphorous and metallic oxides. Polym Degrad Stab 61(3):399–407
23. Levchik GF, Vorobyova SA, Gorbarenko VV, Levchik SV, Weil ED (2000) Some mechanistic aspects of the fire retardant action of red phosphorus in aliphatic nylons. J Fire Sci 18(3):172–182
24. Levchik SV, Weil ED (2000) Combustion and fire retardancy of aliphatic nylons. Polym Int 49(10):1033–1073
25. Wang Z, Qu B, Fan W, Huang P (2001) Combustion characteristics of halogen-free flame-retarded polyethylene containing magnesium hydroxide and some synergists. J Appl Polym Sci 81(1):206–214
26. Liu Y, Wang Q (2006) Preparation of microencapsulated red phosphorus through melamine cyanurate self-assembly and its performance in flame retardant polyamide 6. Polym Eng Sci 46(11):1548–1553
27. Wu Q, Lü J, Qu B (2003) Preparation and characterization of microcapsulated red phosphorus and its flame-retardant mechanism in halogen-free flame retardant polyolefins. Polym Int 52(8):1326–1331
28. Xie R, Qu B (2001) Thermo-oxidative degradation behaviors of expandable graphite-based intumescent halogen-free flame retardant LLDPE blends. Polym Degrad Stab 71(3):395–402
29. Tan Y, Shao ZB, Yu LX, Long JW, Qi M, Chen L, Wang YZ (2016) Piperazine-modified ammonium polyphosphate as monocomponent flame-retardant hardener for epoxy resin: flame retardance, curing behavior and mechanical property. Polym Chem 7(17):3003–3012
30. Pantelaki I, Voutsa D (2019) Organophosphate flame retardants (OPFRs): a review on analytical methods and occurrence in wastewater and aquatic environment. Sci Total Environ 649:247–263
31. Salmeia K, Gaan S, Malucelli G (2016) Recent advances for flame retardancy of textiles based on phosphorus chemistry. Polym 8(9):319
32. Keglevich G, Grün A, Bálint E, Kiss NZ, Bagi P, Tőke L (2017) Green chemical syntheses and applications within organophosphorus chemistry. Struct Chem 28(2):431–443
33. Camino G, Costa L (1988) Performance and mechanisms of fire retardants in polymers-a review. Polym Degrad Stab 20(3–4):271–294
34. Wendels S, Chavez T, Bonnet M, Salmeia K, Gaan S (2017) Recent developments in organophosphorus flame retardants containing P–C bond and their applications. Materials 10(7):784(1–32)
35. Gupta RC (2006) Classification and uses of organophosphates and carbamates. In: Gupta RC (ed) Toxicology of organophosphate and carbamate compounds. Academic Press, Burlington, pp 5–24
36. AbouDonia M, Abou-Donia MB, Salama M, Elgamal M, Elkholi I, Wang Q (2016) Organophosphorus flame retardants (OPFR): neurotoxicity. J Environ Health Sci 2(1). https://doi.org/10.15436/2378-6841.16.022
37. Benin V, Gardelle B, Morgan AB (2014) Heat release of polyurethanes containing potential flame retardants based on boron and phosphorus chemistries. Polym Degrad Stab 106:108–121
38. Dumitrascu A, Howell BA (2011) Flame-retarding vinyl polymers using phosphorus-functionalized styrene monomers. Polym Degrad Stab 96:342–349
39. Su WC, Sheng CS (2005) Method for preparing a biphenylphosphonate compound. U.S. Patent 20050101793

40. Finocchiaro P, Consiglio GA, Imbrogiano A, Failla S (2007) Synthesis and characterization of new organic phosphonates monomers as flame retardant additives for polymers. Phosphorus, Sulfur Silicon Relat Elem 182:1689–1701
41. Ma J, Yang J, Huang Y, Ke C (2012) Aluminum-organophosphorus hybrid nanorods for simultaneously enhancing the flame retardancy and mechanical properties of epoxy resin. J Mater Chem 22:2007–2017
42. Velencoso MM, Battig A, Markwart JC, Schartel B, Wurm FR (2018) Molecular firefighting—how modern phosphorus chemistry can help solve the challenge of flame retardancy. Angew Chemie Int Ed 57(33):10450–10467
43. Carrington CD, Abou-Donia MB (1988) Triphenyl phosphite neurotoxicity in the hen: inhibition of neurotoxic esterase and a lack of prophylaxis by phenylmethylsulfonyl fluoride. Arch Toxicol 62(5):375–380
44. Wolschke H, Sühring R, Xie Z (2015) Organophosphorus flame retardants and plasticizers in the aquatic environment: a case study of the Elbe River. Germany Environ Poll 206:488–493
45. Demchuk OM, Jasinski R (2016) Organophosphorus ligands: Recent developments in design, synthesis, and application in environmentally benign catalysis. Phosphorus, Sulfur Silicon Relat Elem 191:245–253
46. TFRS Fire Death Rate Trends: An International Perspective (2016) vol. 12(8). https://www.usfa.fema.gov/downloads/pdf/statistics/v12i8.pdf. Accessed 1 Sep 2019
47. Fire Analysis and Research Statistical Reports (2014) NFPA. http://www.nfpa.org/News-and-Research/Fire-statistics-andreports/Fire-statistics/Fires-by-property-type/Residential/Home-structure-Fires. Accessed 1 Sep 2019
48. Horrocks AR, Kandola BK, Davies PJ, Zhang S, Padbury SA (2005) Developments in flame retardant textiles—a review. Polym Degrad Stab 88:3–12
49. Lessan F, Montazer M, Moghadam MB (2011) A novel durable flame-retardant cotton fabric using sodium hypophosphite, nano-TiO_2 and maleic acid. Thermochim Acta 520:48–54
50. Gao WW, Zhang GX, Zhang FX (2015) Enhancement of flame retardancy of cotton fabrics by grafting, a novel organic phosphorous-based flame retardant. Cellulose 22(4):2787–2796
51. Alongi J, Carletto RA, Bosco F, Carosio F, Di Blasio A, Cuttica F, Antonucci V, Giordano M, Malucelli G (2014) Caseins and hydrophobins as novel green flame retardants for cotton fabrics. Polym Degrad Stab 99:111–117
52. Bosco F, Carletto RA, Alongi J, Marmo L, Di Blasio A, Malucelli G (2013) Thermal stability and flame resistance of cotton fabrics treated with whey proteins. Carbohydr Polym 94(1):372–377
53. Cheng XW, Guan JP, Kiekens P, Yang XH, Tang RC (2019) Preparation and evaluation of an eco-friendly, reactive, and phytic acid-based flame retardant for wool. React Func Polym 134:58–66

From Smart Materials to Chromic Textiles

Tawfik A. Khattab and Meram S. Abdelrahman

Abstract This chapter presents a selective overview of chromic materials and their application on technical textiles. Most significant chromic materials could be photochromic, halochromic, thermochromic and electrochromic, with the ability to change color depending on the type of the external stimulus. An overview of the major chromic materials and related textile applications is discussed to reflect the progress and significance through ongoing research of high performance textiles. This chapter wraps up with future trends on the development of industrial merchandise from chromic technical clothing.

Keywords Chromism · Colorant · Textile · Coloration · Finish

1 Introduction

Chromic compounds are defined as smart materials that exhibit a distinctive color change upon exposure to one or more external stimuli, particularly when this color variation is controllable and reversible [1–3]. In recent decades, there are a broad range of chromic compounds that have been explored to introduce various textile and non-textile commercial products. Nowadays, chromic compounds are the most commonly applied materials in high-technology non-textile purposes, such as thermometry, biomedicine, electronics and ophthalmic applications [4–8]. Those chromic colorants have not been produced mainly for textile applications. However, there has been a rising interest for their utilization in the production of technical textiles [9, 10]. This chapter presents an overview on the different chemical classes of chromic colorants. Scientific fundamentals of chromic materials and their commercial products particularly technical textiles are discussed. We deal with the four major chromic materials which have been mostly investigated for textile purposes including photochromic (light induced color shift), halochromic (pH induced color

T. A. Khattab (✉) · M. S. Abdelrahman
Dyeing, Printing and Auxiliaries Department, Textile Industries Research Division, National Research Centre, Cairo 12622, Egypt
e-mail: tkhattab@kent.edu

M. Shahid and R. Adivarekar (eds.), *Advances in Functional Finishing of Textiles*, Textile Science and Clothing Technology, https://doi.org/10.1007/978-981-15-3669-4_11

shift), thermochromic (temperature induced color shift) and electrochromic (electric current induced color shift) textiles. Other miscellaneous chromic compounds and their related products is discussed. The chapter concludes the future trends in developing chromic textile products.

2 Chromic Colorants

There is an extensive range of synthetic colorants that have been industrially presented for the coloration of various textile fibers. Conventional textile dyestuffs are required to offer a constant color with high stability to light, washing, heat, perspiration and crocking. Variation in color of a certain fabric, usually known as color fading due to poor colorfastness, upon exposure to washing, crocking, temperature, perspiration or light is highly undesirable [11–13]. However, it was realized recently that there are promising commercial niche applications for chromic colorants that possess a distinctive color shift upon exposure to an external stimulus, particularly when this color shift is reversible and can be controlled (Fig. 1). In general, chromic colorants are those materials with the ability to radiate, replace or erase a color due to stimulation by an external stimulus [14–21].

An extensive range of chromic colorants are currently well known and their fundamental chemical and physical properties have been explored [22–26]. List of different classes of chromic materials and the stimuli involved are summarized in Table 1. Chromic compounds comprising photochromic, halochromic, thermochromic and/or electrochromic properties have been the most extensively investigated smart colorants. The other chromic colorants listed are rather less recognized and either exhibit limited commercial utilization or still an academic curiosity. Chromic colorants provide a considerable potential to offer particular functions in technical smart textiles, which have been designed to sense and response to the surrounding conditions [27–30]. Depending on functional activity, smart clothing can be divided into three different classes including passive, active and ultra smart textiles [31, 32]. Passive smart clothing is only responsive to an external stimulus, such as temperature or light, by changing color, such as presenting a warning signal. On the other hand, active smart textiles respond to an external stimulus as a sensor or as an actuator providing an electrical current from a central control unit, such as electronic textiles. The ultra smart cloth can sense, react and adapt itself to an external stimulus. It is mainly composed of a brain control unit. Ultra smart cloth is a combination between

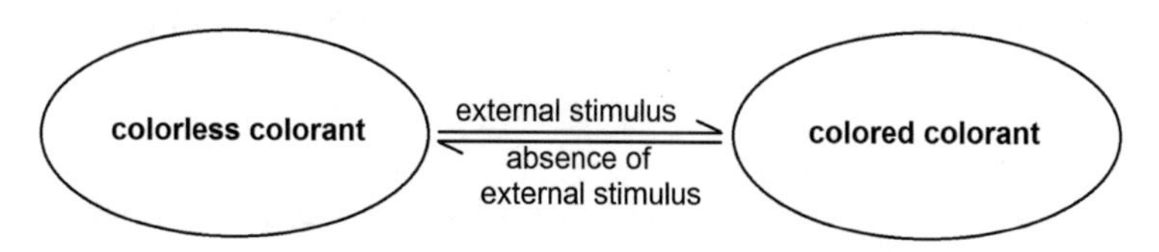

Fig. 1 Reversible performance of chromic colorants

Table 1 Classes of chromic colorants and related external stimulus

Chromic colorant	External stimulus
Photochromic	Light
Halochromic	pH
Thermochromic	Heat
Electrochromic	Electric current
Mechanochromic	Mechanical deformation
Piezochromic	Mechanical pressure
Tribochromic	Mechanical friction
Solavtochromic	Solvent polarity
Hygrochromic	Moisture
Chemochromic	Chemical agents
Ionochromic	Ions
Chronochromic	Time
Gasochromic	Gases
Carsolchromic	Electron beam
Vapochromic	Vapors of organic materials
Biochromic	Biological agents
Aggregachromic	Aggregation of colorants
Crystallochromic	Crystal structure change of a colorant
Magnetochromic	Magnetic field
Cathodochromic	Electron beam irradiation
Radiochromic	Ionizing radiation

conventional clothing technology with other fields of science, such as communication technology, sensors and actuators, materials and biological sciences, artificial intelligence, structural mechanics and advance processing [33, 34].

Thus, chromic textiles could be classified as passive smart clothing. Chromic textile sensors offer a responsive effect to an external stimulus by a visible color change, which offer the benefit of a self-contained responsive action that does not necessitate any electric elements [35]. Chromic colorants currently available are restricted in scope, relatively costly and have been designed for non-textile purposes. Thus, they cannot be employed in exactly the same methodologies as conventional clothing dyestuffs. They may also demonstrate inadequate stability under certain conditions resulting in less durable goods based on chromic clothing [36]. An ambitious motivation anticipated for chromic fabrics is the production of chameleon textiles with highly controlled color shifting, such as textiles providing responsive camouflage characteristics for military applications [37]. These futuristic chameleonic textiles gained their name from reptiles (family Chamaeleonidae) whose skin can change its color depending on the colors of the surrounding environment for attraction and camouflage [38]. Sophisticated clothing that can mimic such natural phenomena is not on the instantaneous horizon. Thus, there have been considerable research efforts to improve our understanding of the behavior of existing chromic colorants employed onto textiles under optimized conditions. Those efforts are needed to deepen into future if the determined targets are to be recognized.

3 Photochromic Colorants

Photochromism is a process in which a substance can go through a reversible color change between two different chemical species of different absorption spectra upon irradiation with ultraviolet or white light. This color change process can be reverted back to the pristine color or colorless state upon the removal of the light source [39, 40]. Irradiation of a photochromic colorless molecule leads to an isomerism process to an intensely colored molecular species. The reversible process may revert back to the pristine colorless molecular form either by removing the light source or by another external stimulus, such as heat. Spiropyrans have been used extensively in commercial products due to their relatively simple preparation and ability to provide reversible deep colors [41, 42]. The photochromic effect in spiropyran is a result of a reversible light-stimulated molecular rearrangement of the colorless spiropyran form via ring opening leading to the generation of the colored photomerocyanine form (Fig. 2). Photomerocyanine typically has violet or blue colors [43].

Due to their moderately low photostability, spiropyrans have been replaced by spirooxazines and naphthopyrans which are characterized by higher durability [44]. Spirooxazines arose as a significant class of organic photochromic colorants owing to their capability to impart a strong visible color, fatigue resistance and relatively simple preparation process. Spirooxazines enclose a spiro sp^3 hybridized carbon atom separating the molecular structure into two moieties comprising orthogonal heterocyclic rings with unconjugated π-systems [45]. The absorption of the localized π-systems are in the ultraviolet range and consequently the spirooxazine molecule is colorless. When the oxazine (C–O) bonding is broken upon exposure to ultraviolet, the spirooxazine was switched to the colored ring-opened photomerocyanine form. The photomerocyanine molecule return back to the colorless ring-closed spirooxazine molecular state as the oxazine-bridge is re-formed as displayed in Fig. 3. Spirooxazine derivatives typically offer red, blue and violet through to turquoise colors upon

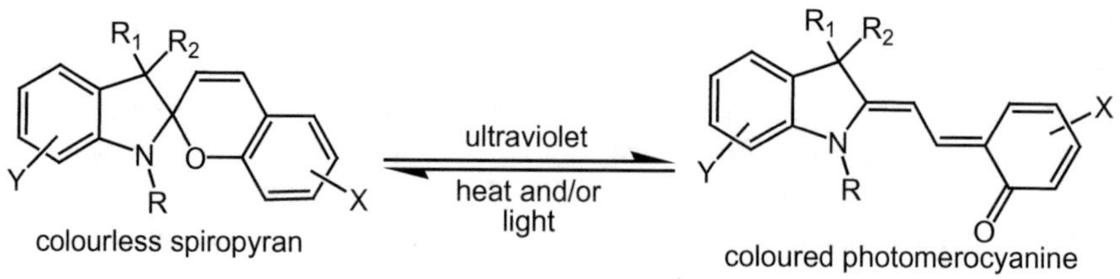

Fig. 2 Photochromic performance of spiropyran chromic colorants; R = alkyl; R', R" = alkyl; X, Y = H, halogen, nitro

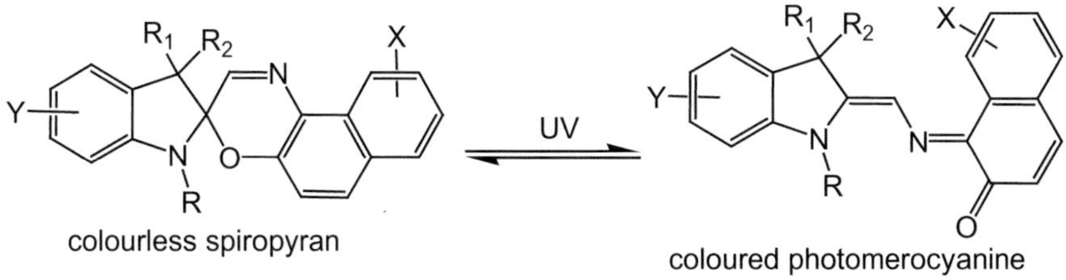

Fig. 3 Derivatives of spirooxazine chromic colorants

ultraviolet irradiation [46, 47]. Naphthopyrans, also known as chromenes, have been widely studied as the most significant group of photochromic colorants. Similarly as spiropyran and spirooxazine, the photochromic behavior of naphthopyrans involves light-stimulated ring-opening to provide the more coplanar colored photomerocyanine as shown in Fig. 4. Naphthopyrans give photochromic colors in the spectrum range between yellow, orange, red, violet and blue depending on the substituents pattern. Neutral colors such as gray and brown which are important for ophthalmic applications are also accessible [19]. Diarylethenes have been also studied in devices using optical switch as photochromic colorants able to switch between colorless or weakly colored ring-opened and colored ring-closed forms (Fig. 5). The reversible

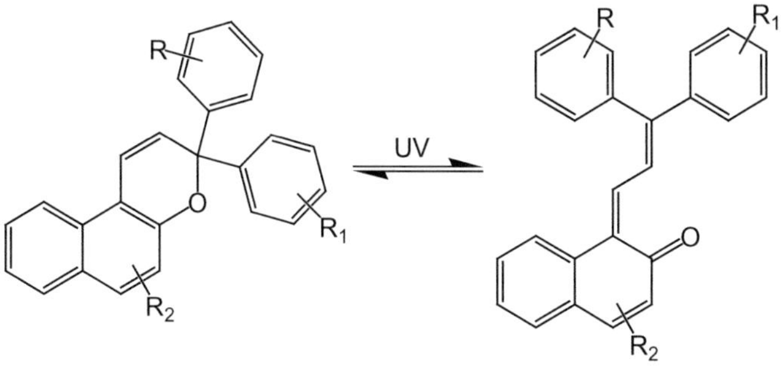

Fig. 4 Naphthopyran derivatives as chromic colorants

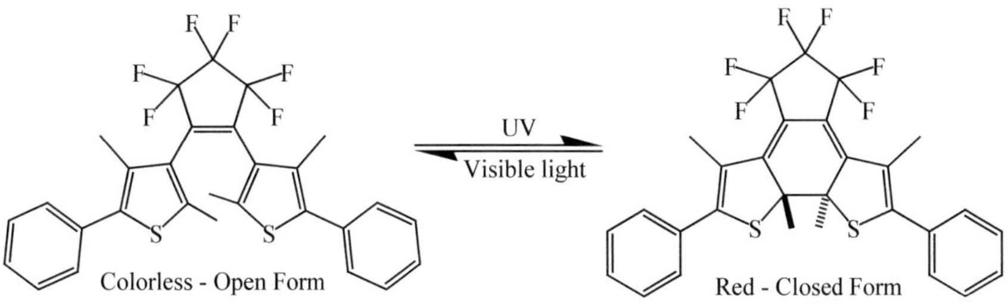

Fig. 5 Diarylethene photochromic molecular switching dye

process only necessitates visible light absorption to revert back to its pristine state [48, 49].

Those organic photochromic colorants generally possess problems upon application in a variety of commercial products, such as the steric hindrance effect arising upon encapsulation in a film matrix leading to inhibiting their optical characteristics. Conversely, inorganic photochromic colorants, such as strontium aluminate, exhibit a better photochromic performance because they do not possess such steric hindrance effect since their photophysical change is not accompanied with molecular structure change [50, 51]. The long time exposure of organic photochromic colorants to ultraviolet radiation can stimulate their degradation and consequently the gradual decrease of their photochromic response leading to low photostability which limits their usage in commercial products for an outdoor environment. Strontium aluminum oxide doped with lanthanides ($SrAl_2O_4:Eu^{2+}/Dy^{3+}$) or strontium aluminate is an inorganic colorant that has been known as long-lasting phosphorescent material. It has been extensively applied as a photochromic colorant for a variety of textile and non-textile applications. It is characterized by highly photostability under ultraviolet irradiation, excellent fatigue resistance and fast reversibility [52–55].

3.1 Photochromism in Textiles

Some early azo disperse dyestuffs, such as azobenzene derivatives, demonstrated a visual color change upon exposure to strong sunlight particularly when applied to cellulose acetate clothing. The process was reversible in the dark. This was attributed to molecular switch from *trans*-azo isomer to the less stable *cis*-azo isomer upon ultraviolet irradiation [56]. Even though recent publications indicate rising interest for smart textile products, there have been relatively few reports on photochromic textiles due to technical difficulties associated with the application process and product performance. There are some reports on studying the exhaustion dyeing of spirooxazines onto synthetic fabrics introducing photochromic clothing able to change from colorless or weakly colored to blue. However, this process was generally characterized by low dyestuff exhaustion. Polyester fibers dyed with phenoxyanthraquinone dyestuffs displayed color variation from yellow to orange after ultraviolet irradiation [57]. Spirooxazine and naphthopyran derivatives were used in the dyeing process of polyester fibers as a commercial photochromic disperse colorants via exhaustion dyeing. Blue spirooxazine offered the most effective results affording a fabric that demonstrated obvious color change upon exposure to ultraviolet light. There have been a few reported research work on photochromic dyestuffs designed for textile purposes by dyeing, such as spirooxazines bearing water solubilizing sulfonate functional groups to introduce acid dyestuffs toward the production of photochromic polyamide fibers [58]. In addition, a spirooxazine bearing a dichlorotriazine moiety has been described as a fiber-reactive dyestuff appropriate for polyamide-based textiles [59]. Although, this molecular modification of photochromic dyes is arguable, further improvement of this modification and dyeing methodologies may be essential to offer commercially usable photochromic textile products.

Long-lasting phosphorescent materials have been applied in various commercial products due to their ability to continue emit light for a longer period of time after excitation. This is usually beneficial to be used in directional safety signs, toys, protective clothing, and other applications where the phenomenon of glow in the dark presents amusement, ornamental and/or safety characteristics [52, 60]. Long-lasting phosphorescent products are composed of a carrier in combination with a photoluminescent synthetic pigment phosphor which can be easily excited by a certain wavelength of light of an external source. The photoluminescent pigment can then slowly discharge the stored light energy after the removal of the external light source. Long-persistence phosphorescent lanthanide-doped pigments have been used in producing various commercial merchandises, such as ornamental decorations, indicators, clothing, guide signs, switches and toys [52, 60–62]. Long-persistent phosphorescent materials are generally consists of crystals and traps. The crystals are photoluminescent collective elements which can be charged via excitation by an external light source. The energy traps are characterized by their high capacity to store this light energy for a long period of time [63]. After excitation, the crystals continue to discharge light and glow in the dark as time proceeds. This is supported by the traps, such as divalent europium and trivalent dysprosium, which extend the time period for light emission. There are various pigment phosphors that have been available for use in affording long-lasting photoluminescence, such as the red emission by $Y_2O_2S{:}Mg^{2+}/Ti^{4+}$, green emission by $SrAl_2O_4{:}Eu^{2+}/Dy^{3+}$, and blue emission $CaAl_2O_4{:}Eu^{2+}/Nd^{3+}$ [64–66]. These long-lasting photoluminescent inorganic pigments are characterized by high stability to chemicals, light and heat, as well as being non-toxic, recyclable, non-radioactive, reversible, bright, long period of phosphorescence time that could be higher than ten hours and high quantum efficiency [67]. Strontium aluminate phosphor has been applied to introduce photochromic and fluorescence functionalities for both textile and paper cellulosic fibers while maintaining their pristine characteristics, such as appearance, handle and mechanical properties. Immobilizing the inorganic Strontium aluminate phosphor at low total content value in an aqueous binder onto on cotton fabric via screen printing was reported recently by Khattab et al. to introduce light-responsive clothing with improved color-exchange performance and colorant stability [68]. The photochromic knitted cotton were firstly dyed conventionally with Reactive Red AEF to introduce a dark red background followed by screen printing with an aqueous paste containing strontium aluminate pigment, binder, diammonium phosphate, synthetic thickener and ammonium hydroxide. The printed cotton was dried under ambient conditions and then exposed to thermofixation at 160 °C. This approach was characterized by fast and simple application, easy to handle and low cost affording photochromic cotton fabric of high durability without affecting its aesthetic features, such as comfort and handling. The color switching of cotton fabric was from red to greenish yellow (Fig. 6). It exhibited excellent fatigue resistance, satisfactory colorfastness, high reversibility, and photo- and thermal stability [68].

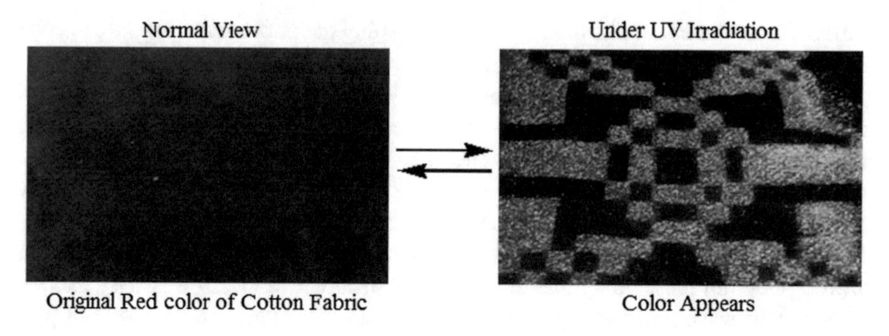

Fig. 6 Screen-printed photochromic cotton fabric before and after ultraviolet irradiation

4 Thermochromic Colorants

Thermochromism is defined as a commonly reversible color change upon exposure to either heating or cooling. Thermochromic materials can be divided into two classes including intrinsic materials, in which the heat is the direct reason for color change, and indirect materials, in which the heat results in changes in an environment enclosing a chromophore, which in turn is affected leading to color change [69]. Various inorganic and metal complex compounds demonstrate intrinsic thermochromism over a wide range of temperature. However, those compounds are usually function at high temperature and consequently are not suitable for textile applications [30]. There are several organic polymers exhibiting reversible intrinsic thermochromic performance, such as poly(alkoxythiophene) which can reversibly alter its color from red/violet to yellow upon increasing temperature as a result of variations in the crystalline form and molecular conformation [70]. There are two classes of thermochromic materials that have been applied to textile products including leuco dyestuff and liquid crystals. Both materials require microencapsulation to enclose the active constituents in a tiny shell to guarantee that those substances are introduced with some protective shielding effect against an environment to which those thermochromic substances could be sensitive [71–74].

The most extensively employed industrial thermochromic material is the leuco class. It can be encapsulated in a composite depends mainly on color generation from the reaction of three components including leuco dyestuff as an organic color former, proton donor as an acid developer such as Bisphenol A and a low-melting/nonvolatile hydrophobic solvent such as aliphatic alcohols. The color former leuco dye is a pH-responsive halochromic dye generally of the spirolactone dye class [71, 72]. Crystal violet lactone has been used as a classic color former which is colorless in its ring-closed molecular species (Fig. 7). Upon decreasing the pH value, the ring-opened protonated molecular form is generated. This ring-opened molecular form exhibits an extended higher conjugated molecular structure, compared to the ring-closed species, leading to a reddish-blue color formation. Leuco-based thermochromic systems vary from colored to colorless upon increasing the temperature [71, 75].

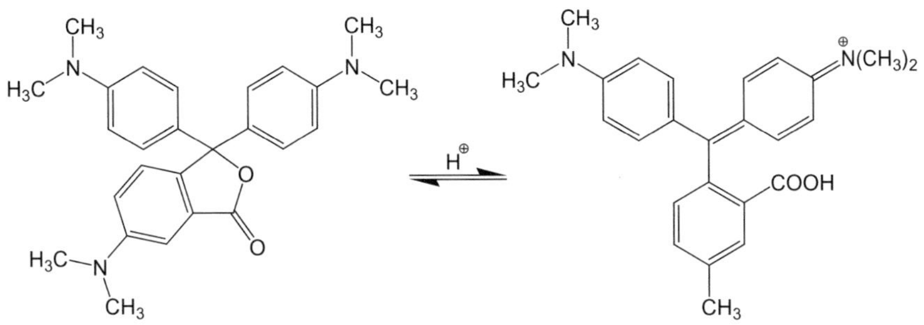

Fig. 7 Protonation of crystal violet lactone

Fig. 8 Molecular structure of chiral nematic liquid crystal; * is an asymmetric centre

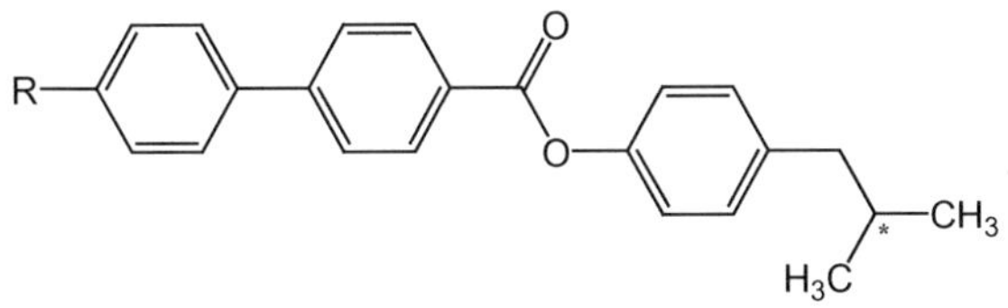

liquid crystals are the second category of materials that can be employed for thermochromic textile products. They offer a continuous varying spectrum of colors over a range of temperatures. Chiral nematic liquid crystalline homologues of esters (Fig. 8) have been applied in commercial thermochromic products [73, 74].

4.1 Thermochromism in Textiles

C.I. Acid Orange 156 has been described as a thermochromic colorant especially for nylon fibers [76]. The thermochromic leuco dyestuffs have been used in apparel and clothing applications. Those thermochromic products were colored to afford color change upon exposure to heat. However, such thermochromic leuco-based products demonstrated insufficient durability particularly to washing [77]. A number of interesting thermochromic apparel goods have been recently described, such as temperature monitoring for baby clothing. There are promising thermochromic textiles that could be designed for future clothing, such as textile thermometer and medical technical fabrics [78]. Thermochromic products of either leuco dyestuffs or liquid crystals have been usually applied by screen print technology onto textiles using the suitable aqueous binder print paste. In contrast to textile print technology, there are few reports on the inclusion of thermochromic materials directly onto textile fibers. Nonetheless, a procedure was reported for the mass dyeing of regenerated cellulose via spinning from a solvent dyeing-bath in which a leuco dyestuff is dispersed and employing a technique similar to that employed to produce lyocell fibers [79, 80].

5 Halochromic Colorants

Halochromism is a reversible color change process depending on a medium pH value. There are a broad range of color variations available which could be either from one color to another or from colorless to colored product [81, 82]. The pH-responsive dyestuffs have been employed as analytical indicators. There are a huge range of colored organic pH-responsive dyestuffs available and documented. The major chemical categories of technically significant pH-responsive dyestuffs are phthalides such as phenolphthalein and triarylmethines [83]. There are a range of simple azo dyestuffs that may undergo halochromism owing to protonation upon decreasing the pH value as shown in Fig. 9 [84].

5.1 Halochromism in Textiles

Halochromic clothing can change color depending on pH levels. The colorants applied for the production of halochromic textiles is the same applied for the establishment of pH sensors. A valuable application of halochromic textiles would be as a wound dress since the pH value of a skin changes considerably during the wound healing progress [83–85]. Halochromic textiles could be also applied as visual indicators protecting workers in a field with materials that is of changing pH levels. There are various publications that have been reported for the development of halochromic textile sensors using pH responsive dyestuffs that can be applied to clothing employing standard coloration techniques [83–87]. Recently, Khattab et al. [88] reported the preparation of halochromic cotton gauze (Fig. 10) coated with microcapsules composed of crosslinked calcium alginate as the shell and tricyanofurane-hydrazone spectroscopic probe together with urease enzyme for recognition of urea in aqueous solutions of either urine or blood serum. The detection mechanism (Fig. 11) depended on an enzymatic catalytic reaction of urease and urea affording alkaline ammonia which in turn abstract proton from the pH sensitive tricyanofuran-hydrazone leading to color change from yellow to purple. The halochromic gauze was prepared simply by padding in an aqueous solution containing sodium alginate, tricyanofurane-hydrazone colorant and urease enzyme under atmospheric conditions [88].

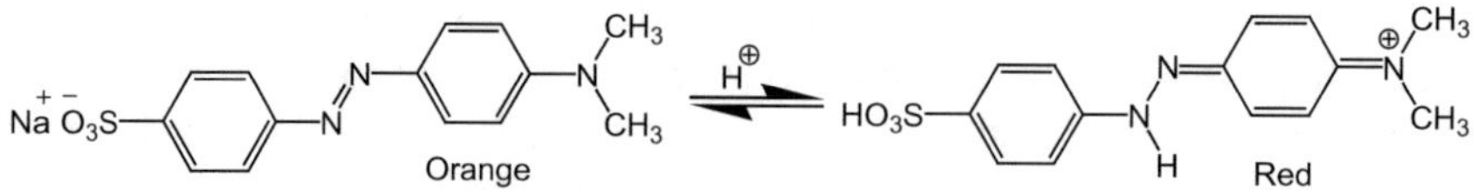

Fig. 9 Halchromic performance of methyl orange

Fig. 10 Scaning electron microscope image of halochromic cotton gauze loaded with calcium alginate microcapsules sensing assay

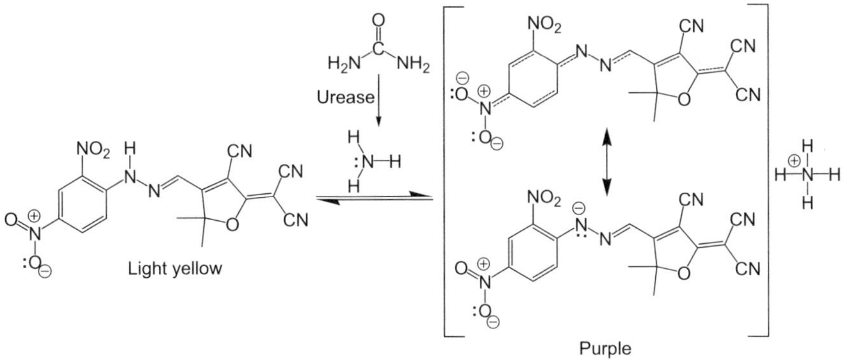

NO_2 H N N CN CN CN O
Light yellow
O H_2N C NH_2
Urease
H :N–H H
NO_2 N N CN CN CN O
NO_2 N N CN CN CN O
Purple
H H–N–H H

Fig. 11 Suggested mechanism for sensing urea using pH responsive tricyanofuran-hydrazone colorant

6 Electrochromic Colorants

Electrochromism is a reversible color alteration due to an electric current which results in electron transfer oxidation/reduction processes [89]. In general, the commercially significant electrochromic colorants are inorganic-based compounds, such as tungsten oxide, metal phthalocyanines and Prussian Blue [89–91]. Pure tungsten oxide has a pale yellow color changing electrochemically to blue owing to partial reduction of Tungsten(VI) oxide to Tungsten(V) oxide at a cathode [89]. Methyl viologen (Fig. 12) is a well known example of electrochromic colorants consisting of

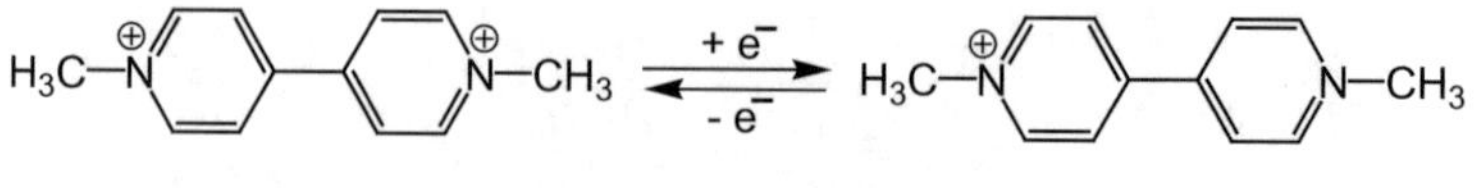

Fig. 12 Electrochromic behavior of methyl viologen

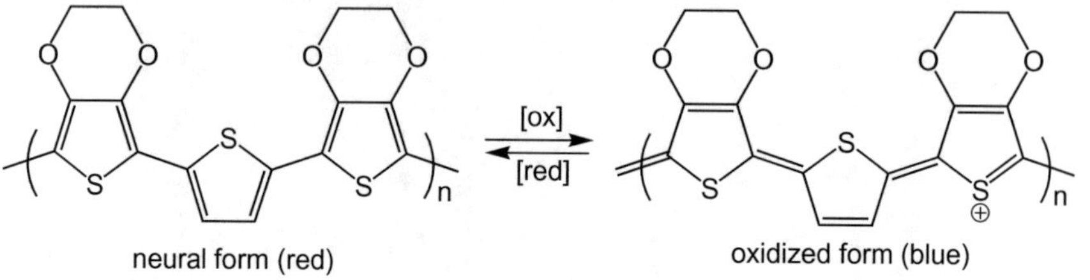

Fig. 13 Chemical structure of poly(3,4-ethylenedioxythiophene)

colorless bipyridylium dication which is able to undergo a reduction process at a cathode to introduce blue radical cation [92]. Other organic electrochromic colorants are also available, such as 1,4-phenylenediamines and thiazines [93, 94]. There has also been an interest in the preparation of electrochromic polymers, such as polyanilines and polythiophenes [95, 96].

6.1 Electrochromism in Textiles

There is a range of applications for eletrochromic textiles, such as biomimicry, flexible displays and camouflage. Developing flexible and stretchable textile-based electrochromic devices introduces serious engineering difficulties [97, 98]. Thus, a highly important prototype of electrochromic clothing was reported recently employing electrodes incorporated in a spandex fabric previously impregnated with poly(3,4-ethylenedioxythiophene)/poly(styrenesulfonate) as an electroactive polymer (Fig. 13). One electrode was coated by a polythiophene derivative as an electrochromic polymer, while the substrates were merged with a transparent organogel electrolyte [99, 100]. The electrochromic textile-based device was able to switch color between red and blue [101]. Such prototype electrochromic textile-based device presents a major initial step to pave the way toward highly developed controllable and chameleon clothing.

7 Future Trends

Smart materials that alter their color according to one or more external stimulus have attracted scientific interest for both academic and commercial purposes. Various chromic compounds have been developed and applied in industry. Researchers have developed novel chromic materials and demonstrated their application on textiles. However, the research field and commercial utilization of chromic clothes has been rather limited owing to technical limitations during application and relatively high cost. Due to these technical difficulties, commercialized chromic textiles are mainly based on photochromic and thermochromic clothing. Chromic materials provide a potential for functional uses in high performance sensing textile-based devices which are designed to respond to external stimuli in the shape of a visible color change. For instance, those textile sensors presents an early warning signal in response to ultra-violet light (photochromism), heat (thermochromism), pH (halochromism) and/or electric current (electrochromism). Chromic textile-based sensors would offer advantages over other sensing devices as they have self-contained response without the necessity for complicated instrumentation, trained personnel or electrical circuitry. There have been research developments to improve the performance of chromic colorants on textiles and the optimal conditions necessary for their application. However, ongoing research of chromic colorants designed for textiles is necessary to broaden the range of those compounds with the enhancement of their properties on textiles. This will be of significance to accomplish further ambitious aspirations. The future may offer developments in chromism based applications which are not unexploited industrially or yet unknown, such as biochromic textiles which might be employed as a medical diagnostic tool.

References

1. Yang XD, Zhu R, Yin JP, Sun Li, Guo RY, Zhang J (2018) Bipyridinium-bearing multi-stimuli responsive chromic material with high stability. Cryst Growth Des 18(5):3236–3243
2. Khattab TA, Dacrory S, Abou-Yousef H, Kamel S (2019) Smart microfibrillated cellulose as swab sponge-like aerogel for real-time colorimetric naked-eye sweat monitoring. Talanta, 120166
3. Jeong J, Min KS, Kumar RS, Mergu N, Son YA (2019) Synthesis of novel betaine dyes for multi chromic sensors. J Mol Struct 1187:151–163
4. Saini A, Christenson CW, Khattab TA, Wang R, Twieg RJ, Singer KD (2017) Threshold response using modulated continuous wave illumination for multilayer 3D optical data storage. J Appl Phys 121(4):043101
5. Chen W, Pan Y, Chen J, Ye F, Liu SH, Yin J (2018) Stimuli-responsive organic chromic materials with near-infrared emission. Chin Chem Lett 29(10):1429–1435
6. Khattab TA, Fouda MMG, Allam AA, Othman SI, Bin-Jumah M, Al-Harbi HM, Rehan M (2018) Selective colorimetric detection of Fe (III) using metallochromic tannin-impregnated silica strips. Chem Sel 3(43):12065–12071
7. Khattab TA, Aly SA, Klapötke TM (2018) Naked-eye facile colorimetric detection of alkylphenols using Fe (III)-impregnated silica-based strips. Chem Pap 72(6):1553–1559
8. Khattab TA, Dacrory S, Abou-Yousef H, Kamel S (2019) Development of microporous cellulose-based smart xerogel reversible sensor via freeze drying for naked-eye detection of ammonia gas. Carbohyd Polym 210:196–203
9. Khattab TA, Kassem NF, Adel AM, Kamel S (2019) Optical recognition of ammonia and amine vapor using "turn-on" fluorescent chitosan nanoparticles imprinted on cellulose strips. J Fluoresc 29(3):693–702

10. Ghosh S, Hall J, Joshi V (2018) Study of chameleon nylon and polyester fabrics using photochromic ink. J TextE Inst 109(6):723–729
11. Khattab TA, Elnagdi MH, Haggaga KM, Abdelrahmana AA, Aly SA (2017) Green synthesis, printing performance, and antibacterial activity of disperse dyes incorporating arylazopyrazolopyrimidines. AATCC J Res 4(4):1–8
12. Tang AYL, Lee CH, Wang YM, Kan CW (2019) Dyeing cotton with reactive dyes: a comparison between conventional water-based and solvent-assisted PEG-based reverse micellar dyeing systems. Cellulose 26(2):1399–1408
13. Khattab TA, Haggag KM, Elnagdi MH, Abdelrahman AA, Aly SA (2016) Microwave-assisted synthesis of arylazoaminopyrazoles as disperse dyes for textile printing. Zeitschrift für anorganische und allgemeine Chemie 642(13):766–772
14. Khattab TA, Allam AA, Othman SI, Bin-Jumah M, Al-Harbi HM, Fouda MM (2019) Synthesis, solvatochromic performance, pH sensing, dyeing ability, and antimicrobial activity of novel hydrazone dyestuffs. J Chem 2019. https://doi.org/10.1155/2019/7814179
15. Rehan M, Khattab TA, Barohum A, Gätjen L, Wilken R (2018) Development of Ag/AgX (X = Cl, I) nanoparticles toward antimicrobial, UV-protected and self-cleanable viscose fibers. Carbohyd Polym 197:227–236
16. Pratumyot K, Srisuwannaket C, Niamnont N, Mingvanish W (2019) Dyeing of cotton with the natural dye extracted from waste leaves of green tea (Camellia sinensis var. assamica). Color Technol 135(2):121–126
17. Rehan M, Barhoum A, Khattab TA, Gätjen L, Wilken R (2019) Colored, photocatalytic, antimicrobial and UV-protected viscose fibers decorated with Ag/Ag_2CO_3 and Ag/Ag_3PO_4 nanoparticles. Cellulose 26(9):5437–5453
18. Basel Y, Ward SC, Christie RM, Vettese S (2019) Textile applications of commercial photochromic dyes: part 7. A statistical investigation of the influence of photochromic dyes on the mechanical properties of thermoplastic fibres. J TextE Inst 110(5):780–790
19. Pinto TV, Cardoso N, Costa P, Sousa CM, Durães N, Silva C, Coelho PJ, Pereira C, Freire C (2019) Light driven PVDF fibers based on photochromic nanosilica@ naphthopyran fabricated by wet spinning. Appl Surf Sci 470:951–958
20. Bao B, Bai S, Fan J, Su J, Wang W, Yu D (2019) A novel and durable photochromic cotton-based fabric prepared via thiol-ene click chemistry. Dye Pigment: 107778
21. De Smet D, Vanneste M (2019) Responsive textile coatings. Smart textile coatings and laminates. Woodhead Publishing, Cambridge, pp 237–261
22. Wang Y, Zhong X, Huo D, Zhao Y, Geng X, Fa H, Luo X, Yang M, Hou C (2019) Fast recognition of trace volatile compounds with a nanoporous dyes-based colorimetric sensor array.". Talanta 192:407–417
23. Faisal S, Farooq S, Hussain G, Bashir E (2019) Influence of hard water on solubility and colorimetric properties of reactive dyes. AATCC J Res 6(2):1–6
24. Lin Hao, Yan Song, Song BenTeng, Wang Zhuo, Sun Li (2019) Discrimination of aged rice using colorimetric sensor array combined with volatile organic compounds. J Food Process Eng 42(4):e13037
25. Liu T, Yang L, Zhang J, Liu K, Ding L, Peng H, Belfield KD, Fang Y (2019) Squaraine-hydrazine adducts for fast and colorimetric detection of aldehydes in aqueous media. SensS Actuators B Chem 292:88–93
26. Sharma DK, Adams ST, Jr KL, Liebmann AC, Miller SC (2019) Sulfonamides are an overlooked class of electron donors in luminogenic luciferins and fluorescent dyes. Org Lett 21(6):1641–1644
27. Cheng CC, Chiu TW, Yang XJ, Huang SY, Fan WL, Lai JY, Lee DJ (2019) Self-assembling supramolecular polymer membranes for highly effective filtration of water-soluble fluorescent dyes. Poly Chem 10(7):827–834
28. Rojas-Sánchez L, Sokolova V, Riebe S, Voskuhl J, Epple M (2019) Covalent surface functionalization of calcium phosphate nanoparticles with fluorescent dyes by copper-catalysed and by strain-promoted azide-alkyne click chemistry. ChemNanoMat 5(4):436–446

29. Kenta N, Morisaki Y, Tanaka K, Chujo Y (2019) Design of thermochromic luminescent dyes based on the bis (ortho-carborane)-substituted benzobithiophene structure. Chemistry–An Asian J 14(6):789–795
30. Okino K, Sakamaki D, Seki S (2019) Dicyanomethyl radical-based near-infrared thermochromic dyes with high transparency in the visible region. ACS Mater Lett 1:25–29
31. Lee W, Someya T (2019) Emerging trends in flexible active multielectrode arrays. Chem Mater
32. Min X, Sun B, Chen S, Fang M, Wu X, Liu YG, Abdelkader A, Huang Z, Liu T, Xi K, Kumar RV (2019) A textile-based SnO_2 ultra-flexible electrode for lithium-ion batteries. Energy Storage Mater 16:597–606
33. Gupta BS, Edwards JV (2019) Textile materials and structures for topical management of wounds. Advanced textiles for wound care. Woodhead Publishing, Cambridge, pp 55–104
34. Wu R, Ma L, Hou C, Meng Z, Guo W, Yu W, Yu R, Hu F, Liu XY (2019) Silk composite electronic textile sensor for high space precision 2D combo temperature–pressure sensing. Small: 15(31): 1901558
35. Promphet N, Rattanawaleedirojn P, Siralertmukul K, Soatthiyanon N, Potiyaraj P, Thanawattano C, Hinestroza JP, Rodthongkum N (2019) Non-invasive textile based colorimetric sensor for the simultaneous detection of sweat pH and lactate. Talanta 192:424–430
36. He J, Xiao G, Chen X, Qiao Y, Dan X, Zhisong L (2019) A thermoresponsive microfluidic system integrating a shape memory polymer-modified textile and a paper-based colorimetric sensor for the detection of glucose in human sweat. RSC Adv 9(41):23957–23963
37. Yu H, Qi M, Wang J, Yin Y, He Y, Meng H, Huang W (2019) A feasible strategy for the fabrication of camouflage electrochromic fabric and unconventional devices. Electrochem Commun 102:31–36
38. Diaz RE Jr, Shylo NA, Roellig D, Bronner M, Trainor PA (2019) Filling in the phylogenetic gaps: induction, migration, and differentiation of neural crest cells in a squamate reptile, the veiled chameleon (Chamaeleo calyptratus). Dev Dyn 248(8):709–727
39. Julià-López A, Ruiz-Molina D, Hernando J, Roscini C (2019) Solid materials with tunable reverse photochromism. ACS Appl Mater Interfaces 11(12):11884–11892
40. Li L, Wang JR, Hua Y, Guo Y, Fu C, Sun YN, Zhang H (2019) "Reversible" photochromism of polyoxomolybdate–viologen hybrids without the need for proton transfer. J Mater Chem C 7(1):38–42
41. Tuktarov AR, Salikhov RB, Khuzin AA, Safargalin IN, Mullagaliev IN, Venidiktova OV, Valova TM, Barachevsky VA, Dzhemilev UM (2019) Optically controlled field effect transistors based on photochromic spiropyran and fullerene C60 films. Mendeleev Commun 29(2):160–162
42. Gan M, Xiao T, Liu Z, Wang Y (2019) Layered photochromic films stacked from spiropyran-modified montmorillonite nanosheets. RSC Adv 9(22):12325–12330
43. Chernyshev AV, Voloshin NA, Solov'eva EV, Gaeva EB, Zubavichus YV, Lazarenko VA, Vlasenko VG, Khrustalev VN, Metelitsa AV (2019) Ion-depended photochromism of oxadiazole containing spiropyrans. J Photochem Photobiol A: Chem 378:201–210
44. Kuroiwa H, Inagaki Y, Mutoh K, Abe J (2019) On-demand control of the photochromic properties of naphthopyrans. Adv Mater 31(2):1805661
45. Fedorov YV, Shepel NE, Peregudov AS, Fedorova OA, Deligeorgiev T, Minkovska S (2019) Modulation of photochromic properties of spirooxazine bearing sulfobutyl substituent by metal ions. J Photochem Photobiol A 371:453–460
46. Tran HM, Nguyen TH, Nguyen VQ, Tran PH, Thai LD, Truong TT, Nguyen LT, Nguyen HT (2019) Synthesis of a novel fluorescent cyanide chemosensor based on photoswitching poly (pyrene-1-ylmethyl-methacrylate-random-methyl methacrylate-random-methacrylate spirooxazine). Macromol Res 27(1):25–32
47. Peng S, Wen J, Hai M, Yang Z, Yuan X, Wang D, Cao H, He W (2019) Synthesis and application of reversible fluorescent photochromic molecules based on tetraphenylethylene and photochromic groups. New J Chem 43(2):617–621

48. Irie M, Mohri M (1988) Thermally irreversible photochromic systems. Reversible photocyclization of diarylethene derivatives. J Org Chem 53(4):803–808
49. Pu S, Zhang F, Jingkun X, Shen L, Xiao Q, Chen B (2006) Photochromic diarylethenes for three-wavelength optical memory. Mater Lett 60(4):485–489
50. Zhang ZJ, Xiang SC, Guo GC, Gang X, Wang MS, Zou JP, Guo SP, Huang JS (2008) Wavelength-dependent photochromic inorganic-organic hybrid based on a 3D Iodoplumbate open-framework material. Angew Chem Int Ed 47(22):4149–4152
51. Luiskutty CT, Ouseph PJ (1973) Valence states of iron in photochromic strontium titanate by Mössbauer effect. Solid State Commun 13(3):405–409
52. Guo X, Ge M, Zhao J (2011) Photochromic properties of rare-earth strontium aluminate luminescent fiber. Fibers Polym 12(7):875
53. Yu SF, Luo WS (2013) Synthesis on spiropyran/SrAl2O4: Eu2+ , Dy3+ photochromic phosphor and its solvent effects. Advanced materials research, vol 805. Trans Tech Publications, Switzerland, pp 1362–1367
54. Nakagawa T, Hasegawa Y, Kawai T (2009) Nondestructive luminescence intensity readout of a photochromic lanthanide (III) complex. Chem Commun 37:5630–5632
55. Pinkowicz D, Ren M, Zheng LM, Sato S, Hasegawa M, Morimoto M, Irie M et al (2014) Control of the single-molecule magnet behavior of lanthanide-diarylethene photochromic assemblies by irradiation with light. A Eur J 20(39):12502–12513
56. Georgiev A, Stoilova A, Dimov D, Yordanov D, Zhivkov I, Weiter M (2019) Synthesis and photochromic properties of some N-phthalimide azo-azomethine dyes. A DFT quantum mechanical calculations on imine-enamine tautomerism and trans-cis photoisomerization. Spectrochim Acta Part A Mol Biomol Spectrosc 210:230–244
57. Wang PY, Wu CJ (1997) Photochromic behaviour of some phenoxyanthraquinone dyes in solution and on polyester substrate. Dyes Pigm 35(3):279–288
58. Christie RM, Chi LJ, Spark RA, Morgan KM, Boyd AS, Lycka A (2005) The application of molecular modelling techniques in the prediction of the photochromic behaviour of spiroindolinonaphthoxazines. J Photochem Photobiol A 169(1):37–45
59. Jang SW, Son SJ, Kim DE, Kwon DH, Kim SH, Lee YH, Kang SW (2005) UV-sensitive photo functional device using evanescent field absorption between SU-8 polymer optical waveguide and photochromic dye. IEEE Photonics Technol Lett 18(1):82–84
60. Khattab TA, Fouda MM, Abdelrahman MS, Othman SI, Bin-Jumah M, Alqaraawi MA, Al Fassam H, Allam AA (2019) Development of illuminant glow-in-the-dark cotton fabric coated by luminescent composite with antimicrobial activity and ultraviolet protection. J Fluoresc 29(3):703–710
61. Khattab TA, Abou-Yousef H, Kamel S (2018) Photoluminescent spray-coated paper sheet: Write-in-the-dark. Carbohyd Polym 200:154–161
62. Khattab TA, Gabr AM, Mostafa AM, Hamouda T (2019) Luminescent plant root: a step toward electricity-free natural lighting plants. J Mol Struct 1176:249–253
63. Qiu J, Shimizugawa Y, Kojima K, Tanaka K, Hirao K (2001) Relaxation of ultraviolet-radiation-induced structure and long-lasting phosphorescence in Eu 2 + -doped strontium aluminosilicate glasses. J Mater Res 16(1):88–92
64. Li W, Liu Y, Ai P (2010) Synthesis and luminescence properties of red long-lasting phosphor Y2O2S: Eu3 + , Mg2 + , Ti4 + nanoparticles. Mater Chem Phys 119(1–2):52–56
65. Yamamoto H, Matsuzawa T (1997) Mechanism of long phosphorescence of SrAl2O4: Eu2 + , Dy3 + and CaAl2O4: Eu2 + , Nd3+. J Lumin 72:287–289
66. Zhang B, Zhao C, Chen D (2010) Synthesis of the long-persistence phosphor CaAl2O4: Eu2 + , Dy3 + , Nd3 + by combustion method and its luminescent properties. Luminescence 25(1):25–29
67. Khattab TA, Rehan M, Hamdy Y, Shaheen TI (2018) Facile development of photoluminescent textile fabric via spray coating of Eu (II)-doped strontium aluminate. Ind Eng Chem Res 57(34):11483–11492
68. Khattab TA, Rehan M, Hamouda T (2018) Smart textile framework: photochromic and fluorescent cellulosic fabric printed by strontium aluminate pigment. Carbohyd Polym 195:143–152

69. Khattab TA, Tiu BD, Adas S, Bunge SD, Advincula RC (2016) Solvatochromic, thermochromic and pH-sensory DCDHF-hydrazone molecular switch: response to alkaline analytes. RSC Adv 6(104):102296–102305
70. Tashiro K, Ono K, Minagawa Y, Kobayashi M, Kawai T, Yoshino K (1991) Structure and thermochromic solid-state phase transition of poly (3-alkylthiophene). J Polym Sci, Part B: Polym Phys 29(10):1223–1233
71. Panák O, Držková M, Kaplanová M (2015) Insight into the evaluation of colour changes of leuco dye based thermochromic systems as a function of temperature. Dyes Pigm 120:279–287
72. Li F, Zhao Y, Wang S, Han D, Jiang L, Song Y (2009) Thermochromic core–shell nanofibers fabricated by melt coaxial electrospinning. J Appl Polym Sci 112(1):269–274
73. Christie RM, Bryant ID (2005) An evaluation of thermochromic prints based on microencapsulated liquid crystals using variable temperature colour measurement. Color Technol 121(4):187–192
74. Wang J, Kolacz J, Chen Y, Jákli A, Kawalec J, Benitez M, West JL (2017) 12–3: Smart fabrics functionalized by liquid crystals. In: SID symposium digest of technical papers, vol 48, no 1, pp 147–149
75. Zhu CF, Wu AB (2005) Studies on the synthesis and thermochromic properties of crystal violet lactone and its reversible thermochromic complexes. Thermochim Acta 425(1–2):7–12
76. Dawson TL (2010) Changing colours: now you see them, now you don't. Color Technol 126(4):177–188
77. Zhang W, Ji X, Zeng C, Chen K, Yin Y, Wang C (2017) A new approach for the preparation of durable and reversible color changing polyester fabrics using thermochromic leuco dye-loaded silica nanocapsules. J Mater Chem C 5(32):8169–8178
78. Christie RM (2013) Chromic materials for technical textile applications. Advances in the dyeing and finishing of technical textiles. Woodhead Publishing, Cambridge, pp 3–36
79. Chowdhury MA, Butola BS, Joshi M (2013) Application of thermochromic colorants on textiles: temperature dependence of colorimetric properties. Color Technol 129(3):232–237
80. Rubacha M (2007) Thermochromic cellulose fibers. Polym Adv Technol 18(4):323–328
81. Khattab TA (2018) Novel solvatochromic and halochromic sulfahydrazone molecular switch. J Mol Struct 1169:96–102
82. AY H, Khattab TA, Youssef YA, Al-Balakocy N, Kamel S (2017) Novel cellulose-based halochromic test strips for naked-eye detection of alkaline vapors and analytes. Talanta 170:137–145
83. Van der Schueren L, De Clerck K (2012) Coloration and application of pH-sensitive dyes on textile materials. Color Technol 128(2):82–90
84. Lai YS, Lu HH, Su YH (2017) High-efficiency water-splitting solar cells with low diffusion resistance corresponding to halochromic pigments interfacing with ZrO2. ACS Sustain Chem Eng 5(9):7716–7722
85. Devarayan K, Kim BS (2015) Reversible and universal pH sensing cellulose nanofibers for health monitor. SensS Actuators B Chem 209:281–286
86. Van der Schueren L, De Clerck K, Brancatelli G, Rosace G, Van Damme E, De Vos W (2012) Novel cellulose and polyamide halochromic textile sensors based on the encapsulation of Methyl Red into a sol–gel matrix. SensS Actuators B Chem 162(1):27–34
87. Rosace G, Guido E, Colleoni C, Brucale M, Piperopoulos E, Milone C, Plutino MR (2017) Halochromic resorufin-GPTMS hybrid sol-gel: Chemical-physical properties and use as pH sensor fabric coating. SensS Actuators B Chem 241:85–95
88. Khattab TA, Fouda MM, Abdelrahman MS, Othman SI, Bin-Jumah M, Alqaraawi MA, … Allam AA (2019) Co-encapsulation of enzyme and tricyanofuran hydrazone into alginate microcapsules incorporated onto cotton fabric as a biosensor for colorimetric recognition of urea. React Funct Polym 142:199–206
89. Shen L, Du L, Tan S, Zang Z, Zhao C, Mai W (2016) Flexible electrochromic supercapacitor hybrid electrodes based on tungsten oxide films and silver nanowires. Chem Commun 52(37):6296–6299

90. Majeed SA, Ghazal B, Nevonen DE, Goff PC, Blank DA, Nemykin VN, Makhseed S (2017) Evaluation of the intramolecular charge-transfer properties in solvatochromic and electrochromic zinc octa (carbazolyl) phthalocyanines. Inorg Chem 56(19):11640–11653
91. Lu HC, Kao SY, Chang TH, Kung CW, Ho KC (2016) An electrochromic device based on Prussian blue, self-immobilized vinyl benzyl viologen, and ferrocene. Sol Energy Mater Sol Cells 147:75–84
92. Li SY, Wang Y, Wu JG, Guo LF, Ye M, Shao YH, Wang R, Zhao CE, Wei A (2016) Methyl-viologen modified ZnO nanotubes for use in electrochromic devices. RSC Adv 6(76):72037–72043
93. Lauw SJ, Xu X, Webster RD (2015) Primary-colored electrochromism of 1, 4-phenylenediamines. ChemPlusChem 80(8):1288–1297
94. Jennings JR, Lim WY, Zakeeruddin SM, Grätzel M, Wang Q (2015) A redox-flow electrochromic window. ACS Appl Mater Interfaces 7(4):2827–2832
95. Zhao Y, Zhang S, Hu F, Li J, Chen H, Lin J, Yan B, Gu Y, Chen S (2019) Electrochromic polyaniline/aramid nanofiber composites with enhanced cycling stability and film forming property. J Mater Sci Mater Electron: 1–11
96. Hu Y, Jiang F, Lu B, Liu C, Hou J, Xu J (2017) Free-standing oligo (oxyethylene)-functionalized polythiophene with the 3, 4-ethylenedioxythiophene building block: electrosynthesis, electrochromic and thermoelectric properties. Electrochim Acta 228:361–370
97. Koc U, Karaca GY, Oksuz AU, Oksuz L (2017) RF sputtered electrochromic wool textile in different liquid media. J Mater Sci Mater Electron 28(12):8725–8732
98. Grancarić AM, Jerković I, Koncar V, Cochrane C, Kelly FM, Soulat D, Legrand X (2018) Conductive polymers for smart textile applications. J Ind Text 48(3):612–642
99. Lang AW, Li Y, De Keersmaecker M, Shen DE, Österholm AM, Berglund L, Reynolds JR (2018) Transparent wood smart windows: polymer electrochromic devices based on poly (3, 4-ethylenedioxythiophene): poly (styrene sulfonate) electrodes. Chemsuschem 11(5):854–863
100. Kai H, Suda W, Ogawa Y, Nagamine K, Nishizawa M (2017) Intrinsically stretchable electrochromic display by a composite film of poly (3, 4-ethylenedioxythiophene) and polyurethane. ACS Appl Mater Interfaces 9(23):19513–19518
101. Chen Y, Zhu X, Yang D, Wangyang P, Zeng B, Sun H (2019) A novel design of poly (3, 4-ethylenedioxythiophene): poly (styrenesulfonate)/molybdenum disulfide/poly (3, 4-ethylenedioxythiophene) nanocomposites for fabric micro-supercapacitors with favourable performances. Electrochim Acta 298:297–304

Plasma Treatment Technology for Surface Modification and Functionalization of Cellulosic Fabrics

Nabil A. Ibrahim and Basma M. Eid

Abstract The present book chapter is devoted to surface modification and functionalization of cellulosic fabrics using atmospheric non-thermal plasma treatment technology as an eco-friendly, efficient and a promising alternative to the conventional wet chemical processing treatments without adversely affecting the bulk properties of treated substrates. In this chapter, conventional wet processing of cellulosic fabrics and its negative impacts on products and environment quality, plasma classification, advantages of plasma-treatments compared with the conventional ones, mode of interaction, modification, and functionalization of the fabric surface are discussed. Moreover, some potential applications of the nominated plasma in chemical processing of cellulosic fabrics, current situation of industrial utilization of plasma technology, as well as future developments are also considered.

Keywords Cellulosic fabrics · Chemical processing · Plasma technology · Surface modification · Functionalization · Future trends

1 Introduction

Growing environmental and energy-savings concerns especially in many environmentally harmful water-based chemical processing of cellulosic fabrics i.e. pretreatments, coloration and final chemical finishes, along with increasing demands regarding production, user as well as disposal ecology have been speeded up the development and adoption of more eco-friendly creative solutions and innovative technologies e.g. nano, bio, plasma etc. [1–6]. Adoption of emerging green technologies in traditional textile finishing industry brings about pollution abatement, materials and energy conservation, environmentally sound processes and textile products, cost reduction, and an enhancement in business performances, which in turn positively affects the competitive edge and the market share [7, 8].

One of the very promising environmentally sound technologies is plasma technology. Recently, potential applications of plasma, as an eco-friendly, dry and economic

N. A. Ibrahim (✉) · B. M. Eid
Textile Research Division, National Research Centre, El-Behouth St., Dokki, Giza 12622, Egypt
e-mail: nabibrahim49@yahoo.co.uk

M. Shahid and R. Adivarekar (eds.), *Advances in Functional Finishing of Textiles*, Textile Science and Clothing Technology, https://doi.org/10.1007/978-981-15-3669-4_12

alternative, to traditional wet chemical processes, to produce desirable effects with minimal environmental negative impacts, and to impart high level of functionalities and new properties without adversely affecting bulk properties of treated substrates, taking in consideration the resource efficient use of both materials and energy, have been practiced [9–12].

The present book chapter primarily focuses on: (i) overview of traditional chemical wet processing of cellulosic substrates and their negative impacts, and (ii) overview of plasma technologies, advantages and shortcomings, factors affecting plasma treatments as well as their potential applications for environmentally sound and green finishing processes. Lastly, recent progress and some future outlooks regarding industrial textile applications of green plasma technology will be discussed.

2 Conventional Wet-Chemical Processing

Wet chemical processing of cellulose-based textiles is a very important stage in production chain due to its positive role in upgrading the aesthetic, performance and comfort properties along with imparting value added functionalities to meet the consumer needs [1, 13]. Textile finishing industry is a very diverse production sector in terms of raw materials, processing stages, available techniques and final products. Conventional wet-chemical processing of cellulosic fabrics is the segment of the textile chain production processes that involves pretreatment (desizing, scouring, bleaching and mercerizing), coloration (dyeing and printing), and chemical finishing (Fig. 1). The traditional wet-chemical processing, especially of cellulose based textiles, is characterized by huge consumption of water, energy and processing chemicals along with creation of highly polluted effluents which in turn negatively impact both the eco-system and human health [9, 12–14].

Textile industry is one of the top 10 most polluting industries [15]. Pretreatment and finishing processes consume the greatest share ($\approx$ 35%) of the thermal energy in form of steam and heat [16]. On the other hand, pretreatments of cotton based textiles consume nearly 38% of the total water consumption for cotton wet-chemical processing [13]. Moreover, nearly 25% of globally produced chemicals are used in textile wet-chemical processing [16]. About 8000 different chemicals are utilized in textile wet-chemical processing, and some of these processing chemicals are harmful, hazardous and/or non-biodegradable which in turn negatively affect the sustainable development of the finishing industry and its environmental performance [17]. Additionally, textile finishing industry significantly affects the carbon footprints as well as greenhouse gas emissions as a direct consequence of high thermal energy consumption [18]. Key factors affecting the generation of various types of waste streams such as large volume of chemical-laden waste water, air emissions, solid and hazardous wastes include type of textile substrate, processing chemicals, available technologies, production chain, availability of on-line monitoring systems as well as the demanded performance and functional properties of the final product

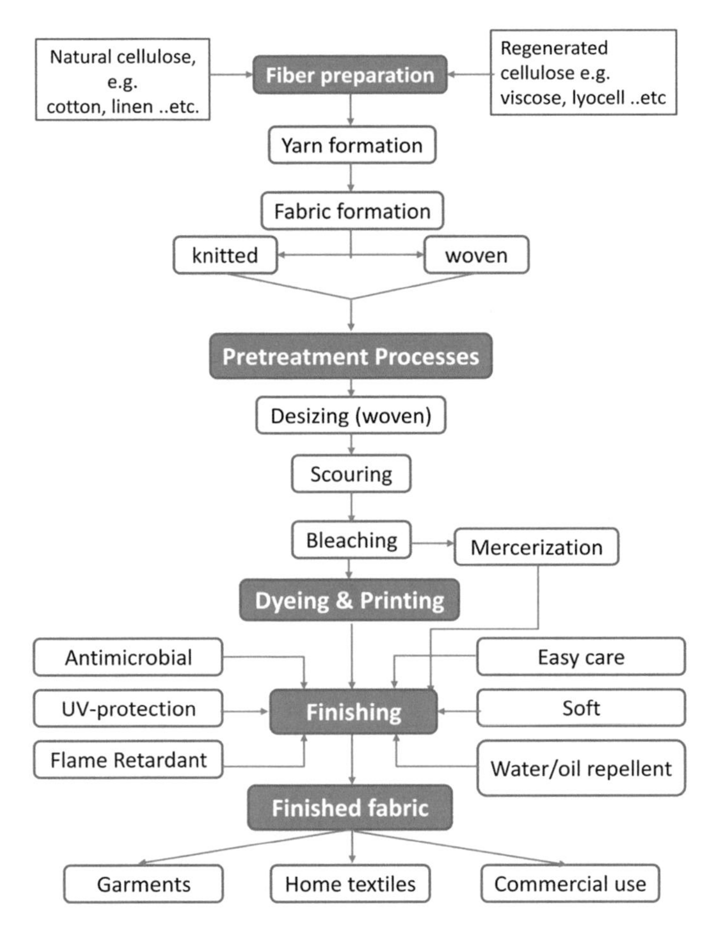

Fig. 1 Cellulosic textile manufacturing chain

[19]. Figure 2 demonstrates negative impacts associated with various conventional we-chemical processes [1, 14].

For sustainable textile processes and products, it becomes mandatory to adopt green technologies, eco-friendly production processes using renewable or recyclable raw materials and fewer amounts of water and energy, taking in consideration social responsibility, economic and environmental concerns [20].

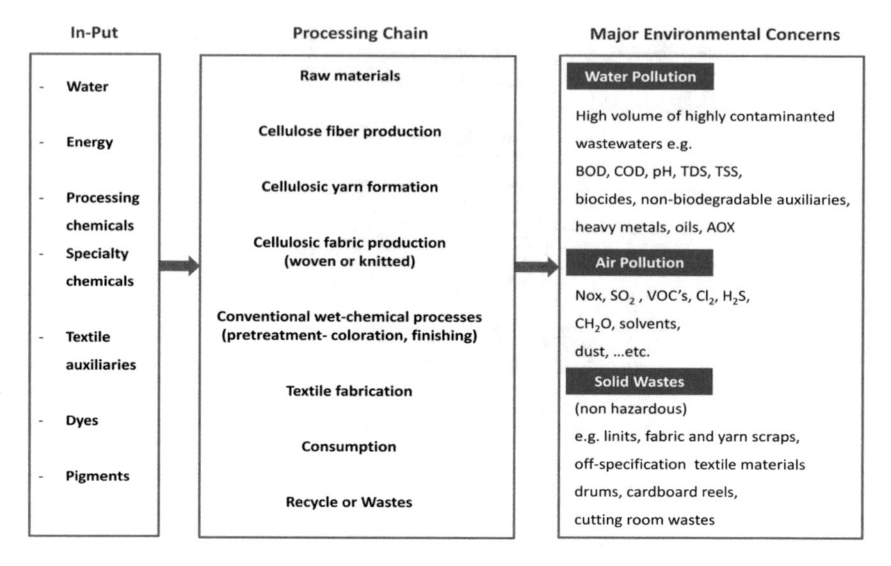

Fig. 2 Negative impacts of conventional wet-chemical processing of cellulosic substrates

3 Plasma Technology

3.1 Some Basic Aspects

Plasma, as an emerging/green/water-free technology, can be utilized for development of nano-scale surface modification of treated textile materials to impart value added new functionalities without adversely affecting their bulk properties [10, 15]. The generations of plasma active species like ions, energetic electrons, radicals, photons, UV-light etc. and the subsequent attack and bombardment of fabric surface result in surface activation modification and functionalization [21–23].

Plasma can be generated by using low frequency (50–450 kHz), radio frequency (13.56 or 27.12 MHz) or microwave (0.915 or 2.45 GHz) power supply. Additionally, the required power range (10–500 W) is selected based on both the reactor size as well as the demanded surface treatment [24].

Plasma can be classified onto thermal and non-thermal. On account of its negative impacts on the treated substrate and its destructive nature, hot plasma is not recommended for surface modification and there is no interest to textile manufactures [23].

On the other hand, applications of cold plasma, as a dry/non energy intensive/eco-sound technique, provides an alternative tools of online pretreatment and finishing treatments of cellulose based textiles, which in turn enable surface modification and functionalization of treated substrates in an environmentally sound manner for wide potential applications [9]. Various types of possible surface modification and functionalization can be obtained by suing cold plasma treatments which include

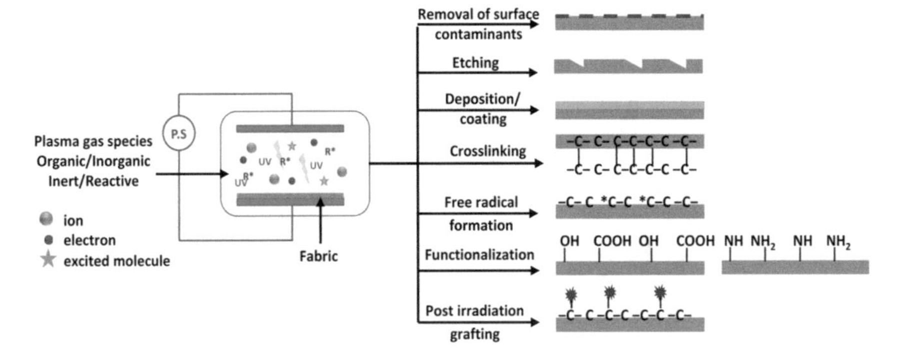

Fig. 3 Possible surface modifications of textile fabric using cold plasma

(i) removal of surface contaminants, (ii) etching and creation of surface roughness, (iii) thin film deposition and coating, (iv) creation of new active sites and functional groups and (vii) post-irradiation grafting (Fig. 3) [9, 21, 25].

The commonly used gasses for cold plasma treatment include: chemically inert, e.g. argon, neon, helium … etc., reactive/non-polymerisable, e.g. air, nitrogen, ammonia etc., reactive/polymerisable, e.g. tetrafluoroethylene, hexamethyldisiloxane etc. [9, 24]. Pretreatment and finishing of textile substrates by cold plasma bring about a versatile surface modification via creation of large variety of active sites and functional groups onto the treated fabric surface thereby enhancing its accessibility, including significant morphological and surface chemical modification as well as finally upgrading both the performance and functional properties of the treated substrates.

Additionally, non-thermal plasma may be classified into atmospheric pressure (APPs), or low-pressure plasma (LPPs). Atmospheric pressure plasmas for surface modification prove to be an efficient alternative, cost-effective methods to (LPPs) and traditional wet-chemical processing of textile substrates as a direct consequence of avoiding the need for costly vacuum equipment as well as allowing continuous and uniform surface treatments [21], i.e. can be utilized as part of the whole chain of continuous textile wet processing [23]. On the other hand, main types of APPs applied to textile materials are: corona discharge, dielectric barrier discharge (DBD), glow discharge (APGD), and atmospheric pressure plasma jet (APPJ) [21, 24]. DBD in air proves to be one of the most effective non-thermal APP source especially for industrial applications most probably due to its scalability to huge systems [21, 26].

3.2 *Potential Applications in Textile Processing*

With the ever-growing ecological and economical restrictions imposed on traditional textile finishing industry, environmentally benign, energy efficient and commercially available technologies have become more demanded taking in consideration production, user and disposal ecology aspects [1, 2]. Amongst the numerous emerging technologies available, atmospheric pressure cold plasma-surface modification is now a first priority for imparting specific desired functionalities to the treated textile materials and prove to be an eco-friendly, economical and effective process to replace or enhance the conventional wet-chemical processing, i.e. pretreatment, coloration and finishing of cellulosic substrates, as well as to meet the increasing needs of textile industry for fast, consistent and low footprint production processes [9, 21, 27].

On the other hand, the extent of modification of chemical, physical and structural surface properties of the treated substrate, taking in consideration both the process evenness and treatment effectiveness is affected by device parameters, plasma gas, treatment time, plasma density and energy as well as type and structure of the textile material [28–30]. Moreover, treatment parameters must be optimized to achieve the desirable surface modification and functionalization for specific applications without adversely affecting the performance and bulk properties.

Cold-plasma treatment can be used as an effective technique for: surface cleaning, surface activation and functionalization, improving adhesion to ensure effective coating, enabling graft-copolymerization/in situ deposition, and imparting hydrophilicity or hydrophobicity, to the treated textile materials [9, 21, 31, 32].

Creation of reactive sites and functional groups at plasma treated substrates is governed by the nature of gas plasma, i.e. polymerising and non-polymerising gas plasma. Only cold plasma either at atmospheric or at low pressure, is proper for surface treatment of heat sensitive textiles [32]. Tables 1, 2 and 3 demonstrate a vast range of some potential applications of non-polymerizing and polymerizing gas plasma for surface modification and functionalization of the cellulosic substrates as well as for imparting the specific and desirable value-added functional properties [9, 20].

4 Future Trends

Recently, potential applications of non-thermal atmospheric plasmas, as an emerging, green, water-free and surface modification technologies, for achieving eco-friendly textile process and products and for developing innovative, multifunctional, performance enhanced/value added cellulosic textile materials have been attracted a great deal of attraction. Replacement of conventional wet-chemical processing with plasma treatments will offer a great promise to develop unique properties, to impart innovative functional and to achieve economic savings and environmental protection. Full industrial exploitation and application of proper plasmas will help to develop sustainable textile process and products, taking in consideration environmental/economic concerns and social responsibility, and will continue in the near future. Moreover, the co-application of emerging technologies such as plasma/nanotechnology, plasma/biotechnology … etc. in the textile industry will offer eco-friendly solutions to impart new functionalities to textile materials and develop environmentally benign products taking in consideration economic concerns, market demands and environment aspects.

Table 1 Some potential applications of non-thermal atmospheric plasma for eco-friendly pretreatment of cellulosic fabrics

Plasma type	Working gas	Cellulosic substrate	Main task	Positive impacts	References
DBD	Air	Grey cotton fabric	Upgrading surface properties	• Removal of various impurities which in turn positively affects the subsequent bleaching and dyeing processes along with reduction in chemical and energy consumption	[33]
DBD	Air/argon	Grey cotton fabric		• Enhancing wettability as well as modification and functionalization of the treated fibers surfaces via removal of impurities, etchings as well as generation of new functional groups as –COOH, –OH … etc.	[34]
APPJ	Helium/oxygen	Grey cotton fabric		• Improving treated substrate wettability as a direct consequence of etching, de-waxing and creation of new active sites	[35]
DBD	Helium/air	Grey cotton fabric	Pretreatment—desizing/scouring	• Enhancing wettability and desizability of the used sizing agent via etching and introducing new functional groups	[36]
APGD	Air/O_2/He and air/He	Cotton sized with polyvinyl alcohol	Desizing	• Facilitate the removal of PVS-size with cold after-wash • A remarkable desizing treatment is achieved by using O_2-plasma	[37]

(continued)

Table 1 (continued)

Plasma type	Working gas	Cellulosic substrate	Main task	Positive impacts	References
DBD	Air	Grey cotton fabric	Scouring	• Improvement in wettability of treated substrate via removal of non-cellulosic impurities	[38]
APPJ	O_2	Grey cotton fabric	Desizing/scouring	• Desizing/scouring in one step, i.e. size and non-cellulosic impurities removal in one step, positively affects the subsequent dyeing process	[39]
DBD	O_2	Grey cotton fabric	Facilitate bio-scouring using pectinases (plasma treatment followed by bioscouring)	• Facilitate removal of wax as well as etching of non-cellulosic impurities thereby enhancing the accessibility of pectin substances to the enzyme attack	[40]
DBD	Air	Grey cotton fabric	Bleaching of grey cotton fabric	• To achieve high degree of whiteness, it is recommended to apply a two stage treatment, i.e. plasma pretreatment followed by H_2O_2-bleaching	[33]
Corona	Air	Grey cotton	Mercerization without using wetting agent	• Plasma treatment followed by mercerization brings about numerous advantages such as low-cost, reasonable environmental impact, better conditions for soda-recovery	[41]

(continued)

Table 1 (continued)

Plasma type	Working gas	Cellulosic substrate	Main task	Positive impacts	References
DBD	Air, oxygen, or nitrogen	Linen containing fabrics	Pretreatment to upgrade properties and dyeability of linen-containing fabrics	• Eco-friendly plasma pretreatment followed by enzymatic treatment using neutral cellulases and alkaline pectinase for enhancing the wettability, degree of whiteness, softness as a well post reactive dyeing	[42]
DBD	Air, oxygen, or nitrogen	Linen containing fabrics	Enhancing hydrophilicity and affinity for H_2O_2 bleaching	• Eco-friendly substitution of conventional scouring process taking in consideration, quality, ecology and economy concerns	[43]

DBD dielectric barrier discharge; *APGD* atmospheric glow discharge plasma; *APPJ* atmospheric pressure plasma jet

Table 2 The positive role of non-thermal atmospheric plasma in enhancing the coloration properties of cellulosic textiles

Plasma type	Working gas	Cellulosic substrate	Main task	Positive impacts	References
DBD	Argon and air	Cotton	Enhancing dyeability of cotton with acid dyes	• Formation of free radicals on fabric surface, grafting with ethylenediamine or triethylenetetramine to create new active/basic sites for subsequent acid dyeing	[34]
Corona	Air	Cotton	Creation of active sites, i.e. –COOH, –OH groups onto the fabric surface for post-dyeing	• Enhancing dyeability with reactive dyes	[44]
APPJ	O_2	Cotton textile	Creation of active sites for pigment coloration	• Surface functionalization for enhancing the extent of post-pigment coloration	[45]
DBD	Air, O_2	Cotton fabric	Surface-roughness and activation for printability with natural dyes	• Pre-activation and production surface-roughness for attaining better coloration properties with natural dyes	[46]

Table 3 Plasma-aided cellulosic textile functional finishing

Plasma type	Working gas	Cellulosic substrate	Main task	Positive impacts	References
APPJ	He/O_2	Cotton	To impart antibacterial properties	• Environmentally sound surface modification of cotton textile for enhancing its surface roughness and functionality followed by loading of ZnPNPs	[47]
APPJ	Nitrogen	Cotton	To enhance regenerable antibacterial finishing of cotton fabric	• Creation of nitrogen-containing active sites onto cotton fabric for enhancing the extent of adhesion and fixation of 5,5-dimethylhydantoin (DMM), which in turn positively affects the imparted antibacterial functionality	[48]
DBD	Oxygen or nitrogen	Linen containing fabrics	Production of linen based textiles with noticeable UV-protection and/or antibacterial functions	• Creation of active sites using eco-friendly pre-plasma treatment for subsequent treatment with various functional materials, e.g. ionic dyes, metal salt, nanomaterials, antibiotic … etc. to impart the desired functional properties	[49, 50]
APPJ	He/O_2	Cotton fabric	Imparting flame retardancy to cotton fabric	• Creation of active sites using plasma for post-treatment with ZnONPs	[51]

(continued)

Table 3 (continued)

Plasma type	Working gas	Cellulosic substrate	Main task	Positive impacts	References
DBD	He/O_2	Cotton fabric	To impart antibacterial flame retardancy/thermal stability functions	• Eco-friendly plasma pretreatment followed by subsequent deposition of nano-TiO_2/SiO_2 onto the modified fabric surface	[52]
APGD	He	Cotton substrate	To impart high hydrophobicity along with thermal stability to the treated cotton	• Grafting of stearyl methacrylate (SMA) onto fabric surface to form a functional coat coped with the desired properties	[53]
DBD	N_2	Cellulosic fabrics	To improve antibacterial functionalization and coloration properties	• Green surface modifications using N_2-plasma to generate –NH_2 groups for subsequent treatment with AgNPs/antibiotic hybrids	[54]
DBD	O_2	Cellulosic substrates	Eco-friendly functionalization of cellulosic substrates	• Surface modification using O_2-plasma for subsequent post-treatment with AuNPs/ZnONPs combination	[55]

References

1. Ibahim NA, Eid BM (2018) Emerging technologies for source reduction and end-of-pipe treatments of the cotton-based textile industry. In: Yusuf M (ed) Handbook of textile effluent remediation. Pan Stanford-Taylor & Francis Group, New York, pp 185–226
2. Shenai V (2001) Non-ecofriendly textile chemicals and their probable substitutes—an overview. Indian J Fibre Text Res 26:50–54
3. Ibrahim NA (2015) Nanomaterials for antibacterial textiles. In: Rai M, Kon K (eds) Nanotechnology in diagnosis, treatment and prophylaxis of infectious diseases. Academic Press, Boston, pp 191–216
4. Noor-Evans F, Peters S, Stingelin N (2012) Nanotechnology innovation for future development in the textile industry. In: Horne L (ed) New product development in textiles. Woodhead Publishing, pp 109–131
5. Vigneswaran C, Ananthasubramanian M, Kandhavadivu P (2014) Bioprocessing of natural fibres. In: Vigneswaran C, Ananthasubramanian M, Kandhavadivu P (eds) Bioprocessing of textiles. Woodhead Publishing India, pp 53–188
6. Stegmaier T, Linke M, Dinkelmann A, Von Arnim V, Planck H (2009) Environmentally friendly plasma technologies for textiles. In: Blackburn RS (ed) Sustainable textiles. Woodhead Publishing, pp 155–178
7. Gulzar T, Farooq T, Kiran S, Ahmad I, Hameed A (2019) Green chemistry in the wet processing of textiles. In: Shahid-ul-Islam, Butola BS (eds) The impact and prospects of green chemistry for textile technology. Woodhead Publishing, pp 1–20
8. Sheikh J, Bramhecha I (2019) Enzymes for green chemical processing of cotton. In: Shahid-ul-Islam, Butola BS (eds) The impact and prospects of green chemistry for textile technology. Woodhead Publishing, pp 135–160
9. Dave H, Ledwani L, Nema SK (2019) Nonthermal plasma: a promising green technology to improve environmental performance of textile industries. In: Shahid-ul-Islam, Butola BS (eds) The impact and prospects of green chemistry for textile technology. Woodhead Publishing, pp 199–249

10. Shahidi S, Ghoranneviss M, Moazzenchi B (2014) New advances in plasma technology for textile. J Fusion Energy 33(2):97–102
11. Verschuren J, Kiekens P, Leys C (2007) Textile-specific properties that influence plasma treatment, effect creation and effect characterization. Text Res J 77(10):727–733
12. Jelil RA (2015) A review of low-temperature plasma treatment of textile materials. J Mater Sci 50(18):5913–5943
13. Saxena S, Raja ASM, Arputharaj A (2017) Challenges in sustainable wet processing of textiles. In: Muthu SS (ed) Textiles and clothing sustainability: sustainable textile chemical processes. Springer Singapore, Singapore, pp 43–79
14. Bhatia SC (2017) Textile industry and its impact on environment. In: Devraj S (ed) Pollution control in textile industry. Woodhead Publishing India in Textiles, India, pp 11–28
15. Hasanbeigi A, Price L (2012) A review of energy use and energy efficiency technologies for the textile industry. Renew Sustain Energy Rev 16(6):3648–3665
16. Roy Choudhury AK (2014) Environmental impacts of the textile industry and its assessment through life cycle assessment. In: Muthu SS (ed) Roadmap to sustainable textiles and clothing: environmental and social aspects of textiles and clothing supply chain. Springer Singapore, Singapore, pp 1–39
17. Kant R (2012) Textile dyeing industry an environmental hazard. Nat Sci 4(1):22–26
18. Bhatia SC (2017) Carbon footprint in textile industry. In: Devraj S (ed) Pollution control in textile industry. Woodhead Publishing India in Textiles, India, pp 223–238
19. Ibrahim NA, Abdel Moneim NM, Abdel Halim ES, Hosni MM (2008) Pollution prevention of cotton-cone reactive dyeing. J Clean Prod 16(12):1321–1326
20. Samanta KK, Basak S, Chattopadhyay SK (2017) Sustainable dyeing and finishing of textiles using natural ingredients and water-free technologies. In: Muthu SS (ed) Textiles and clothing sustainability: sustainable textile chemical processes. Springer Singapore, Singapore, pp 99–131
21. Zille A, Oliveira FR, Souto AP (2015) Plasma treatment in textile industry. Plasma Process Polym 12(2):98–131
22. Pavliňák D, Galmiz O, Pavliňáková V, Poláček P, Kelar J, Stupavská M et al (2018) Application of dielectric barrier plasma treatment in the nanofiber processing. Mater Today Commun 16:330–338
23. Mather RR (2009) Surface modification of textiles by plasma treatments. In: Wei Q (ed) Surface modification of textiles. Woodhead Publishing, pp 296–317
24. Roy Choudhury AK (2017) Various ecofriendly finishes. In: Roy Choudhury AK (ed) Principles of textile finishing. Woodhead Publishing, pp 467–525
25. Desmet T, Morent R, De Geyter N, Leys C, Schacht E, Dubruel P (2009) Nonthermal plasma technology as a versatile strategy for polymeric biomaterials surface modification: a review. Biomacromolecules 10(9):2351–2378
26. Borcia G, Anderson CA, Brown NMD (2006) Surface treatment of natural and synthetic textiles using a dielectric barrier discharge. Surf Coat Technol 201(6):3074–3081
27. Buyle G (2009) Nanoscale finishing of textiles via plasma treatment. Mater Technol 24(1):46–51
28. Ceria A, Rombaldoni F, Rovero G, Mazzuchetti G, Sicardi S (2010) The effect of an innovative atmospheric plasma jet treatment on physical and mechanical properties of wool fabrics. J Mater Process Technol 210(5):720–726
29. Guo L, Campagne C, Perwuelz A, Leroux F (2009) Zeta potential and surface physico-chemical properties of atmospheric air-plasma-treated polyester fabrics. Text Res J 79(15):1371–1377
30. Kamel MM, El Zawahry MM, Helmy H, Eid MA (2011) Improvements in the dyeability of polyester fabrics by atmospheric pressure oxygen plasma treatment. J Text Inst 102(3):220–231
31. Bazaka K, Jacob MV, Crawford RJ, Ivanova EP (2011) Plasma-assisted surface modification of organic biopolymers to prevent bacterial attachment. Acta Biomater 7(5):2015–2028
32. Samanta KK, Basak S, Chattopadhyay SK (2014) Environment-friendly textile processing using plasma and UV treatment. In: Muthu SS (ed) Roadmap to sustainable textiles and clothing: eco-friendly raw materials, technologies, and processing methods. Springer Singapore, Singapore, pp 161–201

33. Prabaharan M, Carneiro N (2005) Effect of low-temperature plasma on cotton fabric and its application to bleaching and dyeing. Indian J Fibre Text Res 30:68–74
34. Karahan HA, Özdoğan E (2008) Improvements of surface functionality of cotton fibers by atmospheric plasma treatment. Fibers Polym 9(1):21–26
35. Tian L, Nie H, Chatterton NP, Branford-White CJ, Qiu Y, Zhu L (2011) Helium/oxygen atmospheric pressure plasma jet treatment for hydrophilicity improvement of grey cotton knitted fabric. Appl Surf Sci 257(16):7113–7118
36. Bhat N, Bharati R, Gore A, Patil A (2011) Effect of atmospheric pressure air plasma treatment on desizing and wettability of cotton fabrics
37. Cai Z, Qiu Y, Zhang C, Hwang Y-J, Mccord M (2003) Effect of atmospheric plasma treatment on desizing of PVA on cotton. Text Res J 73(8):670–674
38. Dave H, Ledwani L, Chandwani N, Chauhan N, Nema SK (2014) The removal of impurities from gray cotton fabric by atmospheric pressure plasma treatment and its characterization using ATR-FTIR spectroscopy. J Text Inst 105(6):586–596
39. Kan C-W, Lam C-F, Chan C-K, Ng S-P (2014) Using atmospheric pressure plasma treatment for treating grey cotton fabric. Carbohyd Polym 102:167–173
40. Wang Q, Fan X-R, Cui L, Wang P, Wu J, Chen J (2009) Plasma-aided cotton bioscouring: dielectric barrier discharge versus low-pressure oxygen plasma. Plasma Chem Plasma Process 29(5):399–409
41. Carneiro N, Souto AP, Rios MJ (2005) Evaluation of cotton fabric properties after mercerization using CORONA as a preparation step. In: 5th international Istanbul textile conference recent advances in innovation and enterprise in textile and clothing, Istanbul-Turquia
42. Ibrahim NA, El-Hossamy M, Hashem MM, Refai R, Eid BM (2008) Novel pre-treatment processes to promote linen-containing fabrics properties. Carbohyd Polym 74(4):880–891
43. Ibrahim NA, Hashem MM, Eid MA, Refai R, El-Hossamy M, Eid BM (2010) Eco-friendly plasma treatment of linen-containing fabrics. J Text Inst 101(12):1035–1049
44. Patiño A, Canal C, Rodríguez C, Caballero G, Navarro A, Canal JM (2011) Surface and bulk cotton fibre modifications: plasma and cationization. Influence on dyeing with reactive dye. Cellulose 18(4):1073–1083
45. Man WS, Kan CW, Ng SP (2014) The use of atmospheric pressure plasma treatment on enhancing the pigment application to cotton fabric. Vacuum 99:7–11
46. Ahmed H, Ahmed K, Mashaly H, El-Halwagy A (2017) Treatment of cotton fabric with dielectric barrier discharge (DBD) plasma and printing with cochineal natural dye. Indian J Sci Technol 10:1–10
47. Kan CW, Lam YL, Yuen CWM, Luximon A, Lau KW, Chen KS (2013) Chemical analysis of plasma-assisted antimicrobial treatment on cotton. J Phys: Conf Ser 441:012002
48. Zhou C-E, Kan C-W (2015) Plasma-enhanced regenerable 5,5-dimethylhydantoin (DMH) antibacterial finishing for cotton fabric. Appl Surf Sci 328:410–417
49. Ibrahim NA, Eid BM, Hashem MM, Refai R, El-Hossamy M (2010) Smart options for functional finishing of linen-containing fabrics. J Ind Text 39(3):233–265
50. Abdel-Aziz MS, Eid BM, Ibahim NA (2014) Biosynthesized silver nanoparticles for antibacterial treatment of cellulosic fabrics using O_2-plasma. AATCC J Res 1(1):6–12
51. Lam YL, Kan CW, Yuen CWM (2011) Effect of zinc oxide on flame retardant finishing of plasma pre-treated cotton fabric. Cellulose 18(1):151–165
52. Palaskar SS, Desai AN, Shukla SR (2016) Development of multifunctional cotton fabric using atmospheric pressure plasma and nano-finishing. J Text Inst 107(3):405–412
53. Li Y, Zhang Y, Zou C, Shao J (2015) Study of plasma-induced graft polymerization of stearyl methacrylate on cotton fabric substrates. Appl Surf Sci 357:2327–2332
54. Ibrahim NA, Eid BM, Abdel-Aziz MS (2017) Effect of plasma superficial treatments on antibacterial functionalization and coloration of cellulosic fabrics. Appl Surf Sci 392:1126–1133
55. Ibrahim NA, Eid BM, Abdel-Aziz MS (2016) Green synthesis of AuNPs for eco-friendly functionalization of cellulosic substrates. Appl Surf Sci 389:118–125

Cationization as Tool for Functionalization of Cotton

Ashwini Patil, Saptarshi Maiti, Aranya Mallick, Kedar Kulkarni and Ravindra Adivarekar

Abstract The textile processing industry is copiously dependent on utilities, chemicals and colourants to improve its aesthetic properties. Cotton continues to be the most important textile fibre inspite of a large increase in the production and availability of manufactured fibres and is best known for its outstanding comfort characteristics. Current estimates for world production of cotton are about 25 million tonnes or 110 million bales annually, accounting for 2.5% of the world's arable land. It is a natural fibre which assumes negative zeta potential when immersed in water. The zeta potential (ζ-potential) is the potential difference across phase boundaries between solids and liquids. It's a measure of the electrical charge of particles that are suspended in liquid. This results in repelling similarly charged ions of substantive dyes such as direct and reactive. It is possible to reduce the repulsion by chemically modifying the cotton surface by depositing cation onto the surface of cotton. This process is well known as the cationization of cotton. The cationizing agents are used to develop cationic sites on the cotton so that anionic dyes or finish get attracted forming ionic linkages. In a simple way, a cationic auxiliary increases affinity and reactivity of anionic dyes and finishes towards the textile substrate, the cotton. This method is used to enhance the dyeing and finishing property of cotton. These auxiliaries create reactive sites to impart functionality such as colour, flame retardancy, water repellency, antimicrobial properties, etc. to the cotton substrates. These auxiliaries can be successfully applied on the cotton substrate by the exhaust, padding, layer by layer, sol-gel and coating method. The aim of the current chapter is to discuss the importance of cationization, cationizing agents and its application on cotton.

Keywords Pigment dyeing · Exhaust method · Cotton · Cationizing agent · Zeta potential

---

A. Patil · S. Maiti · A. Mallick · K. Kulkarni · R. Adivarekar (✉)
Department of Fibres and Textile Processing Technology, Institute of Chemical Technology, Mumbai 400019, India
e-mail: rv.adivarekar@ictmumbai.edu.in

M. Shahid and R. Adivarekar (eds.), *Advances in Functional Finishing of Textiles*, Textile Science and Clothing Technology, https://doi.org/10.1007/978-981-15-3669-4_13

1 Textile Industry

The textile wet processing industry consumes plenty of textile auxiliaries. These auxiliaries boost up the functional properties of the original fibre/fabric. The global market size of textile auxiliaries was estimated to reach USD 23.62 billion in 2018 and was projected to register a compound annual growth rate (CAGR) of 4.5% during the forecast period. This significant growth is credited to high demand from the rapidly rising apparel industry [1].

2 Cotton

The textile fibres are mainly classified as natural (animal and plant), synthetic semi-synthetic and mineral fibres. The natural fibres are derived from animals (silk and wool) and plant (cotton, jute, linen ramie, etc.) sources. The mineral fibres are developed from asbestos and glass whereas synthetic fibres are man-made which includes; nylon, polyester and acrylics. These fibres are pre-treated before any treatments like dyeing and finishing. All these fibres have their own importance and market share, but even today cotton continues to occupy approximately 73% of market share due to its extraordinary properties like strength, breathability, dimensional stability, comfort and moisture absorption ability [2]. The cotton industry is a combination of old and latest technologies [3]. Along with that, high exhausting dyes and functional finishes are used in the industry which gives cotton improved aesthetic and novel functional properties.

2.1 Cotton Fibre

2.1.1 Cotton Morphology

Cotton fibre has main four parts; cuticle, primary wall, secondary wall and lumen. Cuticle is the outer most layer containing waxes, a mixture of fats, oils, pectin and protein materials. Pre-treatment is carried out to remove these hydrophobic materials and render the cotton hydrophilic. The primary wall is the original thin wall including cellulose strands where molecular chains are arranged randomly i.e. no distinct orientation is seen. This layer also contains pectineus materials. The secondary wall consists of many cellulose layers which constitute the main portion of the cotton fibre. Lumen is the hollow canal responsible for the length of the fibre covered by lumen wall. It contains colouring matter and other impurities which are responsible for the characteristic colour of grey cotton. Pre-treatment, mainly bleaching, decolourizes these natural pigments to get a visible and uniform whiteness. These fibres are flat

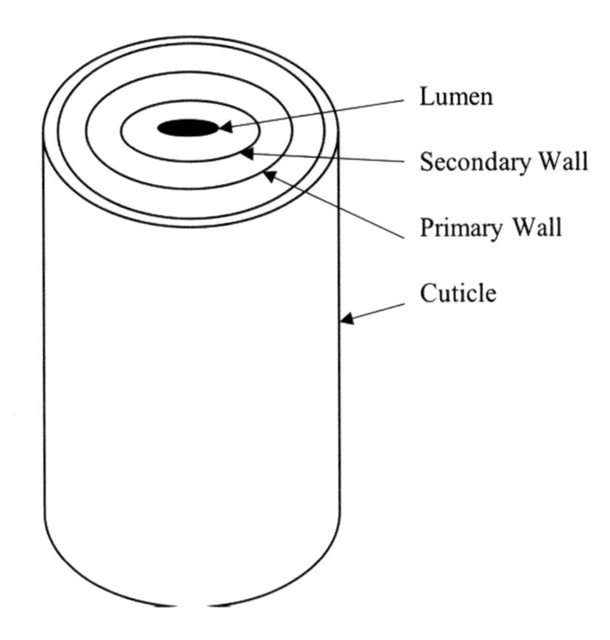

Fig. 1 Cross-section of cotton fibre

twisted ribbons with 50–100 convolutions per inch. The cross-section of cotton fibre shows kidney shaped bean-like structures [4, 5] (Fig. 1).

2.1.2 Chemical Composition

Chemically, cotton is composed of 88–96% of cellulose and remaining are wax, protein, oil, pigment, fat, ash etc. [6]. Cotton is highly crystalline fibre which is long and rigid. The basic building block of cotton is beta-1,4-D (+)glucopyranose. Long cellulose chains are held by 1,4-glycosidic bonds where anhydrous glucopyranose C–O–C link is formed. This anhydrous glucose shows three hydroxyl groups (–OH), one primary on C-6 and two secondary on C-2 and C-3, connected by hydrogen bonding that are chemically active for further additions and also are responsible for ionization in aqueous solution forming cotton with anionic surface characteristics [7–9] (Fig. 2).

2.2 Wet-Processing of Cotton

The harvested and mechanically cleaned, spun and woven cotton fabric goes through various processes like pre-treatment, dyeing, printing and finishing depending upon the end-use of the product. The pre-treatment is the prior process that includes surface preparation for the removal of foreign materials to improve affinity, absorptivity,

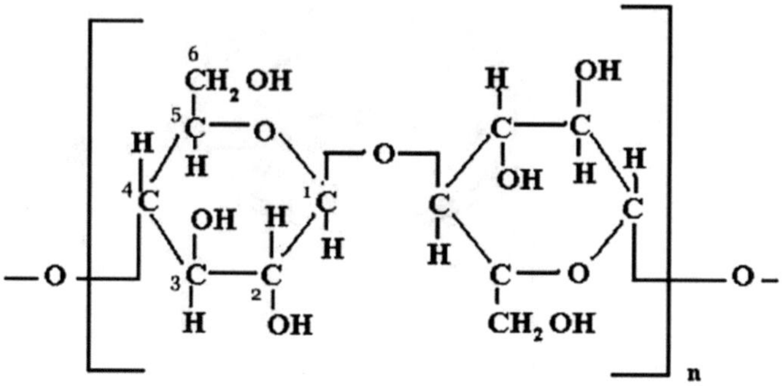

Fig. 2 Chemical structure of cotton

uniformity and strength ensuring to have the right physical and chemical properties before dyeing and finishing.

2.2.1 Pre-treatment

Singeing:

Cotton has hairy appearance due to the presence of loose fibres which are not bound firmly. In singeing operation, such protruding fibres are superficially burned without damaging the yarn or fabric to create a smooth surface feel, to improve lustre and to prevent potential obstructions during wet processing [10, 11].

Desizing:

The size like starch, wax, PVA, etc. is applied on warp yarn before weaving of fabric to increase weave-ability. This size hinders the hydrophilicity of fibre and the removal of such sizes improve its absorbency for further treatments. This process of removal of size is known as desizing. Thus, primarily, added impurities are removed in desizing.

Scouring:

During scouring, natural impurities like dirt, vegetable matter, grease, wax and other hydrophobic materials are removed from the cotton fibres. The scouring process should ensure uniform water absorbency.

Bleaching:

Cotton fibres naturally possess pale yellow colour due to the presence of natural colouring matter and impurities. Bleaching is a chemical treatment involving bleaching agents to decolourize natural colouring matter and obtain a uniform degree of whiteness [12].

Mercerizing:

Cotton is treated with strong sodium hydroxide solution (300 gm/lit) so that cotton fibres swell by breaking hydrogen bonds and Van der Waal forces present between cellulose chains. The expanded chains rearrange and re-orient themselves and when the caustic soda is removed, the chains form new bonds in the reorganized state. The main advantages of mercerization step enhanced lustre, smoothness, strength/elongation and moisture absorption property [10].

2.2.2 Dyeing

It is a process where colouring agents are applied uniformly on textiles to impart attractive looks. These colouring agents are dyes or pigments which are mostly applied in an aqueous medium, although sometimes the solvent system is also used, especially in printing. The selection of the appropriate dye depends on the objective in dyeing and chemistry of the dye and textile material to be dyed. Cotton can be easily dyed with direct, reactive, vat and sulphur dyes. Direct and reactive dyes are water-soluble dyes but vat and sulphur are insoluble dyes and are made soluble using reducing agents and alkalis [13].

2.2.3 Printing

Printing is the localized dyeing process which includes the application of dye or pigment in different designs or patterns. The selection of dye or pigment on textiles remains the same as the dyeing process. Cotton can be easily printed by direct, reactive, azo and vat dyes.

2.2.4 Finishing

This is the last process on the fabric which adds aesthetic and tactile value to the textiles. It should also meet the required demand and secure customer satisfaction by applying such finishes on textiles. Regular finishing is usually carried out using cationic or silicone softeners or a combination of both. Finishing many times involves speciality finishes as well which includes finishes such as, wrinkle-free, soil and stain repellent, water repellent etc.

3 Surface Modification

Finishing is the final step used to improve the appearance, functionality and durability of the textile products. The process of finishing can be classified according to purpose or end result [14, 15]. Generally, the surface of each textile fibre carries its individual properties such as hydrophilicity, hydrophobicity, adhesion and optical

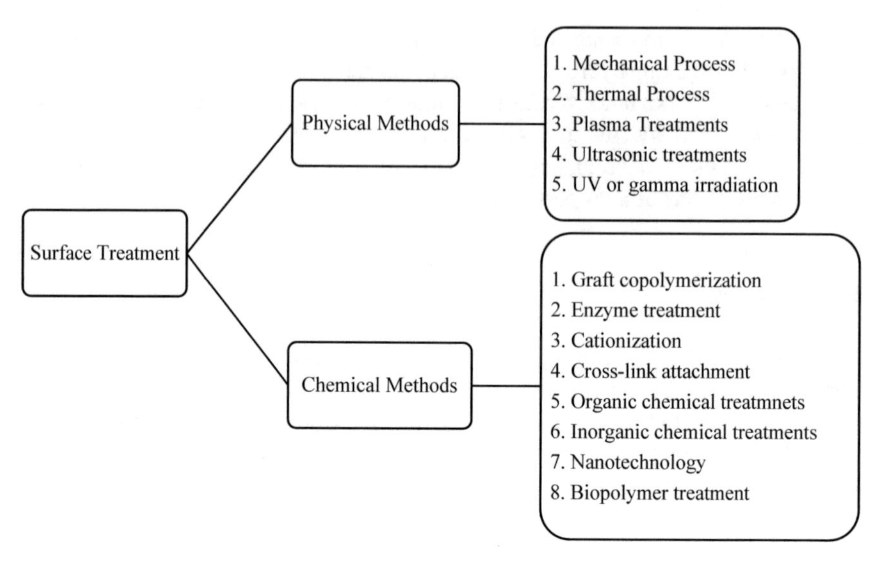

Fig. 3 Types of surface treatments on cotton

appearance depending on the presence of constituent functional groups and their orientation. In order to make desirable changes in given fibres, surface treatment plays an important role, which can be defined as a surface treatment, which improves the surface chemistry of textiles with ease of efficiency for finishing [16].

The following are the ways to modify the textile surface (Fig. 3).

3.1 Types of Surface Treatments

Surface modification of textiles comprises of a wide range of technologies that modifies the surface properties of textiles. The surface modification of textiles can be achieved by physical and chemical methods ranging from traditional solution treatments to biological approaches. Technical and functional surface finishing of textile has attracted a great deal of attention in recent years with the help of new technologies such as plasma treatments, grafting, nanotechnology, enzyme treatment, radiation treatments, organic or inorganic treatments and addition of cationic compound on the surface of textiles. All above surface treatments are used by researcher to enhance the performance of wet processing steps like pre-treatment, dyeing and finishing of natural as well as synthetic textiles [17]. In comparison with physical surface treatment methods, chemical methods are preferred for permanent and durable finish in the industry. The physical methods like mechanical and thermal treatment are dry treatments that do not produce a permanent surface effect and alters the fabric appearance.

3.1.1 Plasma Treatment

In this process, predefined gas is heated under high pressure to produce charged plasma state and fabric is placed between it, the layer formed on textile modify the surface property. The plasma treatment is widely used nowadays for desizing, wettability enhancement, improving dyeing-printing ability and physical property of textiles.

3.1.2 Ultrasonic Application

In a bath, when ultrasonic waves are introduced, it gets absorbed in liquor forming gas bubbles. Due to vibrations these tiny bubbles collapse generating shock waves and if the bubbles are not collapsed then they push water producing streaming. This energy helps in better penetration of dyes and chemicals, certain surface treatment with better efficiency, etc. [18]. The process is cost-effective and eco-friendly but the formation of high-intensity ultrasound in a big vessel is a difficult task [19].

3.1.3 Irradiation Treatments

UV and gamma irradiation treatments are also extensively used in the industry to modify properties especially for synthetics and non-wovens [20]. Due to the high energy source and high penetrability of such rays, they can affect chemical bonding with the deterioration of physical properties of textiles [17].

3.1.4 Grafting Copolymer

The grafting is the method where copolymers are strongly bonded onto the polymer chain, when it comes in contact with dry heat, the mixture of polymers form a coated layer on the surface that alters the physical forces of the treated substrate [21, 22].

3.1.5 Enzyme Treatment

Many enzymes are used for cotton pre-treatment and finishing processes due to their positive properties towards the environment as they are very specific in action, biodegradable and best alternative to toxic chemicals [23].

3.1.6 Cross-Link Modification

In cross-link modification, polymer or fibre are chemically joined forming a covalent or ionic linkage [24]. Such modifications are used to improve certain physical properties of fibres.

3.1.7 Nanotechnology

The application of nanomaterials on textiles offers high-performance functionality, which is vigorously used for finishing, coating, dyeing etc.

3.1.8 Biopolymer Application

The polymers which are produced by living organisms are known as biopolymers. Biopolymers are biodegradable and eco-friendly [25, 26]. Other than starch or sodium alginate, commonly used biopolymer in textiles are chitosan and carrageenan derived from carbohydrate family, they form a covalent linkage with the substrate or other agent offering durable and permanent effect [27].

3.1.9 Chemical Addition

These treatments are commonly used for finishing processes. Examples of this are, cyclodextrins used for fragrance finishes, silicones used for water repellency, fluorocarbons used for flame retardancy, DMDHEU used as cross-linkers, silicone-based softeners for textiles, zinc/titanium oxides for antimicrobial finish and UV protection finish, etc.

3.1.10 Cationic Compound Addition

The process where a cationic compound is applied on the cotton surface is known as cationization. The popularly cationic agents used for the process are amines and/or quaternary ammonium compounds. They chemically modify the surface on cotton introducing permanent cationic sites. Other surface treatment methods used for cationization include organic or inorganic chemical addition, biopolymer application on different textiles etc. Lewis and Lei in early 1930s had reported and developed various cationizing processes and cationic agents to be used on cotton substrates that introduce cationic sites after application. These chemicals are as follows.

Quaternary ammonium chlorides like Glycidyl trimethylammonium chloride (Glytac A) [2], Betaine (*N*, *N*, *N*-trimethyl glycine), 3-chloro-2-hydroxypropyltrimethylammonium chloride (CHPTMAC), polyepichlorohydrin acrylamides, polyamino chlorohydrin quaternary ammonium compound i.e. Cibafix WFF

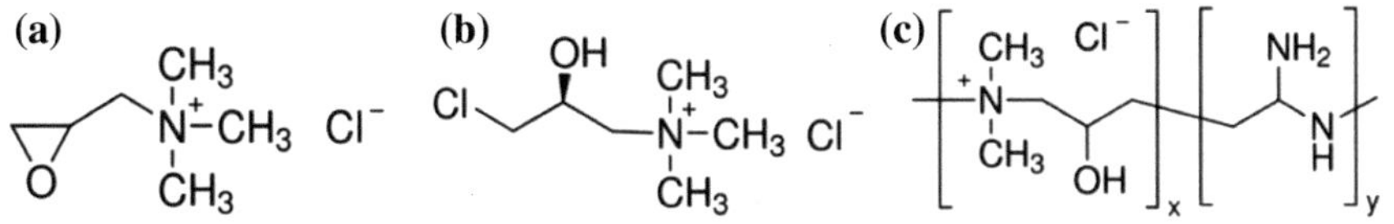

Fig. 4 **a** Glytac A, **b** CHPTMAC, **c** poly-epichlorohydrin acrylamides

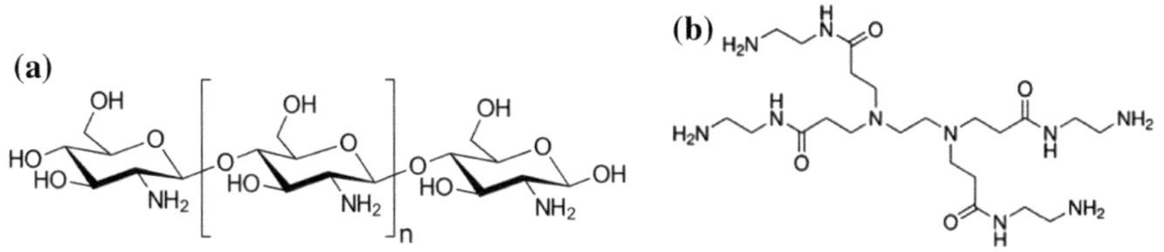

Fig. 5 **a** Chitosan, **b** PAMAM Dendrimer

above all those have been patented as a cationizing agent for cotton. Most of these are amine ($-NH_2$) derivatives. The amines are commonly used cationizing agents as they contain donor nitrogen (N_2) compound in the group that are known for their cationizing power and also the hydrogen (H^+) which are responsible for protonation and improving cationic property [28–41] (Fig. 4).

Nowadays biopolymer like chitosan and its derivatives are in use as cationizing agent on cotton for dyeing and finishing processes. Chitosan is biodegradable, biocompatible, antimicrobial active and non-toxic. It includes three reactive groups, one amino and two hydroxyl in its structure. The other cationizing agent made with dendrimers has amongst these polyamidiamine dendrimer known as PAMAM dendrimer contains four amine groups at terminal end formed by amidation reaction and possess surface modifying property with cationic characteristics found to be highly active (Fig. 5).

3.2 Need for Cationization of Cotton

Common textile wet processes like dyeing and finishing can be divided into steps like adsorption, absorption and fixation of chemicals on textile materials. Due to the presence of various forces existing in the liquor, adsorption of dyestuff and textile auxiliaries on substrate is not easy [42]. The dyestuff and textiles chemicals might be in liquid or solid form and in different ionic nature. The very popular cotton fibre contains reactive hydroxyl groups (–OH) that react with dyes or chemicals. When such cotton is dipped in aqueous solution it shows negative zeta potential [43]. Two mechanisms are involved behind cationization. Firstly, the layer of cationic charges is formed on the cotton surface when applied and the second mechanism involves the application of cationic agents that modifies cotton surface by getting covalently linked. From dyeing perspective, anionic nature of cotton poses a problem while taking up dyes which are also anionic in nature. Because negative dye/chemical and fibre surface repel each other, the electrolyte is added in the dyeing process to neutralize the electronegativity of fibre surface. When alkalinity is introduced functional group of reactive dye forms a covalent bond with fibre [44]. These electrolytes, however, do not get absorbed during the dyeing process and such a high concentration of electrolytes is discharged. It is not acceptable by environmental norms as it disturbs aquatic life. It also has an adverse effect on dye solubility. Lowering or eliminating the concentration of electrolytes including alkali does not give satisfactory fixation of the anionic dyes on the cotton substrates. Similar issues arise in case of finishing process as negative surface nature of cotton also restricts the anionic chemical agents to be applied on its surface. Anionic softeners cannot get attached on the cotton. Hence, it is necessary to perform cationic surface modification of cotton. In textile industry various surface modification techniques like grafting, inorganic chemical addition, biopolymer addition etc. are used but most of these treatments are not economic and effective. They are not durable as there is no strong linkage between surface of substrate and target chemical agent.

The durability of any finishing process is the most important criteria in textile processing, where strong and permanent linkage of cationic chemical on the surface of the substrate is the necessity, so the cationization of cotton is important surface-modifying treatment [45]. The cationization process has a high ecological impact as for processing like dyeing and finishing. By cationization, chemical consumption to dye cotton can be reduced by 50% compared to conventional processing.

3.3 Mechanism

The process of cationization is a simple technique, where a cationic agent is chemically reacted with hydroxyl ions of cotton to develop new cationic sites on the fibre surface. The nucleophilic substitution reaction takes place between nucleophile i.e. hydroxyl group of cotton and a cationic group of cationizing agent. When such cationized cotton is further treated with anionic dyes or finished with a finishing agent, the coulombic attraction takes place between the positive charge on the fibre and the negative charge on the dyes or finish [46].

The efficiency of the cationization is directly proportional to the concentration of the cationizing agent and the capacity of the cotton surface. In a simple way, more is the amount of cationizing agent applied, more is the degree of cationization (Fig. 6).

3.4 Application

Cationization can be achieved on fibre, yarn or fabric form and the reaction depends upon parameters like time, temperature and pH. The temperature and time can be

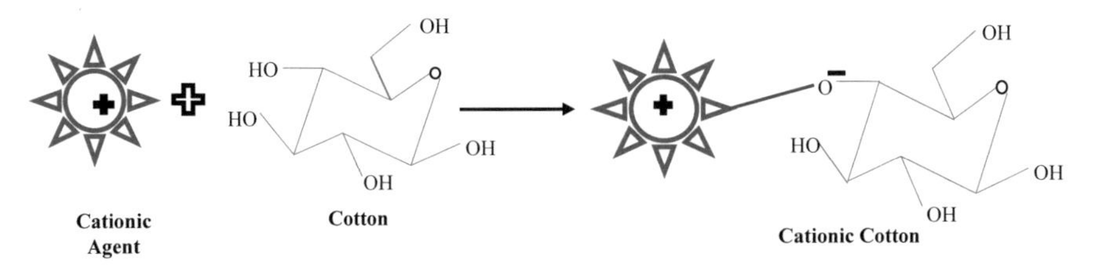

Fig. 6 Cationization mechanism

adjusted according to the process requirement but the desired pH for cationization treatment is set from neutral to slightly acidic because at this pH the protonation of amine groups ($-NH_2$) of cationizing agent occurs and that increases the degree of cationization. The cationizing agents can be applied by the following methods.

3.4.1 Exhaust Method

Exhaust method is also known as batch or discontinuous application method. In this method, dye molecule or finishing agents migrate from solution on to the textile material. The efficiency of the process depends upon the influencing factors like the chemistry of the dye or finishing agent, liquor ratio, pH and temperature required for the process [47, 48]. The process is preferred if the chemical agents have a strong affinity to the substrate [28].

3.4.2 Padding Method

This method is used when chemical agents have a low affinity for substrates. This process involves the application of chemicals through padding solution and then squeezing with the help of pair of rollers as per desired pickup percent, followed by drying and curing for fixation of chemicals on the substrates. The process is also referred to as pad-dry-cure method [49].

3.4.3 Coating Method

In this process, coating material i.e. polymeric material is applied on one or both sides of fabric forming a layer on the surface followed by drying and curing [50].

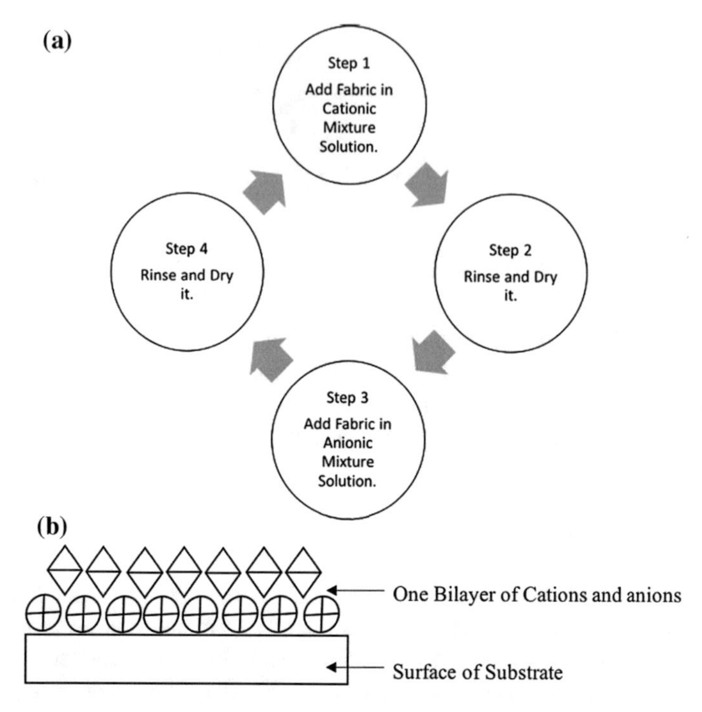

Fig. 7 **a** LBL treatment, **b** formation of bilayer by LBL treatment

3.4.4 Layer by Layer Technique

This process is similar to padding method, but here desired padding solution pots are made with the addition of polyelectrolytes i.e. polycations and polyanions consisting of oppositely charged chemicals. The process is repeated until the satisfactory multilayer polymeric film of cations-anions are developed on the substrate surface [51, 52] (Fig. 7).

3.4.5 Sol-Gel Technique

In this method, the monomer solution is converted to a colloidal solution which is known as sol [53]. The sol is made up of metal salt or metal-organic components in colloidal suspension form due to the hydrolysis and polymerization reaction [54]. In this process, the sol is converted to gel i.e. layer of metal oxide on the surface due to heat treatment. This sol-gel is commonly used in nanoparticles, which are applied by pad-dry-cure method on the textiles [55, 56] (Fig. 8).

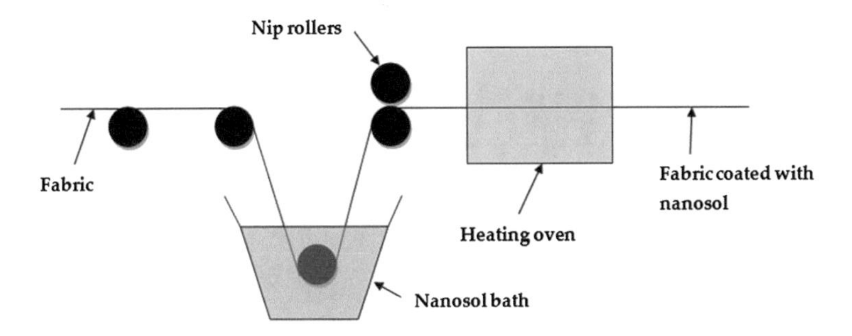

Fig. 8 Sol-gel application

3.5 Recent Developments in Cationization for Wet-Processing Techniques

3.5.1 Pre-treatment

Pre-treatment is the process prior to dyeing, that enhances the physical characteristics of cotton substrate. The surface modification is also a pre-treatment process of cotton which modifies the surface of the substrate according to the required end-use. The application of cationizing agent for dyeing or finishing can be pre-treated or post-treated depending upon the practical parameters by the industry or consumers. This cationic cotton is used for dyeing, printing and finishing.

3.5.2 Dyeing

In the dyeing process, anionic dyes are applied on the surface of cotton. In absence of salt, anionic dye molecules are repelled by the surface of cotton as cotton becomes negatively charged in water. Electrolytes are used to neutralize the surface charge so that dye molecules can reach the fibres. The reactive dye molecule in the presence of alkali reacts with both cotton and water as both contain hydroxyl ions $(–OH)^-$. Dye molecules which react with water are called as hydrolysed dye. This hydrolysed dye cannot chemically react with cotton, but their deposition on the surface affects fastness properties. This is one of the major drawbacks of colouration with reactive dyes. Similar problems are seen in direct and acid dyeing process where dyes are only weakly attached to the fibres. The electrolytes and alkali (in case of reactive dyes) help to neutralize the negative surface of cotton, maintain the pH and mainly increase the affinity of dye to fibre. Utilization of electrolytes and alkali in the dyeing process generates effluent load i.e. salinity of the water. After cationization, the cationic sites present on the cotton surface attract the nucleophilic dye anions directly and gets covalently bonded with cotton when it is dyed with anionic dyes like direct, reactive and acid dyes. The cationization of cotton improves the dyeing ability of salt-free reactive dyeing, acid dyeing, direct dyeing and pigment dyeing

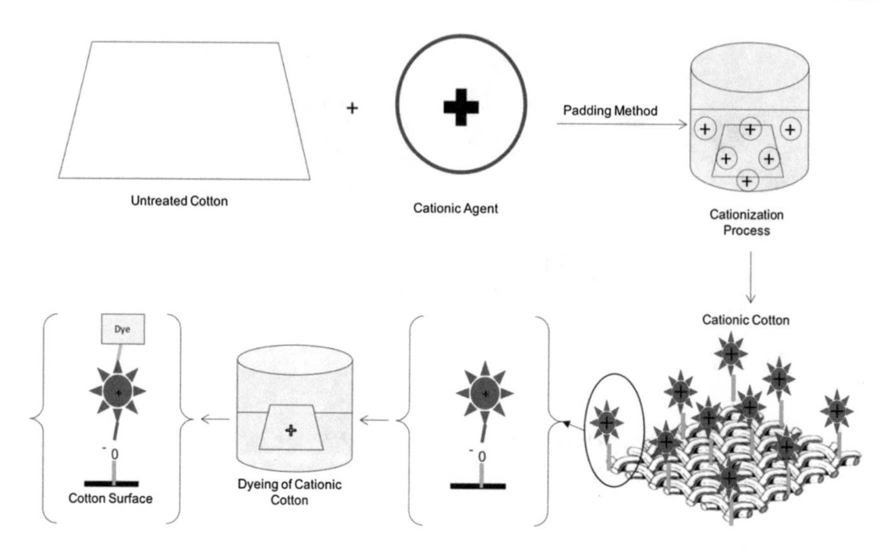

Fig. 9 Padding application of cationizing agent on cotton

with increased dye uptake, colour strength and good wash, light and rubbing fastnesses. The process also helps to reduce or eliminate the use of electrolytes, alkali and other boosting chemicals that are observed in conventional dyeing process. This criterion will ultimately minimize the effluent load. The cationic treatment increases the colour strength, dye fixation and fastness properties of the dyed cotton by enhancing adsorption, exhaustion and fixation of dye molecule on the cotton (Fig. 9).

Aktek et al., broadly describes the modification of cotton by treating with cationizing agents for the purpose of salt-free reactive dyeing. The successful approach was made using cationizing agents like chitosan, cationic starch, cationic monomer and dendrimer [57]. It has been observed that chitosan application on cotton boost up the dyeing ability on cotton by exhaust and padding application, the padding method gave excellent dyeing results [58].

A commercial cationic starch, known as Q-TAC produced by Ali et al. reduce the usage of salt almost totally in reactive dyeing. Owing to the fact that electrostatic attraction with reduced crystallinity of cationic cotton increases the degree of fixation and colour strength [59].

The Polyamido amine dendrimer known as PAMAM was applied on cotton for cationization and its notable achievements for salt-free reactive dyeing and acid dyeing on cotton substrate forming a strong ionic linkage between dye and cationic cotton. The excellent colour strength was achieved for salt-free dyeing with reactive dyes and acid dyes at slightly acidic to neutral pH [60, 61].

The utilization of Chromatech 9414, a commercial cationizing agent for cotton pre-treatment allows direct dye to get easily attached on the cotton surface giving high dye fixation without any after treatment. This can be attributed to strong attraction forces between dye and fibre. It is also observed that the Chromatech 9414 contains high nitrogen content, if the concentration of the cationic agent increases the nitrogen content on the cotton, it also increases its dyeability considerably [62].

A cationizing agent from soya bean hull was synthesized for eco-friendly salt-free dyeing on cotton substrate. This agent contains amino acid compounds in its structure which lowers the negative surface of cotton after modification without affecting its physical property and enhancing the dyeing ability of cationic cotton. The investigation proved that the soya bean hull was found to be effective cationizing agent for salt-free dyeing on cotton [63].

The pigments are insoluble in water and they do not have an affinity for cotton. However, it is observed that cotton can be dyed with pigments if cotton is pre-treated with a cationic auxiliary to provide the necessary affinity [64]. The mixture of chitosan and dendrimer i.e. chitosan-poly(propylene imine) dendrimer hybrid (CS-PPI) was found to be more durable and efficient. This CS-PPI hybrid is biocompatible compound and it can be used for salt-free reactive dyeing to achieve high fixation and excellent fastness property with zero effluent load [65].

Pigments have no affinity for cotton. Hence, the cationic charge is incorporated on cotton by applying a cationizing agent. Cotton is first treated with cationizing agent and dyed with pigment.

In a test, commercial samples of Sandene 2000 and Solidogen NRL were applied to modify the cotton surface and develop cationic sites for pigment dyeing, the result obtained was satisfactory to good for rubbing fastness. As the concentration of the cationizing agent increases, better fixation and level pigment dyeing by exhaust method was achieved [66].

3.5.3 Finishing

The finishing of textiles is the final process, which depends on consumer requirements. This process is used to enhance the aesthetic and functional properties of textiles. The variety of finishing agents or textile auxiliaries used in industry are mostly anionic in nature. The anionic nature of the cotton repelsanionic finishes, resulting in temporary effect. The cationization of cotton was effectively used to develop a new approach for durable and improved finishing effects. This process was studied by many researchers to increase anti-microbial activity, UV-protection effect and for effluent treatment. The one bath treatment using a mixture of cationizing agents along with other finishing chemicals can give multifunctional finish on cotton. The chemical surface modification, cationization plays a vital role in finishing as well. If the cationizing agents have multiple reactive sites or inbuilt multiple properties, then multifunctional finishing is also possible with this treatment. For instance, chitosan shows cationizing property with an anti-microbial effect on cotton (Fig. 10).

In conventional finishing treatments, anionic finishes are fixed on cotton with hydrogen or ionic bonding depending on the type of chemical. In finishing, different auxiliaries are used as a precursor for fixation of chemical on the cotton surface (Fig. 11).

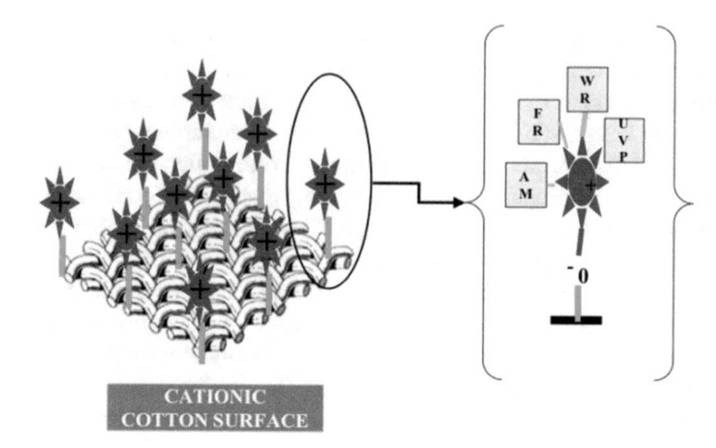

Fig. 10 Desired multifunctional effect on cationic cotton

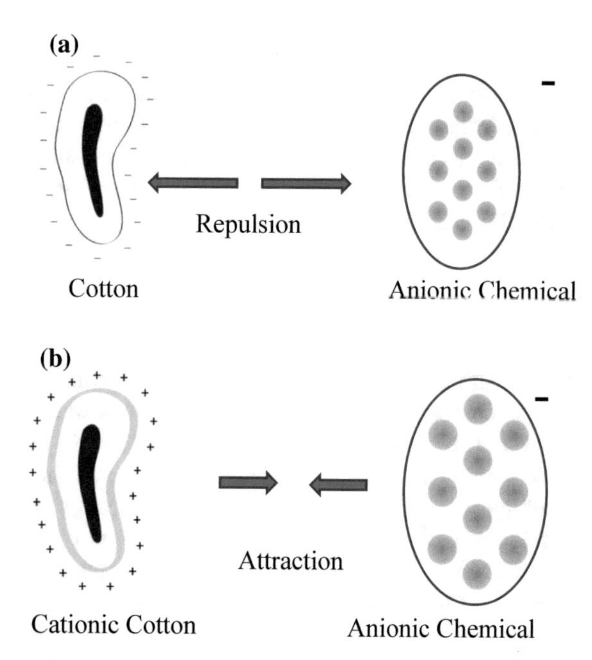

Fig. 11 **a** Repulsion in conventional finishing, **b** attraction after cationic treatment

In the cationized cotton finish, anionic chemical ions show increased affinity towards cationic sites on the fibre and they readily combine to form 'finishing agent-fibre' complex with a strong ionic bond [67].

Thiol–epoxy chemistry showed a permanent and durable antimicrobial property on cotton substrate. In this process, cotton was chemically modified by using 3-mercaptopropyltriethoxysilane (KH-580). This agent developed thiol groups on the cotton surface that helped ease and strong attachment of quaternary ammonium

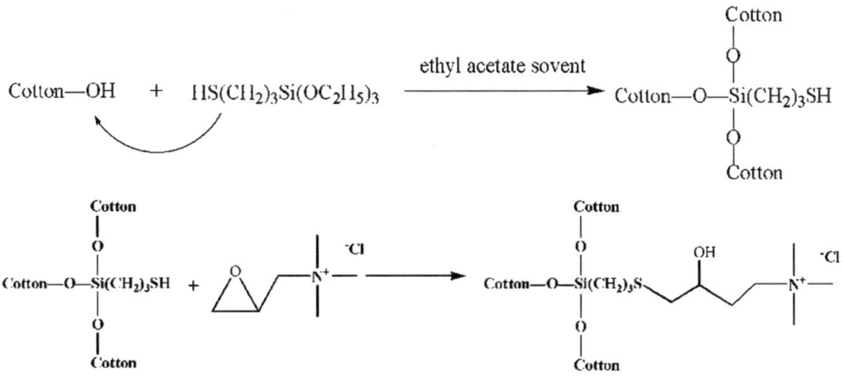

Fig. 12 Thiol–epoxy click chemistry application on cotton

salt (cationizing agent). The presence of thiol groups improved the antimicrobial efficiency of cotton with durable effect [68] (Fig. 12).

The cotton surface was modified by synthesizing carboxymethyl chitosan (CMCS) where chitosan was treated with mono-chloro acetic acid by carboxymethylation reaction. These CMCS possess excellent antimicrobial property due to the presence of chitosan. The CMCS treated cotton showed positive results for salt-free dyeing and proved to achieve good colour fastness properties with levelness dyeing compared to conventional dyeing. The evaluation of CMCS dyed cotton gave notable enhancement in antimicrobial property than untreated cotton [69].

In another work, where water-soluble carboxymethyl chitosan (CMCTS) was used for cationizing, dyeing and finishing of cotton. The evaluation of CMCTS concentration were studied to check the physical properties of cationic cotton. It was illustrated that increasing the concentration improves the antibacterial activity [70]. The cationization process with chitosan and UV absorbers i.e. UV-SUN® (based on Oxalanilides) for multi-protection treatments and dyeing for cotton as well as silk substrates confirmed that both the agents have high nitrogen content (N %) values which further states that as the nitrogen content increases, the antimicrobial property and cationic property also increases. The UV blocking activity on cotton and silk gave remarkable results where some improvement was seen in anti-odour activity with a reduction in microbial activity. The assessment of dyeing ability for cationic cotton and silk showcased durability in treatment as the effect was not affected above 30 wash cycles [71].

In another novel approach for both together salt-free dyeing and functional finishing including UV protection and antibacterial property, the cotton was modified with cationic chemical i.e. 3-chloro-2-hydroxypropyl trimethylammonium chloride. The dyes used were treated with titanium dioxide nano-sol by sol-gel method and then applied on cationized cotton dyeing. The use of salt was eliminated in the dyeing process, still resulting in high colour strength, fixation and good wash, light and rubbing fastness properties. The cationic dyed cotton also gave improved UV blocking, antibacterial activity property during analysis [72].

3.6 Effect of Cationization on Environment

Many of the textile industries cause pollution by the untreated discharge of water and a variety of chemicals. Effluents from textile wet processing industry is causing multiple problems to the environment. Compared with conventional dyeing, cationization process reduces the consumption of water and chemicals like electrolyte, alkali, acid or catalyst etc. Cationic solution can be reused for further processes which result in the saving of water and chemicals. In the dyeing process, cationic dyed cotton bath solution does not contain hydrolysed dye or other auxiliaries because the introduction of electrostatic attraction which increases the substantivity of dye/chemical onto fibre (cotton) easily. The cationic cotton dyed materials also show good fixation and excellent fastness properties. The absence of alkali and electrolyte in dyeing treatment denotes negligible effluent load.

Many studies and researches have claimed that various cationizing agents, cationic salts and cationic biopolymer are used for effluent treatments like removal of anionic dyes of finishes and wastewater treatment [73, 74]. Blackburn and Burkinshaw, in the year of 2003, reported that pre-treatment of cationizing agent on cotton reduces the effluent load and level of water consumption in dyeing and finishing than the normal wet processing methods [75].

Bioflocculants produced by microorganisms which are used for wastewater treatment in effluent plant in the presence of cationic salts like $MnCl_2$, $MgSO_4$ and $CaCl_2$ showed effective removal of dye from textile effluent. The presence of such cationic compounds were also used to determine the effect of salts on dye removal by the bacterial bioflocculants at the optimum temperature and pH [76].

3.7 Advantages of Cationization

- Develops cationic sites on the surface by eliminating negative zeta potential.
- Cationic cotton attracts anionic chemicals and forms strong ionic bonds with durable effect.
- Degree of fixation of anionic dyes and finishes increases giving improved fastness properties.
- Electrolytes and alkali are eliminated from the dyeing process of anionic dyes still giving maximum reactivity and minimum hydrolysis of dyes.
- Simple process with maximum water, chemicals and energy savings.
- Reduced effluent load.
- Ecological and environmental friendly processes [73].

3.8 Limitations

- Cationizing agents are mainly made up of amines so the odour of ammonia on the treated substrate is an issue.
- To control the process of cationization is a laborious task.
- High cost of the commercial cationizing agents.
- Cationizing agents tend to show low fixation rate in the exhaust process and thus higher amount of chemicals are required. Along with it, the normal dyeing procedure requires certain modification for treating cationized cotton.
- Limited number of dyes show good result in dyeing cationized cotton. Therefore, choosing the correct dye is important.

4 Conclusion

Surface modification of cotton by cationization process using a variety of cationizing agents become water, energy-saving and environment-friendly process for anionic dyeing and finishing of textiles [77]. It can be stated that the cationizing agents used for cationization do not affect the physical property of the cotton but chemically develops the cationic sites on its surface. The cationized cotton also denotes higher degrees of dye exhaustion, improved colour strength and strong ionic fixation related to cotton due to the absorption of electrostatic attraction rendered by cationic sites for anionic dyes [78]. It has also been observed that the cationic group of many cationizing agents on the cotton substrate serves a dual functionality like antibacterial activity, UV protection etc. [79]. To highlight, the two main criteria are resolved by the cationization process. First eliminating the requirement of electrolytes i.e. inorganic salt and alkali in few dyeing processes and second to diminish environmental problems. Along with this, many cationizing agents on cotton control the growth of bacteria and protects it from harmful UV rays. The future is bright for the cationization process if the commercialization of cationizing agents is considered. Effluent load can be controlled if the process is introduced in the textile industry.

It can be concluded that cationization can be easily used as dyeing auxiliary and finishing auxiliary for surface modification of cellulosic textiles that will invite novel achievement in future for the textile industry.

References

1. Size, EOM (2019) Share & Trends Analysis Report by aAplication (Cleaning & Home, Medical, Food & Beverages, Spa & Relaxation), By Product, By Sales Channel, And Segment Forecasts, 2019-2025. Report ID, 978-1
2. Patil AA, Maiti S, Adivarekar RV (2018). Salt-free Reactive Dyeing on Surface Modified Cotton Fabric. J Emerg Technol Innov Res 5(11): 628–635

3. Audet D (2001) Structural adjustment in textiles and clothing in the post-ATC trading environment. OECD trade policy working paper
4. Yu, C. (2015). Natural textile fibres: vegetable fibres. In Textiles and fashion. Woodhead publishing (pp. 29–56)
5. Morphological structure of cotton fibre. Textile Academy. https://www.onlinetextileacademy.com/morphological-structure-of-cotton-fibre/
6. Ahsanul IS (2014) Chemical composition of cotton fiber. Textile learner. https://textilelearner.blogspot.com/2014/10/chemical-composition-of-cotton-fiber.html
7. Cotton morphology and chemistry. Cotton Incorporated. https://www.cottoninc.com/quality-products/nonwovens/cotton-fiber-tech-guide/cotton-morphology-and-chemistry/
8. Hsieh YL (2007) Chemical structure and properties of cotton. Cotton: Science and technology, 3–34
9. Chemical structure of cotton (2012) Bdtextileblog-Resource of textile engineering solution. http://bdtextileinfo.blogspot.com/2012/02/chemical-structure-of-cotton.html
10. Al Mottaki (2013) Pretreatment is the heart of wet processing. Slideshare. https://www.slideshare.net/nishohel/wet-processing-23961418
11. Kumar A (2015) All steps of preparation of fabric for dyeing. Slideshare. https://www.slideshare.net/Amitsirohi2/all-steps-of-preparation-of-fabric-for-dyeing
12. Textile bleaching object of bleaching agent recipe for bleaching. Textile Learner (2011) https://textilelearner.blogspot.com/2011/03/textile-bleaching-process_5937.html
13. Cotton: From Field to Fabric, Dyeing, Printing, Finishing. Educational Resources. https://www.cotton.org/pubs/cottoncounts/fieldtofabric/dyeing.cfm
14. Tomasino C (1992) Chemistry and technology of fabric preparation and finishing. Department of textile engineering, Chemistry and science college of textiles, North Carolina State University
15. Dechant, J. (1985). Handbook of fiber science and technology. Vol. II. Chemical processing of fibers and fabrics. Functional finishes: Part B. Hg. von Menachem Lewin und Stephen B. Sello. New York/Basel: Marcel Dekker, Inc. 1984. ISBN 0-8247-7118-4. XX, 515 S., geb. SFr. 283–Acta Polymerica, 36(4)242–242
16. Knittel D, Schollmeyer E (2000) Technologies for a new century. Surface modification of fibres. J Text Inst 91(3):151–165
17. Wei, Q. (Ed.) (2009). Surface modification of textiles. Elsevier
18. Hasib J Textile dyeing process with ultrasonic waves. https://textilelearner.blogspot.com/2016/02/textile-dyeing-process-with-ultrasonic.html
19. Ilangovan R (2015) Ultrasonic waves dyeing. Slideshare. https://www.slideshare.net/raajhashreeilangovan/ultrasonic-waves-dyeing?qid=7e204658-1292-45a1-b72a-4c5f0799c96c&v=&b=&from_search=1
20. Shahidi S, Wiener J (2016). Radiation effects in textile materials. In: Radiation effects in materials, 309–328
21. Bhattacharya A, Misra BN (2004). Grafting: a versatile means to modify polymers: techniques, factors and applications. Prog Polym Sci 29(8):767–814
22. Teli MD, Mallick A (2018) Utilization of waste sorghum grain for producing superabsorbent for personal care products. J Polym Environ 26(4):1393–1404
23. Mallick A (2018) Interconnection Between Biotechnology and Textile: A New Horizon of Sustainable Technology. Handbook Renew Mater Color Finish 549–573
24. Scientific T Overview of crosslinking and protein modification. https://www.thermofisher.com/in/en/home/life-science/protein-biology/protein-biology-learning-center/protein-biology-resource-library/pierce-protein-methods/overview-crosslinking-protein-modification.html
25. Wikipedia contributors. (2020, April 10). Biopolymer. In Wikipedia, The Free Encyclopedia. Retrieved 15:41, April 12, 2020, https://en.wikipedia.org/w/index.php?title=Biopolymer&oldid=950143901
26. Choudhury AKR (2018). Biopolymers in textile industry. Biopolymers and biomaterials. Apple Academic Press, Toronto

27. Knittel D, Schollmeyer E (2006) Chitosans for permanent antimicrobial finish on textiles. Lenzinger Berichte 85(85):124–130
28. Kandasaamy P, Ramasamy M (2005) Effect of cationization of cotton on it's dyeability. Indian J Fibre Text Res 315–323
29. Paul S (1938). U.S. Patent No. 2,131,120. Washington, DC: U.S. Patent and Trademark Office
30. Burkinshaw SM, Lei XP, Lewis DM (1989) Modification of cotton to improve its dyeability. Part 1—pretreating cotton with reactive polyamide-epichlorohydrin resin. J Soc Dye Colour 105(11):391–398
31. Lei XP, Lewis DM (1990) Modification of cotton to improve its dyeability. Part 3–Polyamide–Epichlorohydrin resins and their ethylenediamine reaction products. J Soc Dye Colour 106(11):352–356
32. Lei XP, Lewis DM (1991) New methods for improving the dyeability of cellulose fibres with reactive dyes. J Soc Dye Colour, 107(3):102–109
33. Lewis DM (1993) New possibilities to improve cellulosic fibre dyeing processes with fibre-reactive systems. J Soc Dye Colour 109(11):357–364
34. Hall DM, Leonard, TM, Cofield CD, Barrow HW (1994). U.S. Patent No. 5,330,541. Washington, DC: U.S. Patent and Trademark Office
35. Kanik M, Hauser PJ (2002) Printing of cationised cotton with reactive dyes. Color Technol 118(6):300–306
36. Hauser PJ, Tabba AH (2001) Improving the environmental and economic aspects of cotton dyeing using a cationised cotton. Color Technol 117(5):282–288
37. Hauser PJ (2000) Reducing pollution and energy requirements in cotton dyeing. Text Chem Color Am Dyest Rep 32(6)
38. Lewis DM, Lei X (1989) Improved cellulose dyeability by chemical modification of the fiber. Text Chem Color 21(10)
39. Bhattacharya SD, Agarwal BJ (2001). A novel technique of cotton dyeing with reactive dyes at neutral pH
40. Wu TS, Chen KM (1993) New cationic agents for improving the dyeability of cellulosic fibres. Part 3—the interaction between direct dyes and polyepichlorohydrin-dimethylamine polymers. J Soc Dye Colour 109(09): 365–368
41. Wu TS, Chen KM (1993). New cationic agents for improving the dyeability of cellulose fibres. Part 2-pretreating cotton with polyepichlorohydrin-amine polymers for improving dyeability with reactive dyes. J Soc Dye Colour 109(4)153–158
42. Lewis DM (1992) New methods for improving the dyeability of cellulosic fibres with reactive dyes. In: AATCC international conference and exhibition
43. Cai Y, Pailthorpe MT, David SK (1999). A new method for improving the dyeability of cotton with reactive dyes. Text Res J 69(6):440–446
44. Crangaric AM, Tarbuk A., Warmoeskerken, M (2010) Interface Phenomena of Cationized Cotton with EPTAC. In 22nd IFATCC Conference (pp. B12-1). Associazione Italiana di Chimica Tessile e Coloristica (AICTC).
45. Grancarić AM, Ristić N, Tarbuk A, Ristić I (2013) Electrokinetic phenomena of cationized cotton and its dyeability with reactive dyes. Fibres Text East Eur 21(6):102–106
46. Roy Choudhury AK (2014). Coloration of cationized cellulosic fibers—a review. AATCC J Res 1(3):11–19
47. Wikipedia contributors. (2019, August 9). Batch dyeing. In Wikipedia, The Free Encyclopedia. Retrieved April 12, 2020, https://en.wikipedia.org/w/index.php?title=Batch_dyeing&oldid=910113917
48. Dunning D (2017) Exhaust dyeing process. Sciencing. https://sciencing.com/how-to-regenerate-activated-charcoal-5630591.html
49. Hubbell C (2009) Surface modification and chromophore attachment via ionic assembly and covalent fixation (Doctoral dissertation, Georgia Institute of Technology)
50. Choudhury A KR (2017) Principles of textile finishing. Woodhead Publishing
51. Gebrekiros T (2016) Chemical finishing of textiles. Slideshare. https://www.slideshare.net/Tesfaywasee/chemical-finishing-of-textiles?qid=34578128-ce43-4e5c-863e-c40f691ec351&v=&b=&from_search=12

52. Parmaj O (2016) Textile coating. Slideshare. https://www.slideshare.net/OmkarSParmaj/textile-coating
53. Wikipedia contributors. (2020, April 9). Sol–gel process. In Wikipedia, The Free Encyclopedia. Retrieved April 12, 2020, https://en.wikipedia.org/w/index.php?title=Sol%E2%80%93gel_process&oldid=949958765
54. Stawski D, Simon F, Zielińska D, Połowiński S, Puchalski M (2014) Application of the layer-by-layer technique for knitted fabrics/Aplicarea tehnologiei de depunere strat cu strat pe materiale textile tricotate. Ind Text 65(4):190
55. Mahltig T, Textor B (2008) Nanosol and textiles. World Scientific Publishing Company, New Jercey, USA
56. Camlibel NO, Arik B (2017). Sol-gel applications in textile finishing processes. In: Recent applications in sol-gel synthesis. IntechOpen Limited, UK, 253–281
57. Aktek T, Millat AKMM (2017) Salt free dyeing of cotton fiber-A critical review. Int J Text Sci 6(2):21–33
58. Houshyar S, Amirshahi SH (2002) Treatment of cotton with chitosan and its effect on dyeability with reactive dyes. Iran Polym J (English Ed) 11(5):295–301
59. Ali S, Mughal MA, Shoukat U, Baloch MA, Kim SH (2015). Cationic starch (Q-TAC) pretreatment of cotton fabric: influence on dyeing with reactive dye. Carbohydr Polym 117:271–278
60. Patil AA, Maiti S, Adivarekar RV (2019). The use of poly (amido) amine dendrimer in modification of cotton for improving dyeing properties of acid dye. Int J Cloth Sci Technol
61. Maiti S, Mahajan G, Phadke S, Adivarekar RV (2018). Application of polyamidoamine dendrimer in reactive dyeing of cotton. J Text Inst 109(6):823–831
62. Shahin M (2015) The influence of cationization on the dyeing performance of cotton fabrics with direct dyes. Int J Eng Res Appl 5(8):62–70
63. Dessiea A, Govindanb N (2018). Eco-friendly salt-free reactive dyeing by cationization of cotton with amino acids obtained from soya bean hull. J Text Sci Eng 08(06):1–9
64. Aspland JR (2014) Textile education in coloration and textile wet processing technology. AATCC Review 14(3):34-44
65. Sadeghi-Kiakhani M, Safapour S (2015). Salt-free reactive dyeing of the cotton fabric modified with chitosan-poly (propylene imine) dendrimer hybrid. Fibers Polym 16(5):1075–1081
66. Patra AK, Bhaumik S, Kaur H (2006) Studies on pigment dyeing of cotton by exhaust method. Indian J Fibre Text Res 31(3):450–459
67. Cationic Cotton. Cotton works. https://www.cottonworks.com/topics/sourcing-manufacturing/dyeing/cationic-cotton/
68. Yu D, Xu L, Hu Y, Li Y, Wang W (2017) Durable antibacterial finishing of cotton fabric based on thiol–epoxy click chemistry. RSC Adv 7(31):18838–18843
69. Ibrahim HM, Reda MM (2015) Multi-function Modification of cotton fabrics for improving utilization of reactive dyes. Int J Innov Sci Eng Technol 2(7):501–507
70. El-shafei AM, Fouda MM, Knittel D, Schollmeyer E (2008) Antibacterial activity of cationically modified cotton fabric with carboxymethyl chitosan. J Appl Polym Sci 110(5):1289–1296
71. Kotb RM (2017). Innovative Multi-Protection Treatments and Free-Salt Dyeing of Cotton and Silk Fabrics. J Eng Fiber Fabr 12(3):155892501701200308
72. Alebeid OK, Zhao T (2016) Simultaneous dyeing and functional finishing of cotton fabric using reactive dyes doped with TiO2 nano-sol. J Text Inst 107(5):625–635
73. Ghazi A-U, Lahiri S, Quan H (2018). Preparation and application of Cationic agent on cotton fabric to cut down the amount of salt in conventional dyeing. The 8th cross strait conference of textile technology. 371–381
74. Huang XY, Mao XY, Bu HT, Yu XY, Jiang GB, Zeng MH (2011) Chemical modification of chitosan by tetraethylenepentamine and adsorption study for anionic dye removal. Carbohydr Res 346(10):1232–1240
75. Blackburn RS, Burkinshaw SM (2003) Treatment of cellulose with cationic, nucleophilic polymers to enable reactive dyeing at neutral pH without electrolyte addition. J Appl Polym Sci 89(4):1026–1031

76. Buthelezi SP, Olaniran AO, Pillay B (2012) Textile dye removal from wastewater effluents using bioflocculants produced by indigenous bacterial isolates. Molecules 17(12):14260–14274
77. Karnik PP (2002). Use of cationized cotton for textile effluent color reduction. North Carolina State University
78. Rupin M, Veatue J, Balland, B (1970). Utilization of reactive epoxy-ammonium quaternaries on cellulose treatment for dyeing with direct and reactive dyes. Textilveredlung, 5:829–838
79. Literature review. Shodhganga. 22-91. https://shodhganga.inflibnet.ac.in/bitstream/10603/172177/11/11_chapter%202.pdf

Role of Radiation Treatment as a Cost-Effective Tool for Cotton and Polyester Dyeing

Shahid Adeel, Fazal-ur-Rehman, Tanvir Ahmad, Nimra Amin, Shahzad Zafar Iqbal and Mohammad Zuber

Abstract Currently, the global community is more attentive and alert in using textile products due to widespread awareness of stringent environmental standards. In textile processing effluent load is of much concern, which needs a lot of methods to be treated. Different methods are being used to modify textile processing on the basis of conventional methods, statistical optimization of dyeing variables i.e. response surface modeling but modern methods such as application of radiation in textile dyeing on account of time, energy labor and cost-effectiveness is gaining global fame. The purpose of such radiation is either to activate the dye bath or to modify the fabric surface to decrease the effluents via improved fixation. The current chapter is presenting a brief review of applications of the microwave, ultrasonic and ultraviolet radiation for cotton fabric processing and dyeing of cotton also polyester fabrics using disperse, Vat, reactive and direct dyes using conventional process or via statistical optimization where the improvement in color characteristics of dyed fabrics after irradiation treatment has also been elaborated. This current study has been providing a model of aspiration for academicians, researchers, and traders who intend to utilize the modern tool for developing cost-effective textile processing.

Keywords Cost-effectiveness · Cotton · Microwave · Polyester · Ultraviolet · Ultrasonic · Sustainability · Synthetic dyes

S. Adeel
Department of Chemistry, Government College University, Faisalabad 38000, Pakistan

Fazal-ur-Rehman (✉) · N. Amin · S. Z. Iqbal · M. Zuber
Department of Applied Chemistry, Government College University, Faisalabad 38000, Pakistan
e-mail: furminhas@gcuf.edu.pk

M. Zuber
Department of Chemistry, University of Lahore, Lahore, Pakistan

T. Ahmad
Department of Statistics, Government College University, Faisalabad 38000, Pakistan

M. Shahid and R. Adivarekar (eds.), *Advances in Functional Finishing of Textiles*, Textile Science and Clothing Technology, https://doi.org/10.1007/978-981-15-3669-4_14

1 Introduction

The word textile has been originated from the Latin word known as "texture" which represents braid, to construct or to weave. The textile process comprised of different phases such as the spinning of cotton, manufacturing of fabrics, dyeing and finishing. Since 1947, till today Pakistan has become the 14th largest producer of cotton with yielding capacity of 7.6% which is the total Asian quantity of spinning sector after china [1]. The synthesis of reactive dyes in 1954 and its proper usage in 1956 contracted a major innovation in the dyeing of cotton and other fabric. The oil crises in 1970 rising a costs of raw material used, which in turn, created the drive for extreme energy effective and cost-effective technologies to be used in textile processing. Meanwhile, the world has reached steadily to an estimated cost of 3.5 $\times$ 10^6 tons with the utilization of two major textile fibers i.e. cotton and polyester [2]. During 2010–11, Pakistan has USD10.2 billion textile exports which comprise 3% to the USA and the other 97% to other countries. Despite all this Pakistan textile industry is facing high production cost due to (i) improper implementation of five-year textile policy (ii) an increase in the interest rate, (iii) the devaluation of Pakistan currency, (iv) political unstable atmosphere, (v) removal of subsidy due to shortage of power and gas, (vi) weak internal and financial management, (viii) use of old machinery, (ix) old processing technology, (x) waste of labor and (xi) world environmental standards etc. These problems urged textile researchers to think about the use of such technologies improving production efficiency and make it cost and energy effective [3].

By using such technologies, we can not only save money, labor, power and minimize environmental pollution but also can rise the production rate of textile products. The use of radiation in dyeing and finishing has a great effect which not only fixes the dye onto irradiated fabric firmly but also needs less severs conditions to get excellent color strength. It also helps in reducing the labor and cost by less wastage of dye into the water. It improves water-repelling properties of the fabric, increases the wrinkling and shrinking resistance of the cotton fabric and reduces the pilling of cotton fabric and adds value in the coloration of cotton and polyester via improving wet ability of fabrics [4, 5]. It may also reduce dyeing temperature, dyeing time as well as energy and save the economy previously it has been found that the use of such cost-effective and energy effective technology can help to enhance the rate of export by value-added products which are eco-label. These sustainable products can help in rising foreign exchanges and the economy of Pakistan as well as will improve trade policy in the world textile market.

2 Pretreatment of Raw Greige Cotton

In the textile industry during the manufacturing process, pretreatment is one of the most important procedures. For satisfactory consequential processes, such as printing, textile dyeing, and finishing, and the important purpose of the pretreatment is to remove most of the impurities from the textiles. Generally, the pretreatment process includes chemical and mechanical processing wherein the mechanical pretreatments consist of checking, stitching, cleaning, collecting and distribution, or even singeing processes, etc. For good preparation, wet-chemical pretreatment processes have been done such as desizing, scouring, bleaching, and mercerization, which improve the quality of ultimate products [6, 7].

3 Chemistry of Cotton

The cotton fabric is the most comfortable and easily available fabric found in the textile world. Mostly the cotton fabric is cellulosic and non-cellulosic in nature, where the major part of cotton fibres contains α-cellulose up to 96.5%. The non-cellulosic part is found either on the outer layer or inside the lumen of the fibers [8]. The non-cellulosic part contains proteins (1.9%), wax (1.2%), Inorganic material (1.6%) and other substances (8%). The degree of polymerization of the primary wall ranges 2000–6000 and their distribution is broader similarly for secondary cell wall it is about 14,000 with molecular weight of uniform distribution [9]. Cotton cellulose contains β- 1,4-D(H)- glucopyranose building blocks bonded through 1, 4 glycosidic linkage. Each unit has three hydroxyl group C-6, C-2 and C-3, where due to these hydroxyl group, the chain is packed via intermolecular and intramolecular H-bonding shown in (Fig. 1) [10]. These hydroxyl groups are responsible for the interaction with dye molecules of the cellulosic unit present in the amorphous region [11, 12].

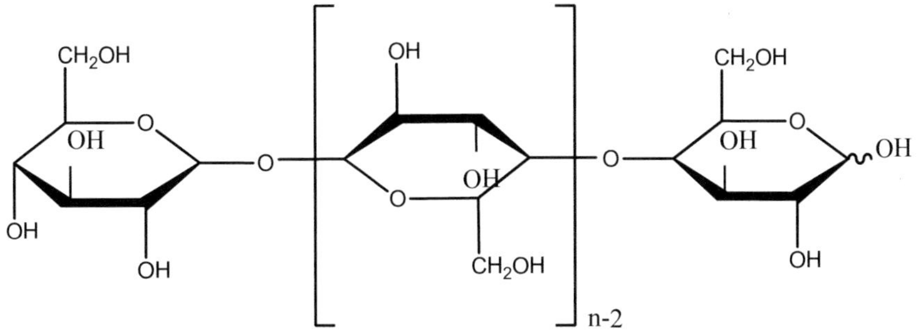

Fig. 1 Structure of cellulose

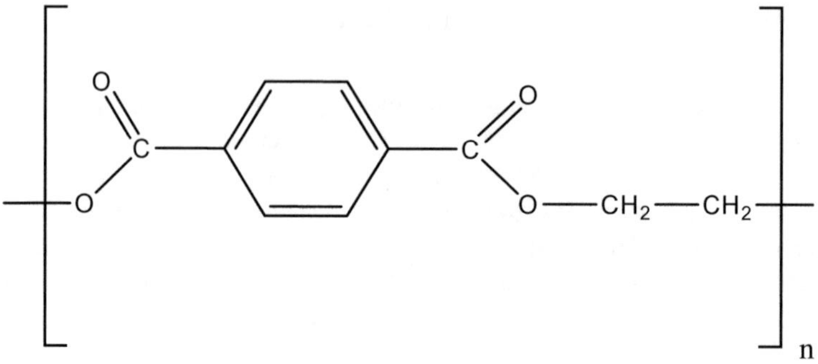

Fig. 2 Structure of polyester

4 Chemistry of Polyester

The term polyester is known as many esters, a combination of different acids and alcohols and represented as –CO–O. The composition of polyester is based on three main groups such as carbonyl groups (–OCO–), methylene groups ($–CH_2$) and ($R—\overset{O}{\overset{\|}{C}}—O—R$) ester group represent in Fig. 2. The interaction with dye molecules takes place due to the presence of weak hydrogen bonding and Van der Waal's forces which has a prominent role in the world of dyeing [13].

5 Classes of Dyes

Dyes are organic molecules which are divided into two main classes according to origin i.e. Natural dyes and synthetic dyes

5.1 Natural Dyes

These dyes are extracted from different sources, as shown in Fig. 3.

5.2 Synthetic Dyes

Synthetic dyes are those colorants that are synthesized from petroleum. The colored organic substances are used to dye textile material like cotton, wool, silk, polyester, leather, nylon [14] and attach themselves by chemical and physical bonding with fibers and develop the colored textiles. Synthetic dyes have a property of reproducible shades and colors due to which the use of natural dyes in textiles is obsolete [15].

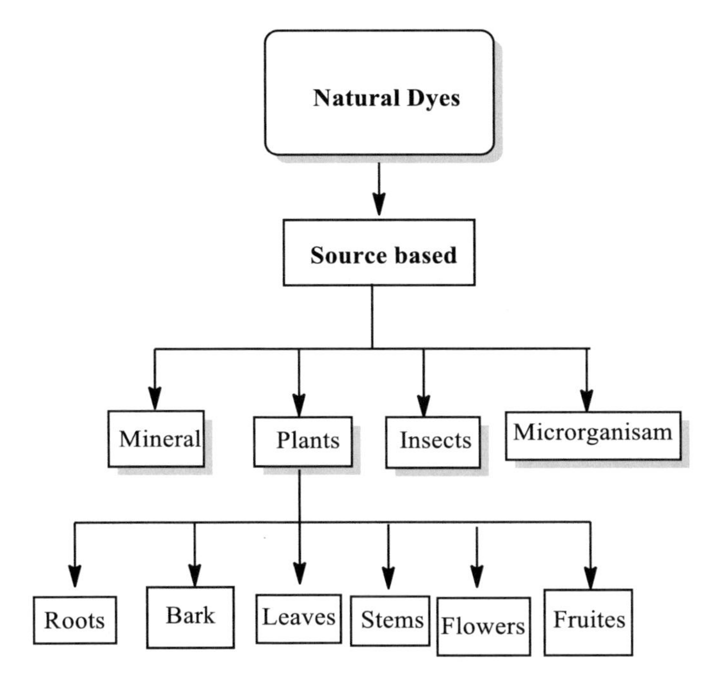

Fig. 3 Classification of natural dyes

6 Classification of Synthetic Dyes

Synthetic dyes are classified into three categories on the basis of color, chemical structure and mode of application. Here we will discuss only the classification depending upon the mode of application as shown in Fig. 4 [16].

7 Methods of Dyeing

In most, textile processing usually conventional method is being used but nowadays radiation assisted dyeing methods are being reviewed an account of saving, cost, labor, energy, modern methods are being used. These methods include:

(i) UV Radiation
(ii) US Radiation
(iii) MW Radiation.

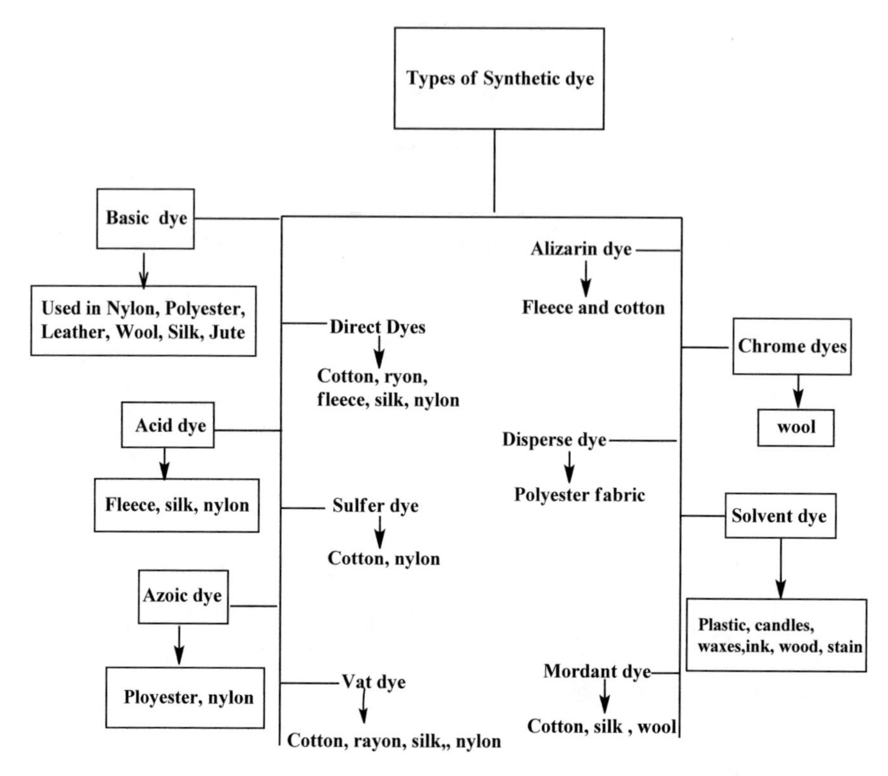

Fig. 4 Schematic diagram of classification of synthetic dyes on the basis of application

7.1 UV Radiations

UV radiation is the type of electromagnetic radiation having wavelength lesser than visible rays (200–380 nm) which is unable to perceive with the naked eye. There are different parts of UV region such as UVA, UVB, and UVC, where UVC has been used in textile processing that ranges from 100 to 280 nm. UV radiation helps in the improvement of color strength (K/S) of dye by increasing fabric uptake ability and also helps to improve wettability. Similarly, the photo variation of external fibers increases the wettability of dye and allows dye uptake at low temperature as well as helps to fix more dye on the fabric surface, but also give excellent color characteristics these radiations have only used to modify the fabric surface, while leaving the other textile material unaffected [17].

7.2 US Radiations

Ultrasound radiations another form of ecofriendly rays in the sound spectrum have frequency under the range of 20–10 MHz. These rays be further divided into two parts, the primary is known as powerful ultrasound (20–2 MHz) and another is named as diagnostic ultrasound (5–10 MHz). Ultrasonic radiation has typically long wavelengths in the millimeter range which is too much greater as compared to the molecule, thus it cannot interact directly with molecules to produce some chemical change in it. The phenomena of powerful ultrasound are acoustic cavitation in liquids which accelerates the different types of chemical and physical reactions. Since, these phenomenon is based on advanced and implosive interruption of microscopic bubbles which are being capable of creating hot spots through which the color yield is enhanced but also process become cost, energy and time effective [18]. In textile dyeing, the main advantage of using ultrasonic energy is to improve the process of mass transfer which can otherwise be attained by the addition of various chemicals, prolonged time and using high temperature. The ultrasound-assisted dyeing (UAD) for textiles is broadly specified in terms of environmental benefits and energy savings. Different researchers have reported the use of these radiations in dyeing on different textiles material by applying, basic dyes, acid dyes and disperse dyes [19–21].

7.3 MW Radiation

Microwave radiation is the type of electromagnetic radiation, where its frequency range is from 30 to 300 GHZ with a wavelength of 1 cm to 1 m. Currently the use of microwaves has been employed in many applied fields on account of energy consumption and decrease treatment time [22]. Conventional heating is surface heating while microwave heating is based on activation of polar molecules known as volume heating which adds value in textile processing by saving energy, cost and labor. These rays have ability to penetrate easily inside the fabric material uniformly to enhance its substantive nature [23–25].

8 Role of Radiation in Textiles

Radiations are mostly used in the textile industry because of their promising nature in saving time, fast treatment time. Microwave irradiation gives better color in a short period as compared to the common pad batch dyeing process. These rays use a small amount of energy as compared to the conventional method. It also reduces the heat transfer problem. By using the microwave, we can save time and get better results and also the process is ecofriendly. It gives dye exhaustion and a better dyeing rate. Ultrasonic (US) are other sustainable source of heating gives sharp color and increase the dye uptake [26]. US radiations have advantages such as energy saving, water saving, reduce process time and increase color depth [27]. These rays have found a widespread industrial application, such as for surface treatment, soldering, degradation, and formation of emulsions [28, 29]. The decolorization of dyes can take place by using ultrasound, while ultrasound alone is not effective to obtain their complete removal. Ultraviolet radiation has also a relatively good potential for modification of textile processing. Previous studies reported that UV-treatment of fabric brings about photo-modification of surface fibers and it can allow internal vicinities of fibers become uptake able to more dye with firm fixation and develops deeper shades and also enhances the coloration value of fabric and garments [5, 30]. The effect of Ultraviolet treatment considerably depends upon the nature of pigment and dye, the existing absorptive groups in the dyestuff, penetration after dyeing, the consistency and additives used in the dyeing processes [4]. Several modifications and treatments were carried out to improve the dyeability and fastness properties of different fabrics using natural and synthetic dyes, but the very few studies exist on the effect of UV radiation treatment on the dyeing of cotton and polyester fabric with reactive and disperse dye [31].

The major objectives of using these modern tools are

1. Reduce the energy, money and time during textile processing
2. To improve fixation of dye onto irradiated cotton and polyester at mild reaction conditions using irradiated dye solution
3. To improve the color depth and colorfastness properties through radiation.
4. To reduce the number of chemicals in dye house effluents to make dyeing process eco-friendly and eco-label.

Following aspects under various dyeing conditions have been discussed in this chapter and their role has been compared on account of the results obtained.

(i) Effect of microwave treatment (MW) in reactive, vat and disperse dyeing of cotton and polyester fabric
(ii) Effect of ultrasonic treatment in reactive, vat and disperse dyeing of cotton and polyester fabric
(iii) Effect of ultraviolet treatment in reactive, vat and disperse dyeing of cotton and polyester fabric.

9 Methodology

The precise methodology of dyeing is given in tabulated form which represent the improvement in dyeing behavior of reactive dyes under the influence of MW, UV, US treatment (Tables 1, 2, 3, 4, 5, 6 and 7).

Table 1 Application of MW radiation on dyeing of cotton using Reactive dyes

Radiation used	Dyeing parameters	Work done
MW (1–4 min.) Reactive Red 195	Curing temperature (150–190 °C), curing time (1:40–2:30 min:s), pH $NaHCO_3$ (25–40 g/100 mL), urea (40–70 g/100 mL)	Muhammad Kamran
MW (2–10 min.) Reactive violet H3R	Volume (60 mL), pH (7.5), salt (8 g/100 mL), temperature (60 °C), time (55 min.)	Maria Jannat
Mw (1–4 min.), Reactive Blue 19	Temperature (25–55 °C), time (20–60 min.), pH (3–10), salt (1–6 g/100 mL)	Tayyaba Ayesha
Mw (2–6 min.) Reactive Green 6	Temperature (30–90 °C), time (10–40 min.), pH (7–11), salt (1–4 g/100 mL	Huma Hira Rehman
MW (1–6 min) Basic violet 3B dye	pH (6–11), salt (3–7 g/100 mL)	Sobia Qadeer
Mw (1–5) Reactive Blue 21 Dye	Temperature (30–80 °C), time (10–40 min.), pH (7–12), salt (1–6 g/100 mL)	Moneeba Mushtaq
Mw (2–6 min.) Reactive Black 5	Temperature (30–70 °C), time (10–60 min.), pH (7–11), salt (1–4 g/100 mL)	Adilla Arshad
Reactive Blue 19 Mw (1–6 min.)	Statistical analysis temperature (30–70 °C), time (10–60 min.), pH (1–9), salt (1–4 g/100 mL), volume 20–60 mL	Muhammad Sultan

Table 2 Application of MW radiation on vat dyeing of cotton at various dyeing conditions

Radiation used	Dyeing parameters	Work done
Mw (1–6) Vat Red 10	Redox conditions (i) $NaHSO_4$ (0.2–2.2 g) (ii) CH_3COOH (0.5–3 mL) (iii) H_2O_2 (0.5–3 g/100 mL), temperature (25–55 °C), time (20–60 min.), pH (3–10), Salt (1–6 g/100 mL)	Maham Majid
MW (1–4) Vat Yellow 46	Temperature (40–70 °C), time (10–25 min.), Redox reagent, NaOH (60–90 g/100 mL), $Na_2S_2O_4$ (80–110 g/100 mL)	Muhammad Kamran
MW (2–6) Reactive red 195	Temperature (30–80 °C), time (2–10 min.), pH = 7–10 salt 10–60 g/100Ml	Farwa Khalid

Table 3 US treatment and reactive dyeing of cotton at various dyeing conditions

Radiation used	Dyeing parameters	Work done
US (10–60 min.) Reactive Red 195	Curing temperature (160–210 °C), curing time (1:40–2:30 min:s), pH NaHCO3 (10–30 g/100 mL), urea (60–90 g/100Ml	Muhammad Kamran
US (20–50 min.) Reactive violet H3R (CI 182130)	Volume (15–65 mL), pH (7–11.5), Salt (2–10 g/100 mL), temperature (20–70 °C), time (15–65 min.)	Maria Jannat
US (10–50 min.) Reactive Yellow 160	Temperature (30–90 °C), time (20–50 min.), pH (1–12), volume (15–50 mL), salt (0.1–0.6 g/100 mL)	Muhammad Umar
US (10–60 min.) Reactive Yellow 145	Temperature (70 °C), time (35 min.), pH (9), Salt (5 g/100 mL)	Muhammad Zeeshan

Table 4 US rays and vat dyeing of cotton at various conditions

Radiation used	Dyeing parameters	Work done
US (15–60 min.) Vat Blue 4	Temperature (55 °C), time (60 min.), volume (50 mL), $Na_2S_2O_4$ (2.5 g/100 mL), acetic acid (2 mL), hydrogen peroxide (2 mL)	Aqsa Abdullah
US (10–60 min.) Vat Yellow 46	Temperature (20–80 °C), Time (1–40 min.), redox conditions, $Na_2S_2O_4$ (9–11 g/100 mL), NaOH (60–80 g/100 mL)	Muhammad Kamran

Table 5 Efficacy of MW rays on disperse dyeing of polyester at given conditions

Radiation used	Dyeing parameters	Work done
MW (1–6 min.) Disperse Red 153	Temperature (20–60 °C), time (20–60 min.), dispersant (1–10 g/100 mL), pH (3–11), volume (20–70 mL)	Hina Tariq
MW (1–6 min.) Disperse Yellow 79	Time (20–60 min.), dispersant (1–10 g/100 mL), pH (3–11), volume (20–70 mL)	Hina Tariq
Mw (1.5–7.5 min.) Disperse Orange 25	Temperature (80 °C), time (25 min.), pH (8), volume (1:75 mL)	Saba Ayub

Table 6 Efficacy of US rays disperse dyeing of polyester at various conditions

Radiation used	Dyeing parameters	Work done
US (20–40 min.) Disperse Blue 56	Time (25–60 min.), pH (5–9), volume (15–45 mL), dispersant (0.6–0.9 g/100 mL)	Rabia Akhter
US (10–50 min.) Disperse Red 13	Volume (15–50 mL), pH (3–12), dispersant (0.1–0.6 g/100 mL), time (25–60 min.), temperature (90–140 °C), time (20–50 min.)	Muhammad Umar
US (10–50 min.) Disperse Red 343	Volume (10–60 mL), pH (5–11), dispersant (1–5 g/100 mL), time (10–60 min.)	Muhammad Asif

Table 7 Efficacy of UV rays on disperse dyeing of polyester at various dyeing conditions

Radiation used	Dyeing parameters	Work done
UV (15–60 min.) Disperse Red 153	Time (20–60 min.), pH (3–11), volume (20–70 mL), dispersant (1–10 g/100 mL)	Hina Tariq
UV (15–60 min.) Disperse Yellow 79	Time (20–60 min.), pH (3–11), volume (20–70 mL), dispersant (1–10 g/100 mL)	Hina Tariq
UV (15–75 min.) Disperse Orange 25	Time (45 min.), pH (10), volume (70 mL), temperature (130 °C)	Saba Ayub
UV (15–60 min.) Disperse Yellow 221	pH (8), volume (70 mL), temperature (70 °C), dispersant (2%)	Sania Shahid
UV (10–50 min.) Disperse Red 13	Temperature (90–140 °C), pH (3–12), volume (15–50 mL), dispersant (0.1–0.6 g/100 mL), time (20–50 min.)	Muhammad Umar

9.1 *Effect of Microwave on Cotton Dyeing*

The effect of pretreatment of fabric with microwave radiation for reactive dyeing using reactive Blue 21 dye have been studied by Mushatq [32]. They found that microwave irradiation of fabric (RC) for 3 min has improved its dyeing behavior using an un-irradiated dye solution (NRS). According to date it is found that good color characteristics was due to the mechanical heating surface of cotton fabrics which in turn facilitated the dye uptake ability more profoundly [33, 34]. For low time treatment the surface is not tuned enough to take more dye while for long time treatment, there may be weakening of fibers which in turn cannot hold dye molecules firmly [35]. Upon washing with hot and cold water most of the unfixed dye is washed and low K/S is obtained. The other factor is the irradiation of dye bath, but it has no significant effect as irradiation may cause degradation of dye which results in low K/S. Hence in this study, she found that fabric treatment by MW for 3 min. is of much important rather than heating of dye bath.

Microwave treatment has always shown promising results as it provides leveled heating and modified the fabric and dyeing process said Kamran [36] found that MW treatment has enhanced the dyeing behavior of treated desized cotton fabric (RC) using a non-irradiated solution of Reactive Red 195 (NRS). From this study, he observed that above optimal time, too much heating has tuned the fabric which during dyeing has sorbed colorant molecules in cluster form. This sudden aggregation of big clusters of dye did not give good color depth because clusters were failed to penetrate into voids and being mainly remained at the surface. After dyeing, once finishing process was carried out resulting color depth is obtained [37, 38]. Hence overall from his experimental work MW treatment for 4 min. before desizing should be given to get enhanced uptake ability of the fabric. It was also found that 3 min. treatment to desized fabric gives good color depth using an un-irradiated dye solution. Whereas, MW treatment for 3 min. after bleaching has also helped to modify the fabric for making firm bonding with an un-irradiated dye solution (NRS). Upon application of MW, irradiation of fabric for 3 min. after mercerization has also given acceptable results but less than of treatment after bleaching. Hence overall from this study, he concluded that irradiation to fabric for 3 min. after bleaching should be given to improve its uptake ability using untreated Reactive Red 195 dye. The results obtained from this study have revealed that, microwave treatment not only reduced the time, labor, and money but also improved color characteristics of fabric dyed with Reactive Red 195 dye.

Another study conducted by Kamran [36] revealed that the radiation treatment for processing of dyeing using Vat Yellow 46 has also given excellent results. The dyeing of irradiated cotton fabric (RC), and non-irradiated cotton (NRC) using irradiated solution (RS), and non-irradiated dye solution (NRS) has shown various results. The effect of MW treatment on greige fabric has been found promising followed by desizing, bleaching, and mercerization using an un-irradiated Vat dye solution. The treatment of fabric for 4 min. has helped to improve the uptake ability before dyeing, due to surface modification. The microwave treatment after desizing for 4 min. using the un-irradiated solution (NRS) has given good color strength (K/S) value, due to the desizing of the fabric after MW has made it more suitable for further processing [39, 40]. After desizing the MW treatment helped to modify the surface and affect the crystalline area of the fabric to enhance the uptake ability.

Kamran [36] studied the role of microwave treatment during the processing of greige fabric to make it viable for dyes. After every step from desizing to mercerization, MW treatment was carried out up to 6 min. and dyed with reactive and Vat dyeing. MW treatment for 4 min. of bleached fabric gives a maximum K/S value by using the un-irradiated dye solution (NRS). Because after bleaching the fabric surface becomes more prone to dye sorption and upon irradiation, the surface becomes modified enough to sorb more dye. Treatment for a high time due to continuous heating may weaken the fabric to hold the dye aggregates firmly. Hence 4 min. MW treatment is the optimum time for fabric after bleaching. After mercerization, the irradiation for 2 min. to fabric gives more K/S value. This is because mercerization not only helps to improve whiteness as well as the orientation of fibers but also enhanced the dye uptake ability. After mercerization, the fabric upon irradiation becomes more prone to dyeing of fabrics resulting in good color depth (K/S). Hence overall it is recommended that for reactive dyeing after bleaching of fabric (RC) the MW treatment for 3 min. is recommended using un-irradiated dye (NRS). Hence it is proved that MW treatment is not only a cost-effective tool also time, energy and labor effective tool for fabric processing. This experiment has improved the bleaching process of fabric after irradiation and enhance the uptake ability of dye. During these experiments, it is observed that radiation of desized fabric before bleaching for 4 min. should be given followed by further processing to improve its substantive behavior. However, dye bath after and before irradiation for dyeing is not significant.

Jannat [41] reported the substantive improving behavior of fabric using Reactive Violet H3R dye under the influence of microwave treatment. In this work, MW treatment employed to both dye bath and fabric up to 10 min. It is found that microwave irradiation for 8 min. at high power has given good color strength when 60 mL of irradiated dye bath of 7.5 pH containing 8 g/L of salt as an exhausting agent was used to dye irradiated fabric at 60 °C for 55 min. It is concluded that, at these optimum conditions the shade made, upon the exposure to fastness agencies. It has been revealed that microwave irradiation has improved the rating of fastness properties. It is reported that at a greyscale on compassion these characteristics were improved and microwave irradiation has an excellent potential to improve the color strength and dyeing properties below insignificant conditions.

Khalid [42] work revealed that, the dyeing after irradiation has a great impact on dye uptake ability for fabrics. Microwave treatment for 6 min. onto fabric has given excellent results for reactive dye in dyeing bath (Fig. 5). Irradiation for short time does not tune the fabric whereas for high exposure of time the cellulosic fabric is tuned physically in such a way that dye molecules rush towards fabric to give good tint via forming firm bonding [34, 43]. Hence for better color strength both fabric and dye solution should be irradiated for 6 min. by dyeing fabric at 80 °C for 50 min. using a dyeing bath of pH 7 keeping material-to-liquor ratio 1:75 containing 40 g/L of salt as exhausting agent (Fig. 6). It is also found that color fastness properties assessed after dyeing the fabric at optimal conditions were improved fair to good under the influence of MW treatment (Table 8).

Majid [44] described, microwave treatment in textile dyeing has been the topic of research since the last decade as it has effected the dielectric and thermal properties of fabric which in turn helped to enhance uptake of dye. In this study the role of MW in Vat dyeing of cotton using Vat Red 10 dye and found that irradiation of dye bath and cellulosic fabric for 5 min (RS/R C) has shown maximum color strength (Fig. 7). It was observed that less exposure of MW treatment has no effect while for a long exposure time may degrade the colorant due to which low color strength is observed [33, 45, 46]. The result displayed in (Fig. 8) shows that the microwave radiation has reduced the amount of redox reagent, where 1.4 g of $NaHSO_3$, 1.5 mL of CH_3COOH and 0.5 mL of H_2O_2 have given good color strength during the vat dyeing of irradiated cotton using irradiated dye solution (RS/RC, 5 min). The reduction in redox reagent

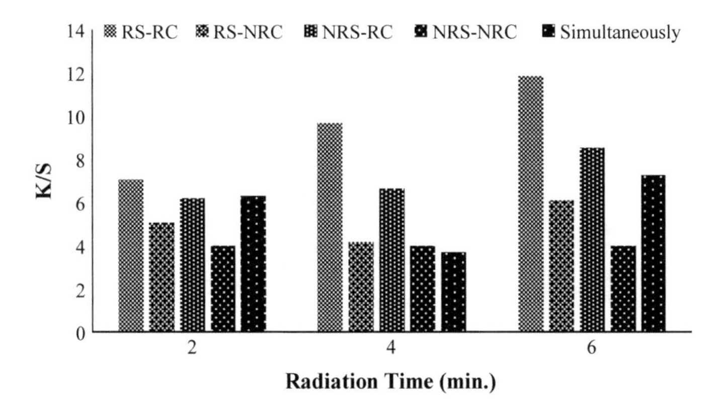

Fig. 5 Effect of microwave radiation on dyeing of cotton with Reactive Red 195 dye

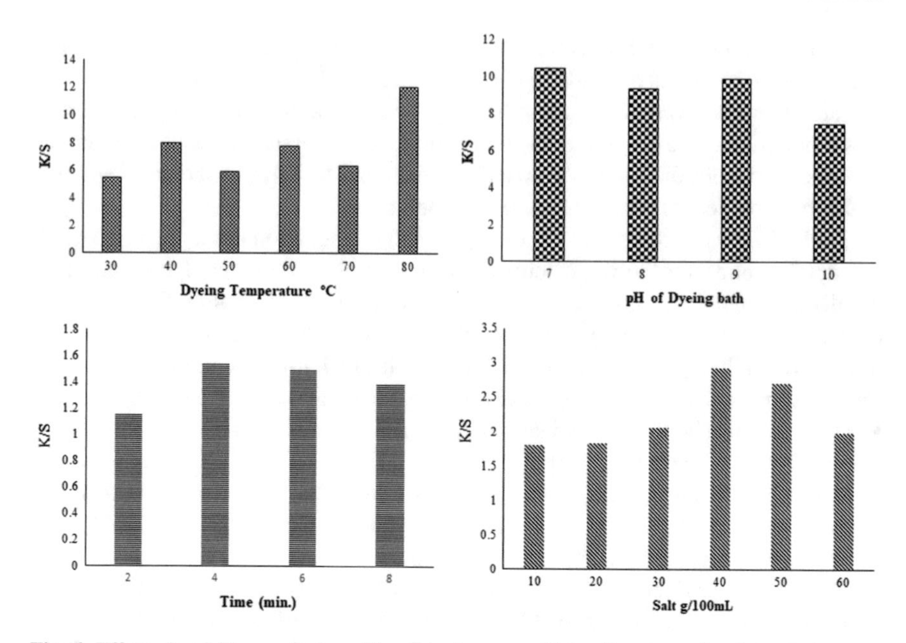

Fig. 6 Effect of variables on dyeing of irradiated cotton with irradiated reactive dye solution

Table 8 Effect of microwave treatment on light and wash fastness of dyed fabrics

Shades (%)	Wash fastness		Light fastness
	c.c	c.s	
0.5	4	3–4	4–5
1	4–5	4	4
1.5	4–5	4	4–5
2	4–5	4	4–5

c.c color change, *c.s* color stain

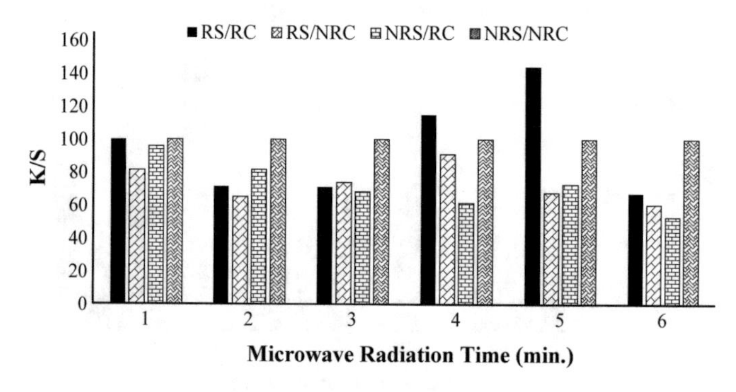

Fig. 7 Effective of microwave radiation on dyeing of cotton fabric using Vat Red 10 dye

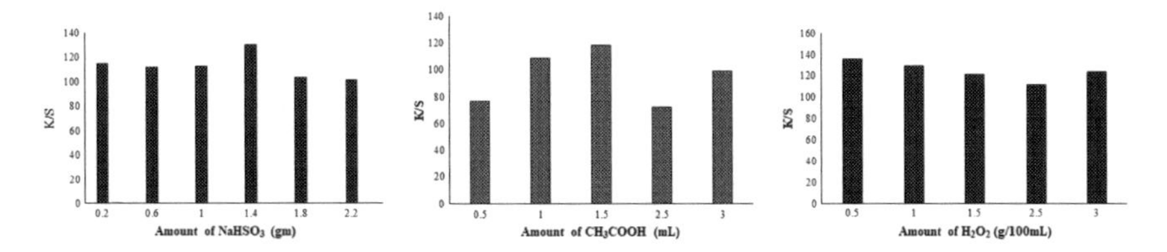

Fig. 8 Effect of redox reagents on vat dyeing of irradiated cotton fabric using irradiated vat dye solution

amount shows that MW treatment is a cost-effective tool for the dyeing process using Vat Red 10.

Arshad [47] designed the study to observe the effect of different parameters on color strength of cotton fabric which is dyed with Reactive Black 5. It was observed that MW treatment for 6 min. to dye solution and fabric has given good color depth (K/S) (Fig. 9). Similarly, un-irradiated cotton used for dyeing gives good and darker shades as compared to irradiated fabric. Because after employing irradiation the large size of the dye molecule become disintegrate into smaller ones [48, 49]. Upon dyeing the molecules rush towards the fabric evenly and diffused regularly to make covalent bonding, where after dyeing washing with hot and cold water did not fade the color. Hence microwave irradiation of dye solution for 6 min. is recommended to get darker shades and acceptable fastness characteristics. Good color strength was obtained at 60 °C for 50 min. using a dyeing bath of pH 9 keeping material-to-liquor ratio 1:75 (Fig. 10). Colorfastness properties assessed after dyeing the fabric at optimal conditions were improved fair to good by using microwave radiation (Table 9).

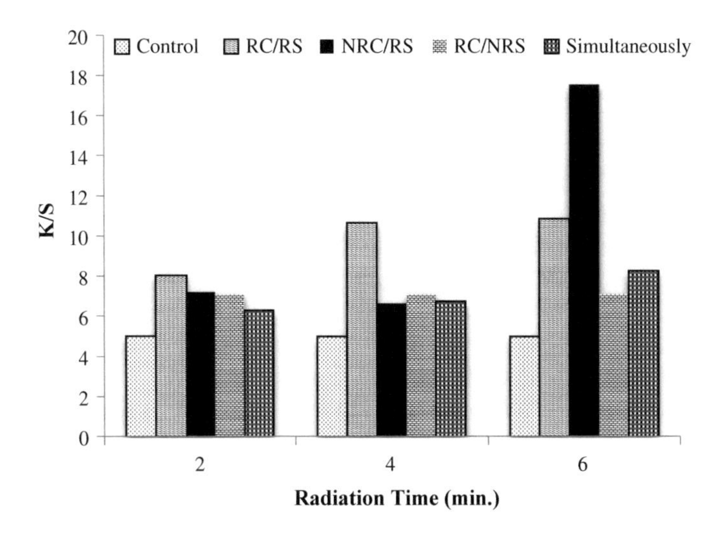

Fig. 9 Effect of irradiation time on the color strength of cotton fabric dyed with Reactive Black 5 dye

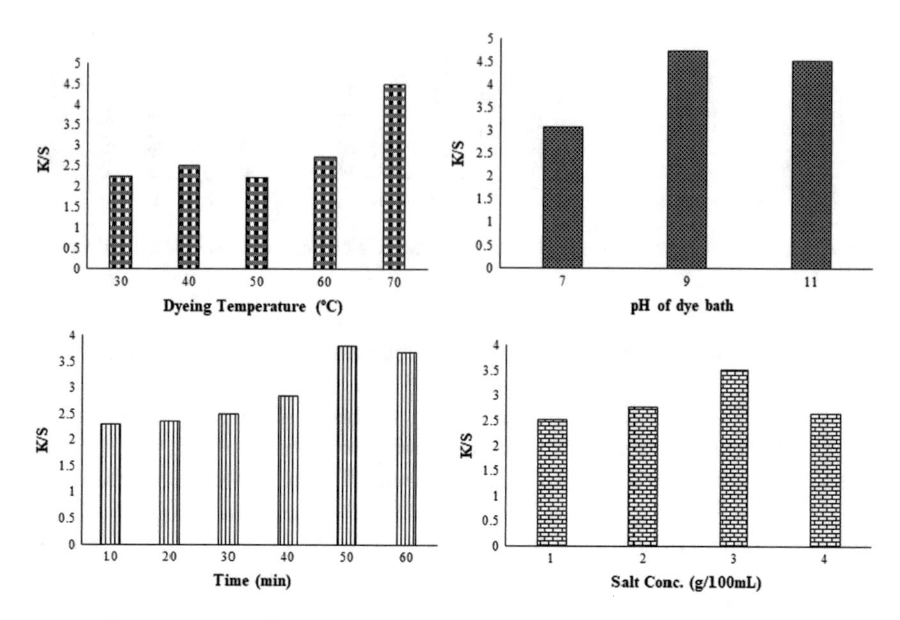

Fig. 10 Effects of variables on dyeing of un-irradiated cotton fabric using irradiated dye solution

Table 9 Rating of fastness properties of various shades made at optimum conditions

Shade (%)	Light fastness	Rub fastness		Washing fastness	Perspiration fatness
		Wet	Dry		
0.1	4–5	3–4	5	3	4–5
0.5	4	3–4	4–5	3–4	4–5
1	3–4	4–5	4–5	4	4
1.5	3–4	3–4	4–5	4–5	4
2	3	3–4	5	3–4	3–4

The behavior of Turquoise blue dye after the application of microwave treatment was also investigated onto the cotton fabric by Abbas [50]. His studies revealed that the microwave treatment of fabric using un-irradiated dye solution has given good color strength (RC/NRS = 2 min.). Irradiation for low time does not tune the fabric surface as well as does not activate dye molecules to rush towards fabric, while for long time irradiation, the dye bath may face degradation as well as over tuning of the fabric, due to which the dye molecules rapidly sorb onto fabric by forming aggregates and upon washing a lot of colors are stripped off and low color strength is observed. It was concluded that, irradiation of fabric for 2 min. is recommended to get good color strength using an un-irradiated solution (RC/NRS, 2 min.) for Turquoise blue dye.

Rehman [51] conducted a series of experiment on dyeing of cotton with Reactive Green 6 dye under the influence of MW treatment. It is found that irradiated dye

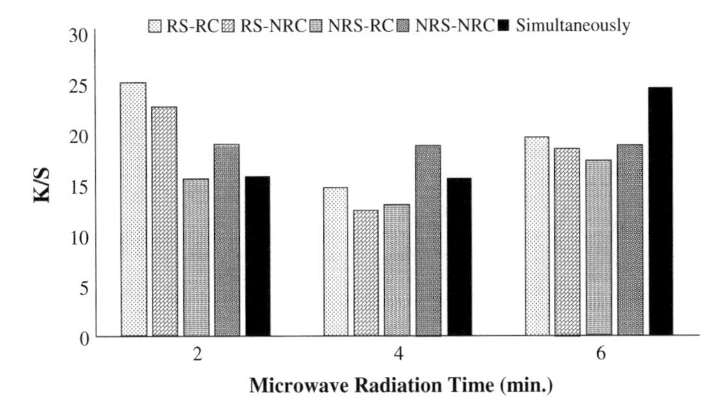

Fig. 11 Effect of microwave treatment on dyeing behavior of cotton using Reactive Green 6 dye

solution (RS, 2 min) applied for dyeing of irradiated fabrics (RC, 2 min) has given good K/S value (Fig. 11). This is because microwave treatment of cotton fabric doesn't affect the chemical structure of the cellulosic unit, where it only causes the physical tuning of fabric in such a way that its dye uptake ability has been enhanced and favored the good color depth. The irradiation causes the disintegration of large dye molecules into small size for making promising penetration more easily into voids of silk fibers after surface modification via MW treatment to give excellent color strength (K/S) [33, 34]. Hence microwave treatment for 2 min is the optimal time for reactive dyeing of cotton using Reactive Green 6 at optimum irradiation condition. Good color strength was obtained at 80 °C for 30 min. using a dyeing bath of pH 10 keeping material-to-liquor ratio 1:25 containing 3 g/100 mL of NaCl as exhausting agent (Fig. 12). The fastness properties given in Table 10 revealed that the shade made at optimum conditions of irradiation and then dyeing had given good to excellent ratings.

Qadeer [52] found that the microwave treatment has shown its effect to improve the color characteristics of fabric. This study reveal that fabric dyed with basic violet 3B dye the irradiation of dye bath (RS) for 5 min. has given excellent results onto irradiated fabric (RC). This is because the big dye molecules are broken into the small molecule which finds their way easy to penetrate fabric thereby resulting in good color strength [43, 45]. Radiation treatment time above optimal level may cause hydrolysis of dye molecules and resulted a slow color strength. So using basic violet 3B dye irradiation of dye bath and fabric should be treated for 5 min. to get good results.

Furthermore, microwave treatment has played a potential role for its surface modification to uptake more reactive dyes. The work reported by Afzal [53] revealed that the nature of dye also has a significant role to make firm interaction with surface-modified fabrics resulted by improving the color characteristics of fabric dyed with basic violet 3B dye. In this studies, it is found that microwave treatment

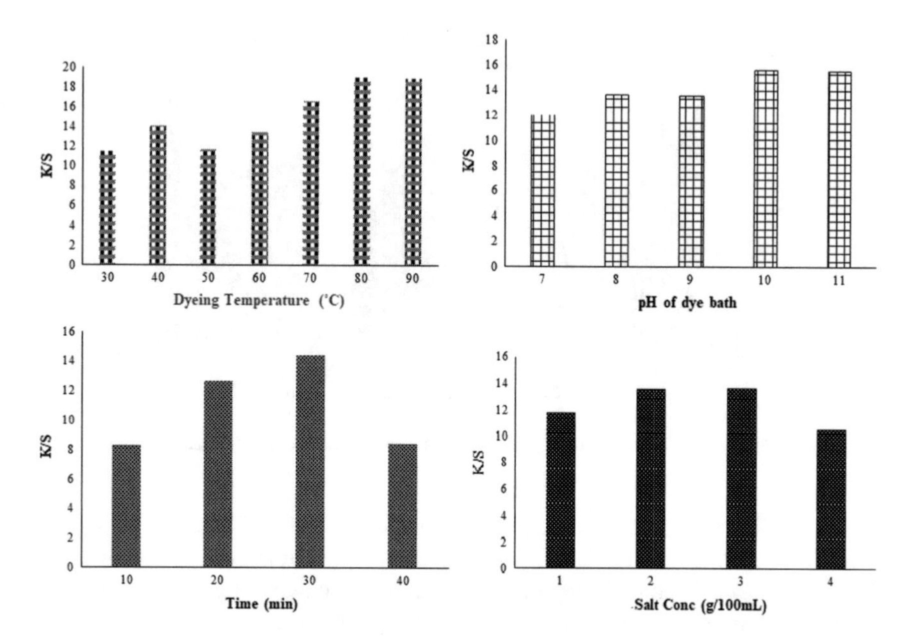

Fig. 12 Effect of variables on dyeing of irradiated cotton using the irradiated dye bath

Table 10 Fastness properties of irradiated fabric dyed at optimal conditions using irradiated dye solution

Shade (%)	Light Fastness	Rubbing test		Washing fastness	
		Dry	Wet	c.c	c.s
0.1	4–5	4	5	4–5	4
0.5	4	4	4–5	3–4	3
1	3–4	4	4–5	3	3
1.5	3–4	3	4–5	3	3
2	3	4	5	4–5	4

Here control is un–irradiated fabric, *c.c* color change, *c.s* color stain

for 6 min before desizing followed by pre-treatment to mercerization had given excellent results (Fig. 13a). It was because the greige fabric has a lot of impurities and it has taken high heating level for surface modification, which upon dyeing has shown excellent, results [48, 49]. Low microwave treatment did not add value by tuning the greige fabric surface which may make easy method for removal of impurities upon the pretreatment process. Hence before desizing MW treatment for 6 min should be given and for dyeing dye bath should also be MW treatment for 6 min. On the other hand when the fabric was irradiated and subjected to pre-treatment and dyed using an un-irradiated dye solution, the low color strength is obtained. Microwave treatment for low time (2–4 min.) less color depth had been eveluated. These results

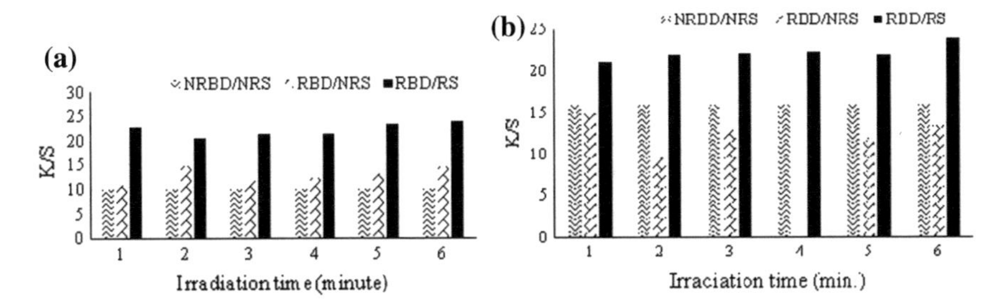

Fig. 13 Effect of MW treated on before resizing **a** and after desizing **b** followed by dyeing using Reactive Black WNN (Black 5)

had was also show that microwave treatment after desizing for 6 min followed by pretreatment upon mercerization had given high color strength when irradiated dye solution was used to dye (Fig. 13b). Again in the other experiment i.e. fabric irradiated after desizing, dye solution was also irradiated. The purpose of scouring is to remove any oil, fat, waxes, etc. from fabric and also to increase its wettability which was enhanced after microwave treatment thereby resulting in good results. Hence after desizing irradiation of fabric for 6 min. good results but the results obtained are poor as compared to irradiation before desizing. Thus change of dye bath, the molecular structure can affect the coloration process.

Bleaching of cotton fabric is carried out to improve its whiteness. During bleaching, the natural coloring matters present in cotton are bleached and improves the uptake of dye. The microwave treatment adds values when the fabric is irradiated before bleaching up to 4 min. (Figure 14a) which may enhance its substantivity towards Reactive Black (WNN black 5). So before bleaching, the results are more significant than before scouring. However, it is found that up to 4 min, microwave treatment of irradiated dye solution was used, high color strength was obtained. Therefore irradiation of fabric for 4 min. was subjected before and after bleaching followed by mercerization has represents excellent results using irradiated solution (Fig. 14a, b). Again it has been observed that irradiation of both fabric and dye solution is significant. Overall MW treatment before bleaching for 4 min. had given excellent results than that of irradiation before desizing and scouring. By applying MW up to 6 min. It was found that irradiation of both bleached fabric and dye solution for 4 min. had given good results. It is because upon irradiation the surface might be modified enough to sorb colorant molecules in a greater amount which may find difficulty to diffuse well into voids upon washing, the dye molecules are stripped and the low tint is observed. The result shows that irradiation of fabric before mercerization for 4 min. has given good results using irradiated dye solution. The fabric from desizing to mercerization when treated with microwave treatment almost all of the impurities are removed and the fabric becomes ready to dye. Overall the results show that irradiation of mercerized fabric for 4 min. had given excellent results using irradiated dye solution. However, treatment time to tune fabric is less, whereas above optimal time, over tuning of the fabric causes many difficulties in

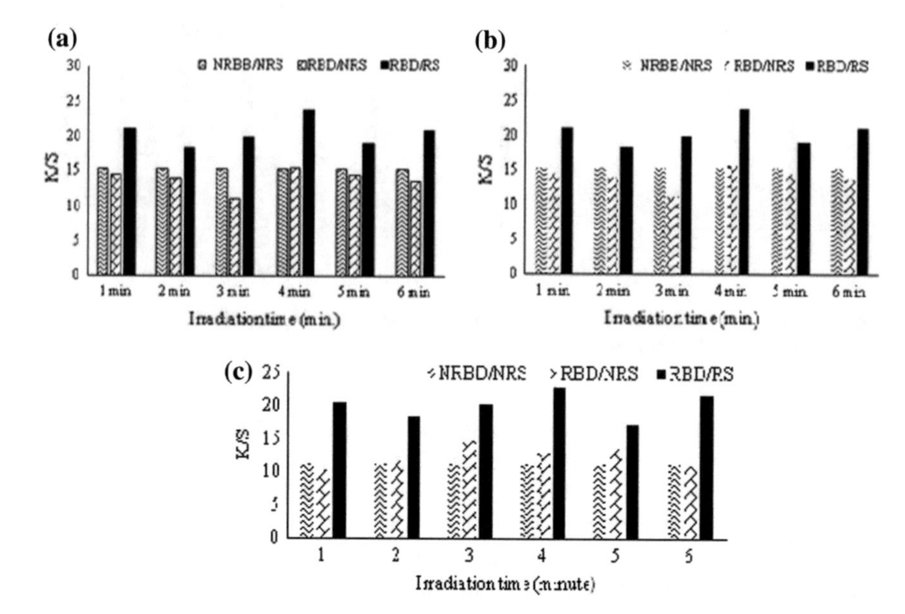

Fig. 14 Effect of MW treatment before **a** and after bleaching **b** and mercerization **c** followed by dyeing using Reactive Black WNN (Black 5)

penetration of molecules into the fibers which are gathered in form of clusters [54]. Upon washing with cold and hot water most of the dye molecules are stripped off and low K/S observed. So after mercerization 4 min. of MW treatment should be given to the fabric using irradiated dye solution to give good results (Fig. 14c). In this comparison, it had been found that before and after bleaching the MW treatment has improved its dyeing behavior and should be used for dyeing. This fabric pretreated by microwave treatment can not only reduce the time, cost energy as well as labor but also give acceptable results. The colorfastness agencies of fabric has been assessed after the dyeing at optimal conditions, enhanced good to excellent due to the effect of microwave radiation treatment (Table 11).

10 Effect of Microwave Treatment on Disperse Dyeing of Polyester Fabric

Muneer et al. [55] studied the effect of MW treatment on dyeing of the polyester fabric using Disperse Orange 25 dye at 80 °C for 25 min. This study revealed that good tinctorial strength was obtained by dyeing irradiated polyester fabric with irradiated disperse dye solution. Fastness properties evaluated onto shade developed at optimized condition the microwave treatment has excellent potential to improve the substantive behavior of polyester fabric using disperse dye.

Table 11 Effect of microwave radiation on colorfastness properties of fabric dyed with reactive black 5

Shade (%)	Light fastness	Washing fastness		Rubbing		Dry cleaning	Perspiration	
		c.c	c.s	Dry	Wet		Alkaline	Acidic
1	4/5	4	4/5	5	5	4/5	4	4
2	5	4/5	4	5	4/5	4	3/4	4
3	4/5	4/5	4	4/5	4	4/5	3	3/4
4	5	4	4/5	5	4/5	5	3/4	3/4
5	4/5	4	4/5	4/5	4	4/5	3	4

Here control is un−irradiated fabric, *c.c* color change, *c.s* color stain

Tariq [56] found that Microwave treatment plays a very significant role in the dyeing of polyester fabric because it makes the dyeing procedure very environment-friendly as well as time, labor and cost-effective [57]. The graphical representation of data is given in (Fig. 15) shows that high color strength (K/S) by dyeing un-irradiated polyester fabric (NRP) using an irradiated solution of Disperse Red 153 (RS) after the microwave treatment for 3 min. Microwave treatment being a volumetric heating source absorb energy directly and internally which leads to controlled, selective, uniform and rapid heating of dye bath [26]. Previously it has been observed that it modifies the fabric in such a way that significant interaction with dye has shown good color strength. The surface of polyester fabric is evenly tuned after the microwave heating due to which good color strength is obtained [58]. After microwave treated for 3 min, the irradiation cause ruptured the cell wall and upon isolation, it causes the breakdown of large size of colorant molecules into smaller size thereby making their

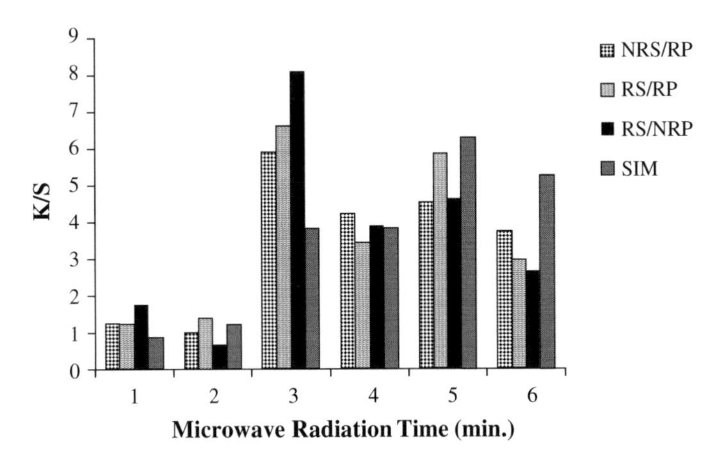

Fig. 15 Effect of microwave radiations on color strength of polyester fabrics using Disperse Red 153. *RP* irradiated polyester fabric, *NRP* un-irradiated polyester fabric, *RS* irradiated dye solution, *NRS* un-irradiated dye solution

diffusion into voids of polyester fabric more promisingly [59, 60]. Therefore, after the process of dyeing due to, washing, fewer dye molecules are detached and good color strength is obtained. For long-time heating, the equilibrium of the dye bath is disturbed and dye molecules rushed towards dye bath from fabric that results in low color depth. In this study depending upon the nature of the dye, fabric irradiation is not required as it did not give good results. In the case of low microwave heating, dye molecules are big that cannot penetrate into the fabric because the voids are unable to open through low treatment time to hold dye molecules tightly [61]. Thus 3 min. is the optimal radiation time for stimulation of dye bath for dyeing of un-irradiated fabric (NRP). Good color strength was obtained by dyeing fabric at 65 °C for 40 min. using 60 mL dyeing bath of pH 3 containing 5 g/100 mL of dispersant material-to-liquor ratio 1:25 dye bath containing 5 g/100 mL dispersant (Fig. 16). It was concluded that the colorfastness properties of the dyed fabric were assessed by using different optimal dyeing conditions, consequently, these fastness properties were enhanced from fair to good due to the effect of microwave radiation for both the polyester fabric as well as dye solution (Table 12).

As discussed earlier, MW treatment is the clean uniform and levelled heating and it can save energy and processing time [25]. Tariq [56] dyed polyester fabric with Disperse Yellow 79 under MW treatment. Through its unique mode of action of microwave not only given excellent colorant yield, but also via rapid mass transfer kinetics decrease the solvent consumption [34]. Dye molecules not only interact with the surface of polyester fabric but also show interaction with the inner parts of which it was observed that irradiation of dye bath for 2 min. (RS) has given excellent results onto un-irradiated fabric (Fig. 17). Above optimal irradiation time, the low color strength (K/S) was found. This is because more irradiation, more is the reduction

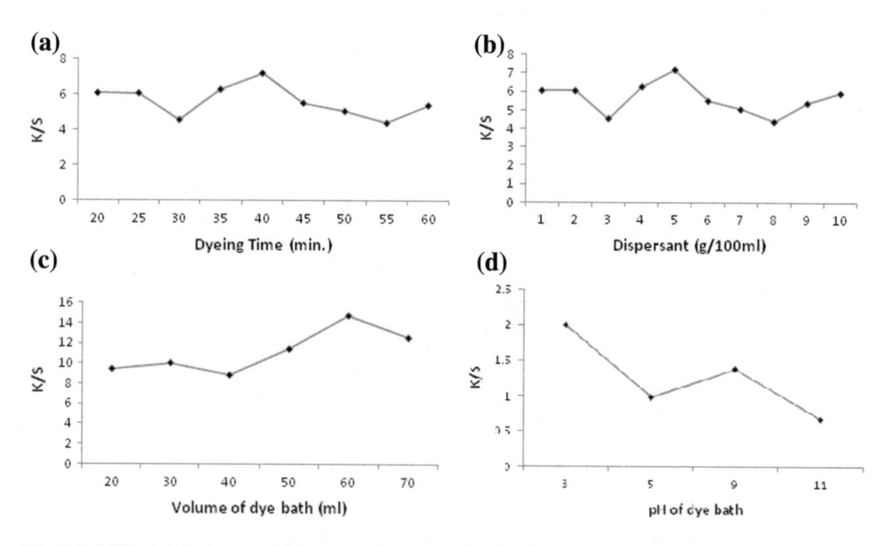

Fig. 16 Effect of dyeing variables on color strength of un-irradiated polyester fabric (NRP) using irradiated Disperse Red 153 solution

Table 12 Fastness properties of fabrics dyed with dispersed Red 153 under microwave treatment at optimal conditions

Shade (%)	Light fastness	Wash fastness		Rubbing fastness	
		c.c	c.s	Dry	Wet
Control	4/5	3/4	5	4	4
0.1	4/5	3/4	4	5	4/5
0.5	4/5	3/4	4/5	5	5
1	4/5	3	5	4/5	5
2	4	3/4	4/5	5	4/5
3	4/5	3	5	4/5	4/5
4	4/5	3/4	4/5	5	4/5

Here control is un−irradiated fabric, *c.c* color change, *c.s* color stain

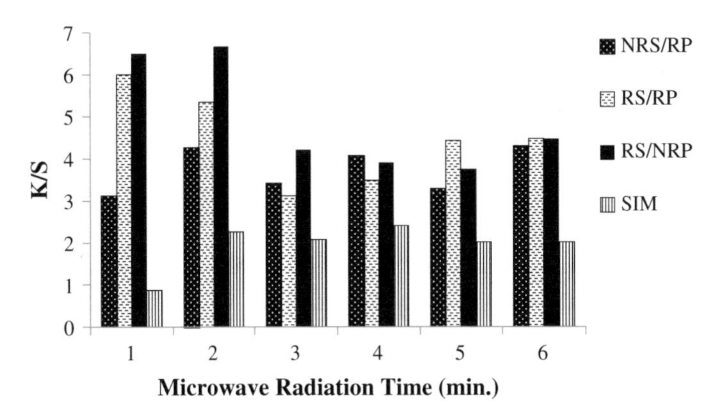

Fig. 17 Effect of Radiation time on color strength of polyester fabric using Disperse Yellow 79 (DY 79)

of size and more is the even penetration of molecule but an aggregation of small molecules to great extent may cause desorption upon washing with hot and cold water, the unfixed dye is removed and low K/S is observed. Good color strength was obtained by dyeing un-irradiated fabric (NRP) at 65 °C for 55 min. using irradiated dye bath of pH 11 heating 3 g/100 mL of dispersant under optimum irradiation and dyeing conditions (Fig. 18). The colorfastness properties of shades were assessed, which found good to excellent (Table 13). Hence microwave radiation has improved the substantive behavior of disperse dyeing of polyester fabric using Disperse Yellow 79.

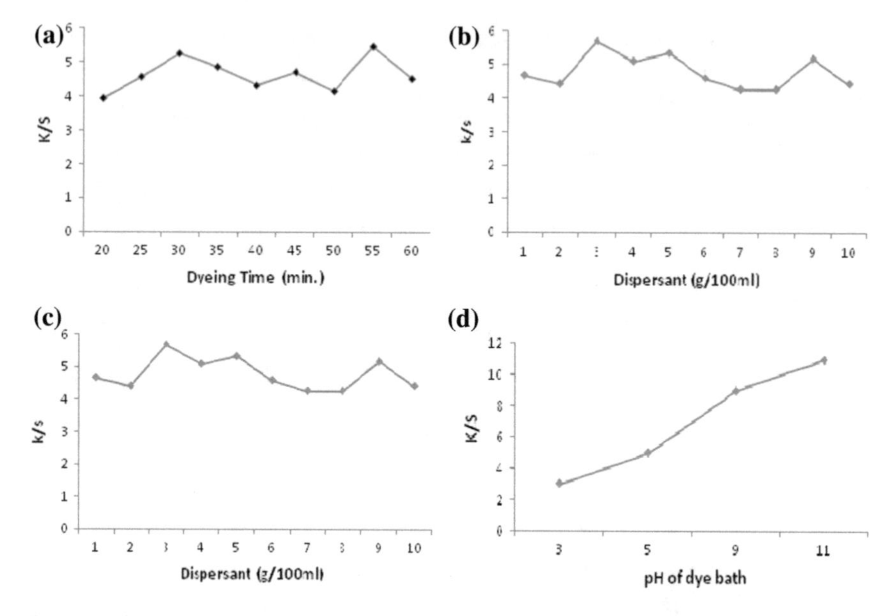

Fig. 18 Effect of dyeing variables on color strength of polyester fabric using Disperse Yellow 79 (DY 79)

Table 13 Fastness properties of polyester fabrics dyed with Dispersed Yellow 79 (DY 79)

Shade (%)	Light fastness	Wash fastness		Rubbing fastness	
		Colour change	Color stain	Dry	Wet
Control	4/5	3/4	5	4	4
0.1	4/5	4	4/5	5	5
0.5	5	3/4	4/5	4/5	5
1	5	4	4/5	5	4/5
2	5	4	4/5	5	5
3	5	4	4	¾	5
4	5	3/4	4/5	4	4/5

Here control is unirradiated fabric

11 Effect of Ultrasonic Radiation on Reactive Dyeing of Cotton

Ultrasonic treatment is another sustainable source of heating for modification of fabric and improvement in the reactive dyeing of cellulosic fabric. The results of Kamran work (2016) reveals that irradiated cotton (RC) and non-irradiated cotton (NRC) using irradiated solution (RS) and non-irradiated dye solution (NRS) valuable results depending upon fabric pretreatment process. Results show that, the irradiated fabric (RC, 30 min) using irradiated dye solution gives maximum K/S value. This is due to US treatment to the greige fabric which helps to remove the dust and other foreign particles. Thereby making viable the surface for more dye uptake ability. The other fabric is the irradiation of dye solution, where US treatment for 30 min. helps to decrease the particle size of dye molecules and stimulate the colorant to move toward treated fabric [62]. The acoustic cavitation also helps to raise the diffusion coefficient which helps to enhance K/S value [63]. The irradiation for low time does not reduce the size while for a long time may cause hydrolytic degradation due to which low (K/S) color strength is observed [38, 64, 65]. Hence US treatment for 30 min. is recommended if applied on to Greige fabric followed by further pretreatment for reactive dyeing using Reactive Red 195. The Ultrasonic treatment after desizing for 20 min using the un-irradiated solution (NRS) gives good color strength (K/S) value. This is due to the desizing of the fabric, where the sizing material through conventional treatment and make it suitable for further processing [63]. After desizing the US treatment helps to modify the surface and affect the crystalline area of the fabric to enhance the uptake ability. Thus, using an un-irradiated dye solution, the US treatment to desized fabric has given good color strength. The US treatment for 30 min. of bleached fabric gives a maximum K/S value by using the un-irradiated dye solution (NRS). After bleaching the fabric surface becomes more prone to dye sorption and upon irradiation, the surface becomes modified enough to sorb more dye. Treatment for a high time due to continuous agitation and cavitation may weaken the fabric to hold the dye aggregates firmly, where for low treatment time, the fabric surface is not tuned enough to interact with dye and tightly give poor results. US treatment for 30 min. is the optimum time for fabric after bleaching which increase uptake of dye to a good extent. After mercerization, the irradiation for 30 min. to fabric gives more K/S value using Reactive Red 195. This is because mercerization not only helps to improve whiteness; orientation of fibers has also enhanced the dye uptake ability. After mercerization, the fabric upon irradiation becomes more voids wider to enhance the sorption of dye molecules resulting in good color depth (K/S) [63]. So overall it is recommended that for reactive dyeing US treatment for 30 min. after bleaching of fabric (RC) is recommended using the un-irradiated dye solution (NRS) to get desired results. Hence it is proved that US treatment is not only a cost-effective but also time, energy effective tool for fabric processing.

For this study show that cellulosic fabric and dye solution was irradiated with ultrasonic (US) radiation for different times (10–60 min.). Umar [66] studied that dyeing was performed using non-radiated and irradiated cellulosic fibers with non-radiated and irradiated Reactive Yellow 145 dye. Different dyeing parameters such as temperature, pH, material to liquor ratio and time were optimized using irradiated dye and irradiated cotton. After evaluation of dyed fabrics in CIE Lab system using spectra flash SF 650, It is found that US treatment of dye solution for 20 min gives good color strength by dyeing un-irradiated fabric at 70 °C, for 35 min using dye bath of pH 9 in the presence of 5 g/100 mL of salt as exhausting agent keeping 40 mL of dye volume. Colorfastness properties of fabrics dyed at the optimum condition of various shades has shown that US treatment improved the grading of fastness from fair to good. It is found that US irradiation had not only enhanced the strength of dye on irradiated fabric but also improved the dyeing properties. It is also concluded that US treatment has not only reduced the dyeing time, temperature but also reduced the amount of salt and dye volume used, which is proof that this tool is cost, time and energy effective. It is inferred that US radiation technology being eco-friendly tools can also be successfully employed for other dyeing processes to get good color strength using different fabrics [3].

The need for eco-friendly tools in textile processing is the major concern of consumers and traders.

Adeel et al. [35] studied the role of ultrasonic radiation (US) on Vat dyeing by using cellulosic fibers. They found that excellent color strength was obtained by using un-irradiated cotton fabric with neutral dye bath by using optimum dyeing conditions. They also found that colorfastness agencies assessed on the shade made at optimum conditions reveal that the ultrasonic treatment have improve the grading of colorfastness good to excellent on cotton fabric dyed using Vat Blue 4.

12 Effect of Ultrasonic Radiation (US) on Disperse Dyeing

Umar [66] studied that the role of Ultrasonic radiation on Disperse dyes by using dye bath radiated for 30 min. has a promising effect on improving color depth onto un-irradiated polyester fabric (NRP). When there is due to the ultrasonic treatment, which produces the cavitation effect and breaks the cluster of dyes into small molecules via mass transfer kinetics which help to make firm bonding with fabric while irradiation of polyester may not cause opening of voids to such extent due to polymeric nature of fabric where big molecules fail to diffuse and upon finishing processes, the unfixed dye is removed and low color strength is observed [27, 63]. He found that US treatment to a solution (RS. 30 min.) should be given to a dye bath for achieving good color depth onto polyester fabric (Fig. 19). The role of US treatment has been found cleaner and more ecofriendly as irradiation of both polyester fabrics and the solution has given good color strength. He found that the US treatment for 50 min. irradiated polyester has given excellent results onto irradiated polyester fabric. As the time of irradiation rises, the reactivity of the polyester fabric for dispersing dyeing rises. This is because more US irradiation provides more cavitation which results in the transfer of energy towards colorant clusters to break into smaller molecules that upon, dyeing finds easy penetration into fiber. Hence US treatment to both polyester and dye solution for 50 min. has given significant results. Good color strength was obtained at 140 °C for 50 min displayed in (Fig. 20a, b). using a dyeing bath of pH 3 keeping material-to-liquor ratio 1:30 by using 0.5 g/100 mL dispersant (Fig. 20c, b). Colorfastness properties of dyed fabric at optimum dyeing conditions assessed, which found fair to good fastness agencies of due to the influence of microwave radiation (Table 14).

Ultrasonic (US) energy has revolutionized the dyeing process by introducing the cost and energy effective tools for textile processing. The objective of this research was to use ultrasonic waves as an eco-friendly technique to enhance the dyeing

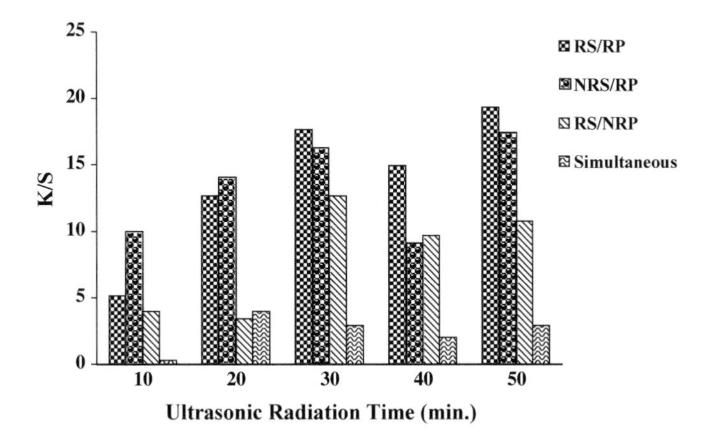

Fig. 19 Effect of ultrasonic treatment on the dyeing of polyester fabrics using Disperse Red 13 dye

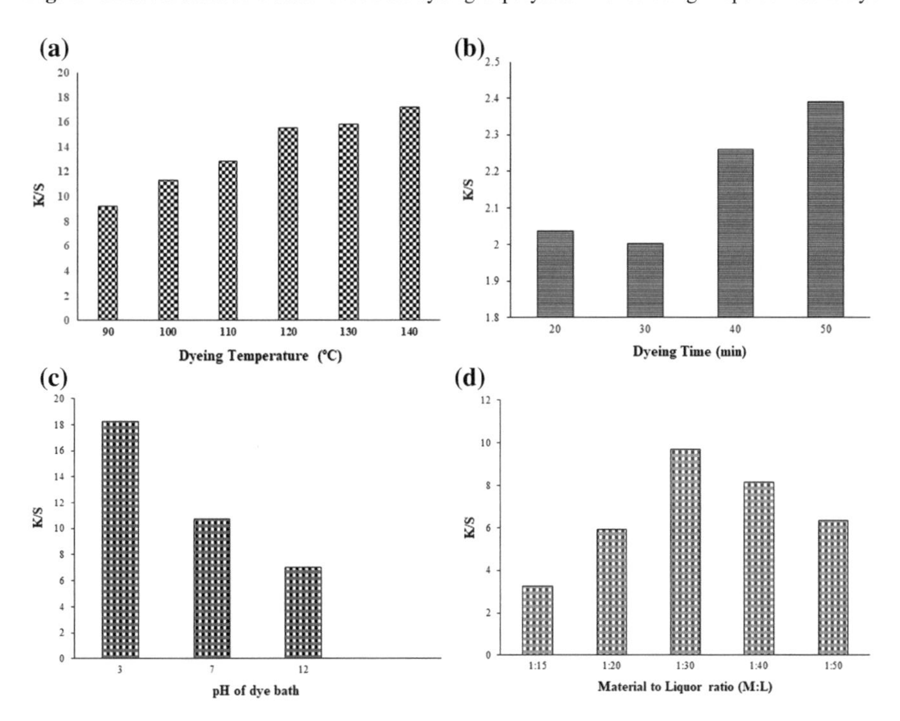

Fig. 20 Effect of dyeing temperature (**a**) and time (**b**) on the dyeing of US treated polyester with the US treated Disperse Red 13 solution. Effect of dyeing pH (**c**) and Volume (**b**) on the dyeing of US treated polyester with the US treated Disperse Red 13 solution

Table 14 Fastness properties of irradiated polyester fabrics dyed with different conc. of irradiated Disperse Red 13 dye solution

Shades (%)	Wash fastness	Lightfastness	Dry rubbing fastness	Wet rubbing fastness
0.5	4–5	3–4	3–4	3–4
1	3	4–5	3	3
1.5	4	4	3–4	3–4
2	3–4	4–5	4	4
2.5	4–5	4–5	3–4	3–4

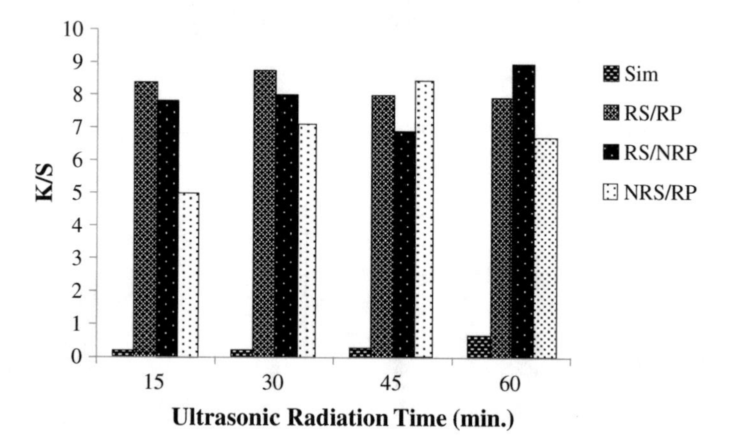

Fig. 21 Effect of ultraviolet treatment time on color strength of polyester fabric using Disperse Red 153

process of polyester fabric using disperse red. The fabrics and dye solutions were exposed to ultrasonic waves for 10–60 min. at 60 °C. In polyester dyeing, surface modification of polyester was done for 40 min. followed by dyeing at 130 °C for 30 min. keeping M: L using 1 g/100 ml dispersant having dyeing bath, of pH 10. At these conditions when different shades are applied, good color fastness has been rated. The current study infers that the use of US energy is a more efficient rapid tool than conventional heating. In terms of energy, cost, efficiency, its use in the textile sector will gain more favor to reduce effluent. So, it is concluded that US energy can successfully be applied in improving dyeing behavior, of other fabrics using various classes of dyes.

13 Effect of UV Radiation on Polyester Fabric

Tariq [56] studied, ultraviolet radiation also played an important role in disperse dyeing. It is found that that (Fig. 21) for the dyeing of polyester fabric, UV irradiation of dye solution of Disperse Red 153 (DR 153) for 60 min. gives high K/S utilizing un-irradiated polyester fabric. Irradiation for a long time and short time do not create any changes at the surface of polyester fabric, to improve its uptake ability. Thus, depending upon the nature of the dye bath, irradiation of solution for 60 min. should be employed for the coloration of un-irradiated fabric. Good color strength was obtained for 50 min. using a dyeing bath of pH 3 keeping material-to-liquor ratio 1:70 by using 2 g/100 mL of dispersant (Fig. 22). The Colorfastness properties was assessed, for cotton fabric dyed at optimum conditions revealed that good to fair fastness agencies were obtained due to the influence of microwave radiation (Table 15).

Tariq [56] studied the coloring behavior of UV treated polyester fabric using irradiated Disperse Yellow 79 (DY 79). It is found that UV treatment of fabric for 15 min. has given good results using un-irradiated disperse dye (Fig. 23). This is because irradiated polyester fabric makes the dyeing procedure evenly tuned (Bhatti et al. 2016). Diffusion of dye molecules becomes easy after the ultraviolet treatment for 15 min. depending upon the nature of the dye, the sorption of dye on irradiated polyester (RP, 15 min.) gives even color with acceptable color characteristics. Above optimal time, the surface is much more tuned to sorb dye molecule to a greater extent, which may face failure to diffuse all molecules. Upon washing with hot and cold water a lot of dye is bleached and less K/S is observed. Again irradiation of dye

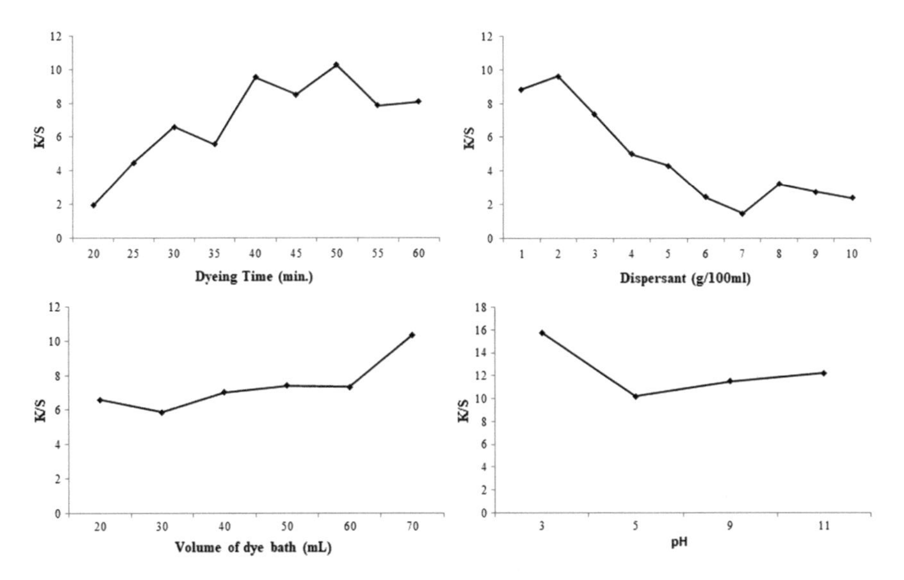

Fig. 22 Effect of different variables on dyeing of polyester fabric using Disperse Red 153 dye

Table 15 Fastness properties of fabrics dyed with different conc. of Dispersed Red 153 under ultraviolet treatment

Shade (%)	Light fastness	Wash fastness		Dry rubbing fastness	Wet rubbing fastness
		c.c	c.s		
Control	4/5	3/4	5	4	4
0.1	4/5	3	4	5	5
0.5	4/5	3/4	3	4	5
1	4/5	3/4	3/4	4/5	5
2	5	3	3	4	5
3	4/5	3/4	3/4	5	4/5
4	4/5	3/4	3	5	4/5

Here control is un−irradiated fabric

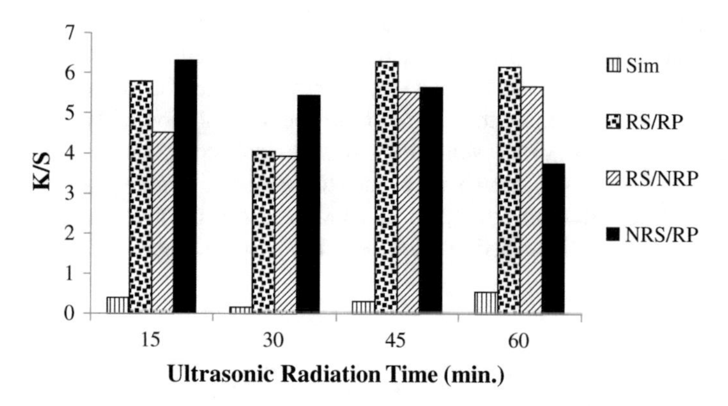

Fig. 23 Effect of ultraviolet treatment time on dyeing of polyester fabric using Disperse Yellow 79 dye

bath does not affect. Hence for UV irradiation an exposure for 15 min. of polyester fabric should be given to get desired results. Good color strength was obtained for 60 min. using a dyeing bath of pH 7 keeping material-to-liquor ratio 1:70 by using 6 g/100 mL of dispersant (Fig. 24). The colorfastness agencies of shade made at optimal condition reveal that good to fair fastness agencies were obtained after the treatment of microwave radiation (Table 16).

Adeel et al. [17] observed the role of UV radiation on dyeing of polyester fabric using Disperse Orange 25. Their studied that good color strength was obtained by using irradiated dye solution on un-irradiated polyester using disperse dye at different optimized dyeing conditions. Fastness characteristics of dyed fabric reveal that shade made at optimized dyeing conditions enhanced the fastness agencies from good to excellent under the influence of the treatment of UV radiation. They concluded that

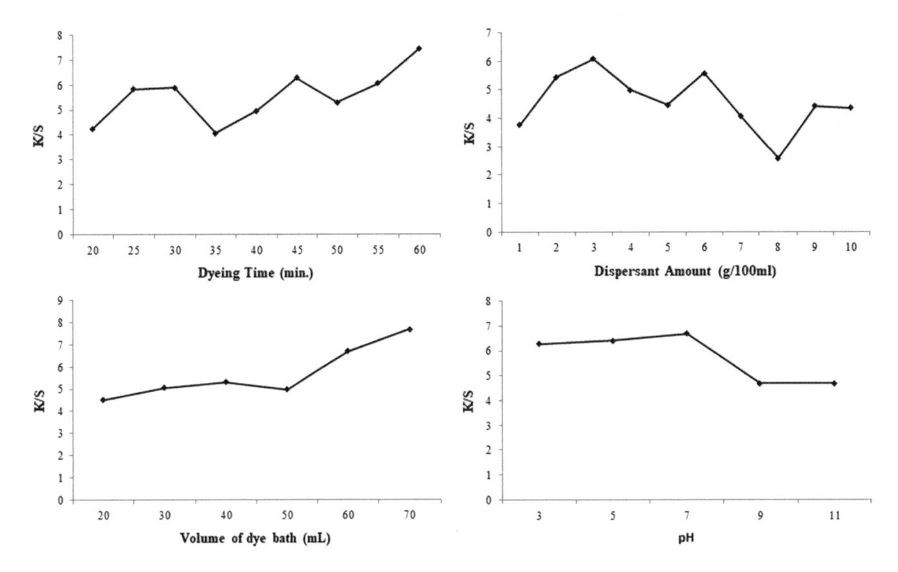

Fig. 24 Effect of different variables on dyeing of polyester fabric using Disperse Red 153 dye

Table 16 Fastness properties of polyester fabric dyed with different conc. of Dispersed Yellow 79

Shade (%)	Light fastness	Wash fastness		Dry rubbing fastness	Wet rubbing fastness
		c.c	c.s		
Control	4/5	3/4	5	4	4
0.1	5	4	3/4	4/5	4/5
0.5	5	3/4	3/4	4/5	4/5
1	5	4	3	4/5	5
2	5	3/4	2/3	4/5	5
3	5	3/4	3/4	4	5
4	5	4	4/5	4/5	4/5

Here control is un−irradiated fabric, *c.c* color change, *c.s* color stain

UV irradiation is useful tool to improve the dyeing behavior of polyester fabric using disperse dye [17].

A uniform and eco-friendly heating source, based on ultraviolet radiation has auspiciously improved the disperse dyeing of polyester fabrics said Shahid [67]. She found that good color strength and fastness properties could be obtained. If Disperse Yellow 211 (DY 211) is used to dye polyester fabric under the influence of ultraviolet treatment up-to 80 min. So, ultraviolet-treated polyester (RP) for 30 min. gives good color strength by dyeing fabric at 70 °C using 70 mL of un-irradiated dye solution (NRS) of pH 8 in the presence of 2% of dispersant. While for the dye bath of pH 11, good color strength obtained by dyeing irradiated polyester (RP) for 45 min. using a 90 mL irradiated dye solution (RS) 100 °C in the presence of 1% of dispersant. The obtained fastness grading revealed that ultraviolet radiation has improved the rating of fastness when shade is made at optimum conditions. Ultraviolet treatment not only reduced time, labor, and money but also improved color characteristics.

14 Conclusion

Modern tools are gaining a lot of popularity on account of their high treatment speed, particularity in fabric processing. In this chapter the role of radiation has been discussed during processing of fabric to make it sustainable for dyeing. The utilization of MW, US and UV rays as modern heating tools not only make fabric processing more viable but also add value in coloration of cotton and polyester fabric using various class of dyes. However, the role of MW treatment is much more prominent as compared to US and UV rays for fabric processing as well as for getting desired results, using MW treatment not only dyeing variables have been reduced but also excellent color characteristics of fabric dyed with reactive Vat and disperse dyes.

Acknowledgements We are highly thankful to Higher Education Commission of Pakistan for funding the research work presented in this chapter under project No. 20-2724/NRPU/R&D/HEC/12-6828. We are also thankful to Government College University Faisalabad for funding the research project under research support program through project No. GCUF-RSP/ACH-09. We are also thankful to Mr. Zafar Iqbal Manager of Noor Fatima Fabric Faisalabad, Mr. Muhammad Iqbal Director of Harris Dyes and Chemical Faisalabad Pakistan for providing us the technical support during this research work.

References

1. Shuli F, Jarwar AH, Wang X, Wang L, Ma Q (2018) Overview of the cotton in Pakistan and its future prospects. Pak J Agri Res 31(4):396–406
2. Aftab AK, Khan M (2010) Pakistan textile industry facing new challenges. Res J Int Stud 14(5):21–29
3. Adeel S, Saif MJ, Khosa MK, Anjum MN, Kamran M, Zuber M, Asif M (2020) Ultrasonic assisted improvement in dyeing behaviour of polyester fabric using Disperse Red 343. Pol J Environ Stud 29(1):261–265
4. Bhatti IA, Adeel S, Siddique S, Abbas M (2012) Effect of UV radiation on dyeing of cotton fabric with Reactive Blue 13. J Saudi Chem Soc 18(5):606–609
5. Migliavacca G, Ferrero F, Periolatto M (2014) Differential dyeing of wool fabric with metal-complex dyes after ultraviolet irradiation. Soc Dyers Color Color Technol 130:327–333
6. Sekhri S (2014) Textbook of fabric science: fundamental to finishing. PHI Learning Private Ltd., Dehli
7. Wang L, Xiang Z, Bai Y, Long J (2013) A plasma aided process for grey cotton fabric pretreatment. J Clean Prod 54:323–331
8. Mahmood N, Tusief MQ, Iqbal D, Khan MA, Ishaque W (2012) Effect of different locations, varieties and micronaire values upon the non-cellulosic and metal contents of cotton. J Chem Soc Pak 34(1)

9. Devi S, Rathinamala J, Jayashree S (2017) Study on antibacterial activity of natural dye from bark of *Araucaria columnaris* and its application in textile cotton fabrics. J Micro Biotech Res 4(3):32–35
10. Hseih YL (2007) Chemical structure and properties of cotton in cotton science and technology. Wood Head Publisher, Cambridge
11. Jose S, Gurumallesh Prabu H, Ammayappan L (2017) Eco-Friendly dyeing of silk and cotton textiles using combination of three natural colorants. J Nat Fibers 14(1):40–49
12. Zia KM, Adeel S, Khosa MK, Aslam H, Zuber M (2019) Influence of ultrasonic radiation on extraction and green dyeing of mordanted cotton using neem bark extract. J Ind Eng Chem 77:317–322
13. Zhang L, Nederberg F, Pratt RC, Waymouth RM, Hedrick JL, Wade CG (2007) Phosphazene bases: a new category of organo-catalysts for the living ring-opening polymerization of cyclic esters. Macromol. 40(12):4154–4158
14. Sinha K, Saha PD, Datta S (2012) Extraction of natural dye from petals of flame of forest (Buteamono sperma) flower: process optimization using response surface methodology (RSM). Dyes Pigments 94(2):212–216
15. Flint I (2007) The eucalyptus dye'. Turk Red J 13(1):1–6
16. Hunger K (2003) Industrial chemistry: chemistry, properties and application, 1st edn. Wiley, Weinheim
17. Adeel S, Muneer M, Ayub S, Saeed M, Zuber M, Iqbal M, Haq E, Kamran M (2017) Fabrication of UV assisted improvement in dyeing behavior of polyester fabric using Disperse Orange 25. Oxid Commun 40(2):925–935
18. Merdan N, Akalin M, Kocak D, Usta I (2004) Effects of ultrasonic energy on the dyeing of polyamide (microfibre)/Lycra blends. Ultrason Sonochem 42:165–168
19. Ferrero F, Periolatto M (2012) Ultrasound for low-temperature dyeing of wool with acid dye. Ultrason Sonochem 19:601–606
20. Kamel MM, Helmy HM, Mashaly HM, Kafafy HH (2010) Ultrasonic assisted dyeing: dyeing of acrylic fabrics C.I. Astrazon Basic Red 5BL 200%. Ultrason Sonochem 17:92–97
21. McNeil SJ, McCall RA (2011) Ultrasound for wool dyeing and finishing. Ultrason Sonochem 18(1):401–406
22. Zhang Z, Shan Y, Wang J, Ling H, Zang S, Gao W, Zhao Z, Zhang H (2007) Investigation on the rapid degradation of congo red catalyzed by activated carbon powder under microwave irradiation. J Hazard Mater 147:325–333
23. Al-Mousawi MS, El-Apasery AM, Elnagdi HM (2013) Microwave assisted dyeing of polyester fabrics with disperse dyes. Mol. 18:11033–11043
24. Hashem M, Taleb AM, El-Shall FN, Haggag K (2014) New prospects in pretreatment of cotton fabrics using microwave heating. Carbohyd Polym 103:385–391
25. Kiran S, Adeel S, Rehman FU, Gulzar T, Jannat M, Zuber M (2018) Ecofriendly dyeing of microwave treated cotton fabric using reactive violet H3R. Glob Nest J. 21(1):43–47
26. Kale MJ, Bhat NV (2011) Effect of microwave pretreatment on the dyeing behavior of polyester fabric. Color Technol 127:365–371
27. Khatri M, Ahmed F, Jatoi WA, Mahar RB, Khatri Z, Kim SI (2016) Ultrasonic dyeing of cellulose nanofibers. Ultrason Sonochem 31:350–354
28. Khajavi R, Atlasi A, Yazdanshenas EM (2013) Alkali treatment of cotton yarns with ultrasonic bath. Text Res J 83(8):827–835
29. Zhou T, Lim T, Wu X (2011) Sonophotolytic degradation of azo dye reactive black 5 in an ultrasound/UV/ferric system and the roles of different organic ligands. Water Res 45:2915–2922
30. Bhatti IA, Adeel S, Taj H (2014) Application of Vat Green 1 dye on gamma-ray treated cellulosic fabric. Rad Phy Chem 102:124–137
31. Adeel S, Shahid S, Khan SG, Rehman F, Muneer M, Zuber M, Akthar N (2018) Eco-friendly disperse dyeing of Ultraviolet treated polyester fabric using Disperse Yellow 211 dye. Pol J Environ Stud 27(5):1935–1939
32. Mushatq M (2015) Bio-processing of cellulosic fiber for efficient ecofriendly textile dyeing. M.Phil. thesis GCUF, Pakistan

33. Haggag K, El-Molla MM, Ahmed KA (2015) Dyeing of nylon 66 fabrics using disperse dyes by microwave irradiation technology. Int Res J Pure App Chem 103–111
34. Mustapa AN, Martin A, Gallego JR, Mato RB, Cocero MJ (2015) Microwave-assisted extraction of polyphenols from *Clinacanthus nutans* Lindau medicinal plant: Energy perspective and kinetics modeling. Chem Eng Proc Proc Intensi 97:66–74
35. Adeel S, Khan SG, Shahid S, Saeed M, Kiran S, Zuber M, Akhtar N (2018c) Sustainable dyeing of microwave treated polyester fabric using Disperse Yellow 211 Dye. J Mexi Chem Soc 62 (1)
36. Kamran M (2016) Microwave and ultrasonic assisted pretreatment of cotton fabric and dyeing using reactive red and Vat dye. M.Phil. thesis GCUF, Pakistan
37. Clark M (2011) Handbook of textile and industrial dyeing: Principles, processes, and types of dyes, vol 2. Woodhead Publishing, Cornwall, pp 425–446
38. Kamel MM, El-Shishtawy RM, Youssef BM, Mashaly H (2006) Ultrasonic assisted dyeing. IV. Dyeing of cationised cotton with lac natural dye. Dyes Pigment 73(3):279–284
39. Adeel S, Ghaffar A, Mushtaq M, Yameen M, Rehman FU, Zuber M, Kamran M, Iqbal M (2016) Bio-processing of surface-oxidised cellulosic fibre by microwave treatment for eco-friendly textile dyeing. Oxid Commun 39(3):2396–2406
40. Buyukakıncı BY (2012) Usage of microwave energy in the Turkish textile production sector. Energy Proced 14:424–431
41. Jannat M (2017) Influence of ultrasonic and microwave treatment on the dyeing behavior of cotton fabric using reactive violet H3R. M.Phil. thesis GCUF, Pakistan
42. Khalid F (2016) Microwave assisted radiation on cotton fabric using reactive dye. Master thesis
43. El-Molla MM, Haggag K, Ahmed KA (2013) Dyeing of polyester fabrics using microwave irradiation technique. Int J Sci Res 4:2319–7064
44. Majid M (2015) Effect of microwave treatment on the dyeing of cotton fabric using VAT red 10. Master thesis
45. Adeel S, Saeed M, Abdullah A, Rehman FU, Salman M, Kamran M, Zuber M, Iqbal M (2018) Microwave assisted modulation of vat dyeing of cellulosic fiber: improvement in color characteristics. J Nat Fibers 15(4):517–526
46. Ghaffar A, Adeel S, Habib N, Jalal F, Munir B, Ahmad A, Jahangeer M, Jamil Q (2018) Effects of microwave radiation on cotton dyeing with Reactive Blue 21 dye. Pol J Environ Stud 28(3):1687–1691
47. Arshad A (2016) Microwave assisted dyeing of cotton fabric using reactive black 5 B.S Hons Report. M.Phil. thesis GCUF, Pakistan
48. Gorjanc M, Bukosek V, Gorensek M, Vesel A (2010) The influence of water vapor plasma treatment on specific properties of bleached and mercerized cotton fabric. Text Res J 80(6):557
49. Kocak D, Akalin M, Merdan N, Şahinbaskan BY (2015) Effect of microwave energy on disperse dyeability of polypropylene fibers. Marmara Fen Bilimleri Dergisi 27:27–31
50. Abbas A (2016) Microwave assisted dyeing of cotton fabrics using Turquoise blue dye. Master thesis
51. Rehman HH (2016) Microwave assisted dyeing of cotton fabric using reactive green 6 dye. Master thesis
52. Qadeer S (2016) Effect of microwave on the dyeing of cotton fabric using basic violet 3B dye. Master thesis
53. Afzal A (2018) Effect of microwave radiations on pre-treatment and its effect on dyeing using reactive black WNN. Master thesis
54. Sakran MA (1996) Effect of ionizing radiation on the spectral and electrical properties of NH3-mercerized cotton fabric strips. J Rad Nuclear Chem Letter 213(1):51–63
55. Muneer M, Adeel S, Ayub S, Zuber M, Rehman FU, Kanjal MI, Kamran M (2016) Dyeing behavior of microwave-assisted surface-modified polyester fabric using Disperse Orange 25: improvement in color strength and fastness properties. Oxid Commun 39(2):1430–1439
56. Tariq H (2018) The influence of radiations on dyeing behavior of polyester fabric using disperse red 153 (DR 153) and Disperse Yellow 79 (DY 79). Master thesis

57. Chakraborty JN (2013) Fundamentals and practices in the coloration of textiles. Woodhead Publishing, Cambridge, pp 222–232
58. Kappe CO, Dallinger D (2009) Controlled microwave heating in modern organic synthesis: highlights from the 2004–2008 literature. Mol diver 13(2):71
59. Bhatti VN, Kale JM, Gore VA (2009) Microwave radiation for heat setting of polyester fibers. J Eng Fiber Fabric 4(4):1–6
60. Haggag K, Hanna HL, Youssef BM, El-Shimy NS (1995) Dyeing polyester with microwave heating using disperse dyestuffs. Amer Dyestuff Rep 84(3):22–37
61. Kamel MM, Allam OG, El-Gabry LK, Helmy HM (2013) Surface modification methods for improving dyeability of acrylic fabric using natural biopolymer. J App Sci Res 9(6):3520–3529
62. Nidheesh PV, Gandhimathi R (2012) Trends in the electro-Fenton process for water and wastewater treatment: an overview. Desalination 299:1–15
63. Gotoh K, Harayama K, Handa K (2015) Combination effect of ultrasound and shake as a mechanical action for textile cleaning. Ultrason Chem 22:412–442
64. Oakes J (2001) Photofading of textile dyes. Rev Prog Color Relat Top 31(1):21–28
65. Yang Y, Huda S (2003) Comparison of disperse dye exhaustion, color yield, and colorfastness between polylactide and poly (ethylene terephthalate). J App Polym Sci 90(12):3285–3290
66. Umar M (2017) Effect of UV and US radiations on cotton and polyester fabrics by using reactive and disperse dyes. M.Phil. thesis GCUF, Pakistan
67. Shahid S (2018) Influence of radiation on the dyeing behavior of polyester fabric using Disperse Yellow 211 (DY 211) Dye. M.Phil. thesis GCUF, Pakistan

Developments in Textile Continuous Processing Machineries

Kedar S. Kulkarni and Ravindra Adivarekar

Abstract In the last century, due to increased industrialization, textile industry has undergone huge changes from technical as well as technological aspects. Particularly textile wet processing machineries have experienced tremendous improvements. From earlier mechanical machines, it has advanced to today's ultra-modern machines which are controlled by electronic components. In this chapter, we have covered commonly used continuous wet processing machineries in the textile industry primarily to process woven fabrics. These machines are explained by detailing various sections and accessories used by different machine manufacturers. The explanations are based on actual work experience and knowledge accumulated during the industrial tenure.

Keywords Wet processing · Continuous machineries · Semi continuous processing · Batch processing

1 Introduction

Textile is considered as one of the necessities of humankind, second only to food. Recorded history of textiles can date back to as early as 6000 years ago. However, industrialisation of textiles production was started only in the nineteenth century.

Improvement in technology changed the customer demands and to fulfil these demands there was lot of research and it resulted in advancement of technologies. During same time, lot of other areas were also improved drastically, and these opened a vast new avenue where textiles could find applications. Some of the examples are, automotive industry, aviation industry, space technology, military applications etc.

Textile wet processing is most important link in the textile supply chain management. Typically, wet processing consists of pre-treatment, coloration and finishing of textiles. These operations can be carried out by means of discontinuous, continuous and semi-continuous systems.

K. S. Kulkarni (✉) · R. Adivarekar
Department of Fibres and Textile Processing Technology, Institute of Chemical Technology, Mumbai 400019, India
e-mail: ks.kulkarni@ictmumbai.edu.in

M. Shahid and R. Adivarekar (eds.), *Advances in Functional Finishing of Textiles*, Textile Science and Clothing Technology, https://doi.org/10.1007/978-981-15-3669-4_15

1.1 Discontinuous or Batch-Type Systems

All the operations are carried out batchwise. It usually involves loading the machine, carrying out the treatments following a predetermined cycle, unload the machine and finally wash it thoroughly before starting a new cycle. This working process is extremely flexible and is suitable for processing small lots. For producing larger lots, the discontinuous process becomes labour-intensive and uneconomical. It also entails long processing times and results may vary from batch to batch.

1.2 Continuous Systems

The operations are carried out by means of a series of machines; every machine carries out always and solely the same process. Every machine is assembled according to specific production requirements. A system like this entails high start-up costs and a complex setup but once the system has started, it requires a smaller staff and grants excellent repeatability and high output rates; continuous systems are therefore suitable for manufacturing large lots of products with the highest cost-efficiency.

1.3 Semi-continuous Systems

In these mixed systems, several operations are carried out with both continuous and discontinuous machines. For example, a continuous pad-batch machine is used to wet the fabric and a discontinuous system is then used for other treatments. These mixed systems are suitable for processing small and medium lots; they require reasonable start-up costs and grant quite good reproducibility.

The wet processing of textiles involve pre-treatment, colouration and finishing. Pre-treatment involve removing of natural as well as added impurities from the textile substrate and making it uniform in terms of absorbency with good degree of whiteness. Colouration involves dyeing and/or printing, while finishing consists of applying various finishes, which can be divided into chemical and physical finishes. Different softeners, speciality finishes like wrinkle free, oil and stain repellent etc. are chemical finishes and calendar, sanforizer etc. are physical finishes.

2 Continuous Processing Machineries

2.1 Inspection Machine

Once a grey fabric is received from weaving/knitting section, it is inspected on inspection machines, which consists of light source and glass. Main aim is to confirm the quality of fabric before wet processing is started. Usually, two different systems are used to decide the quality of fabric known as four point and ten point. Any defects from weaving or knitting can be identified and marked before the fabric is processed. Inspected fabrics are classified based on number of defects found. The machine has options to either batch or roll the fabric. Observed defects such as spots, knots etc. are marked and the operation is called perching. In the next operation, known as burling most of the imperfections are removed.

2.1.1 Recent Innovations

Recent innovation in inspection machine is use of laser beam to detect stains, holes, thick and thin places and thread breaks etc. The system can be used in-line and it helps reduce human interference ensuring consistent quality. Different instruments are available which work on different principles such as coaxial, remission and transmission technique. Scanner light beam is reflected to its source in coaxial method while in remission method, the emitted light is received by a light guide system and directed to a converter. In transmission system light can be received by a receiver system located at the other side and it is used to detect the defects in the fabric.

2.2 Shearing and Cropping

Shearing is an operation where loose ends of fibres from the fabric surface are cut using sharp edged razors. In processing of pile fabrics, cropping operation can be used to control and adjust pile design and pile height. This can be used to reduce pilling tendency of fabrics.

Figure 1 shows one of the earliest 4 cutter shearing machine (a) and line diagrams of 2 cutter Fig. 1b and 4 cutter Fig. 1c machines. The earlier machines consists of a strong frame carrying four revolving cutters fitted with left and right hand spiral blades. A cutting point is a contact line between shearing cylinder and the ledger blade, over which the fabric passes during operation. After completion of cropping, the fabric is passed through revolving brushing rollers which removes the protruding fibres which are cut during operation.

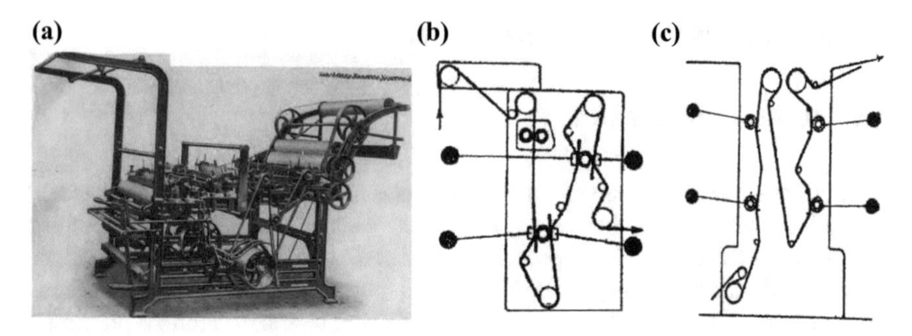

Fig. 1 Photo of four cutter shearing machine (**a**) and line diagram of two cutter (**b**) and four cutter (**c**)

2.2.1 Advanced Shearing and Cropping Machine

It has following features:

- Electronically controlled fabric feeding device to ensure uniform and even fabric transport
- Specially designed soft bed under the cutter to avoid damage to the fabric in process
- Electronic seam joint sensors to protect rollers from fabric joints
- Metal detectors which protects the rollers from iron particles which might be embedded in the fabric. The sensor, when detects iron particles, stops the machine by activating limit switch. The machine restarts once the metal particles are removed
- Gear systems are replaced by DC drives. This enables the modified machines to run at higher speeds.

2.2.2 Main Components of Shearing Machine

Shearing Unit

It is composed of two sharpened elements, that are working like scissors. The unit is equipped with spiral cylinder and high-quality sharpened ledger blade. These spiral blades are installed on solid cylinder. the number of blades varies from 10 to 24, which are fixed with a helical displacement. The speed of shearing cylinder can be continuously adjusted within the range of 600–1200 rpm. The rotating shearing cylinder on the sharpened ledger blade creates a continuously cutting effect. To achieve better cutting performance, the height of ledger blade should match at the centre of shearing cylinder (called as cutting point). The fibres cut during the operation are removed by a vacuum system. Shearing quality directly depends on quality of suction applied (Fig. 2).

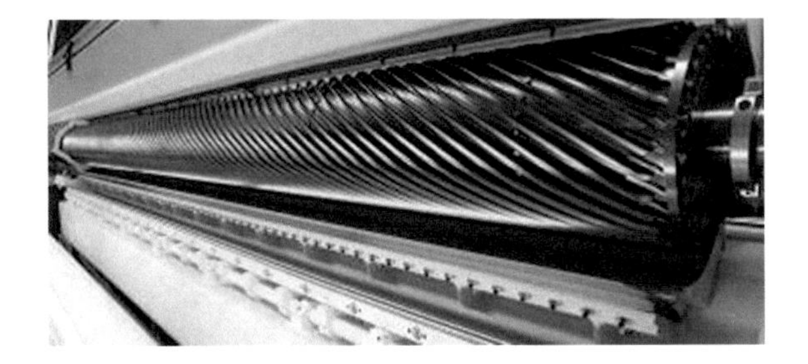

Fig. 2 Shearing unit

Fig. 3 Metal detector

To and from Device

The unit generates alternate axial movement of the shearing cylinder in relation to the ledger blade during operation. It is designed to automatically sharpen the helical blades and ledger blade and to provide more regular shearing.

Inlet Section

The fabric is to be fed to the machine uniformly without any creases. The fabric unwinds from the batch by the un-winder roll drive. For synchronisation and ensuring even fabric tension, a dancer is provided between un-winder and draw roll.

Metal Detector

It is fixed on inlet frame. It detects foreign metal particles and stops the machine instantly. This is to avoid damage to the shearing cylinder and ledger blades (Fig. 3).

Seam Detector

Fabric seam detector is designed to sense seam between fabrics, so that blades are protected.

Fig. 4 Brushing roller

Expanding Rollers/Curve Bar

The expanding rollers make sure that the fabric stays unfolded and does not roll up in the machine. The speed of these rollers can be adjusted with the help of variable frequency inverter.

Brushing and Beating Rollers

Brushing rollers raise the pile to be shear, while beating rollers clean the fabric before shearing. Speed of brushing rollers can vary between 50 to 300 rpm. Distance between fabric and beating rollers is adjustable (Fig. 4).

Exit Section

Used for winding fabric on big batcher carriage. Roller pressure and fabric tension are adjustable (Fig. 5).

2.3 *Singeing Machine*

Singeing operation is carried out to remove protruding fibres from the fabric surface.

2.3.1 Advantages of Singeing Operation

- It improves end use and wearing properties of textiles
- Gives clean surface due to burning off of protruding fibres, which gives clean and brighter appearance of the treated fabric
- Effectively reduces pilling tendency of fabric, especially in case of blends

Fig. 5 Parts of exit section

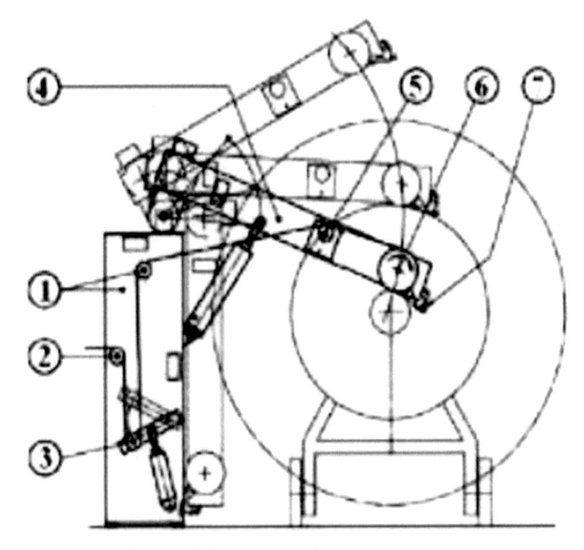

1. Machine stand
2. Overrun roller
3. Dancer
4. Winding arm LH/RH.
5. Guide & Expander roller
6. Transport roller
7. Operator finger protection device

- Closely singed fabric is essential for printing FME intricate patterns
- The singed fabric can be desized using the fabric surface temperature.

2.3.2 Major Disadvantages of Singeing

- Uneven singeing results in streak marks which appear prominent after dyeing
- Singeing of blended fabrics containing synthetic fibres may result in deposition of small globules of melted synthetic fibres. This melt absorbs more dye which gives fabric specky appearance post dyeing
- Temperature sensitive fibres like polyester can get thermally damaged if flame intensity and distance is not managed properly
- Machine stoppages causes heat bar marks on fabrics
- Heat sensitive sizing agents when exposed to heat results in firmly binding of sizing agents with fibres which makes desizing difficult
- When singeing operation is done after dyeing, heat in singeing causes colour loss in case of low sublimation fast dyes
- It may result in loss of tear strength.

2.3.3 Gas Singeing Machine

The latest advancement in singeing machine is gas singeing. There are machines which uses different technique to remove protruding fibres from the fabric surface. Some of the earlier machines use heated plate or roller for singeing, however, incase of gas singeing machine, as the name suggests, gas flame is used for this purpose. Heart of the gas singeing machine is its burner, which is the most advanced part of the machine. Usually, pre-mixed air and gas mixture prepared using proportional mixing valve is supplied to the burner which ensures uniform flame across the width. Ratio of gas to air always remains in accordance with the stoichiometric value so that irrespective of intensity of flame, its height remains same (Fig. 6).

The machine can be divided into three main sections—inlet, singe unit and exit section. The inlet section of the machine may be divided into two main sub-sections—pre-cleaning and inlet. Various sub-assemblies attached in both these sections include,

- Beating and brushing unit
- Cyclone dust collector
- J-scray
- Fabric guider.

Actual singing takes place in Singeing unit, which have number of sub-assemblies such as,

- Burner assembly with adjustments
- Gas train and intensity control
- Hairiness tester.

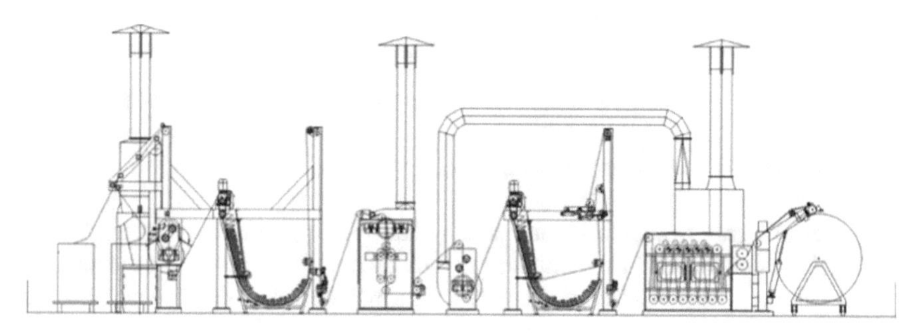

Fig. 6 Line diagram of modern singeing and desizing range

Exit section contains,

- Spark extinguishing device
- Beating unit
- Batcher-optional.

Let us see all the sub-assemblies:

Inlet Section

This unit is used for feeding the fabric to the machine in a uniform manner. The fabric may be plaited, batched on rolls or on horse-back. The fixed bars on infeed section ensures smooth feed of fabric. The centering device ensures alignment of fabric with the machine. Dancer ensures uniform tension along the length and width of fabric.

J-Scray

Is used to store the fabric, which helps in uninterrupted running of machine. It also helps in detecting weak stitch identification, re-stitching and leader cloth passing etc. Inverter controlled take up roller is used to convert fabric from batch to J-scray. Compensator equipped with load cell is used to ensure even fabric tension by adjusting speed of take up roller and stopping the machine when fabric exits (Fig. 7).

Beating and Brushing Unit

It is also called as Vibra plus. Main function of the unit is to pre-cleaning of fabric. The fabric passes between driven brushing and beating rollers. The cloth runs from top to bottom, while cleaning brushes rotates in opposite direction. Action of beaters remove the dust particles. This removed dust is collected in a cyclone dust collector (Fig. 8).

Cyclone Dust Collector

The exhaust air passes through the system from top to bottom. Flying dust, fibre residues and lints are removed by brushes, is effectively sucked away via bottom ducts to entire width. The contaminated air from pre-cleaning chamber is passed into the cotton bag, so that dust is settled into the bags and normal air goes out. This system prevents contamination of outside air (Fig. 9).

Singeing Unit

This unit is a heart of the machine. The recent features of the singeing unit give various advantages over earlier machines such as,

Fig. 7 J scray

Fig. 8 Beating roller

- Even effect of singeing across the length and width of the fabric.
- Gas and air mixture is used to ensure uninterrupted homogeneous concentrated flame.

The machine can be used to singe various textile materials such as, natural, animal, regenerated and synthetic fibres and their blends. To achieve even singeing effect on all these substrates, various parameters that can be adjusted in the machine includes,

- Brushing intensity
- Flame intensity

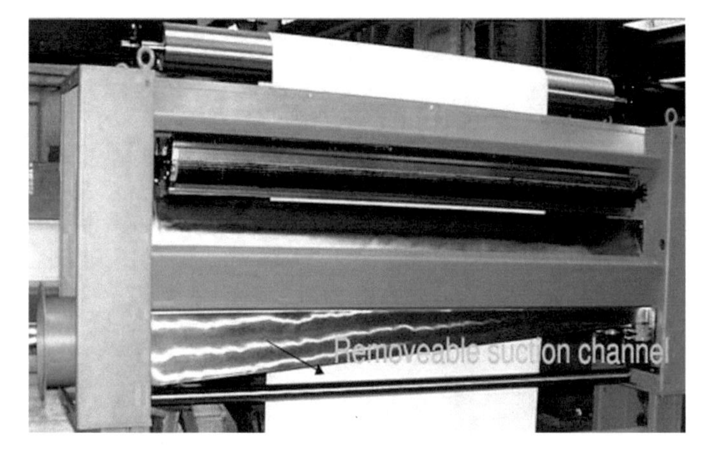

Fig. 9 Cyclone dust collector

Fig. 10 Singeing unit

- Speed of the machine
- Singeing positions
- Burner to fabric distance
- Selection of burner
- Burner width adjustment.

Latest advances in singeing machine is the accessory such as hairiness tester which is and optional accessory (Fig. 10).

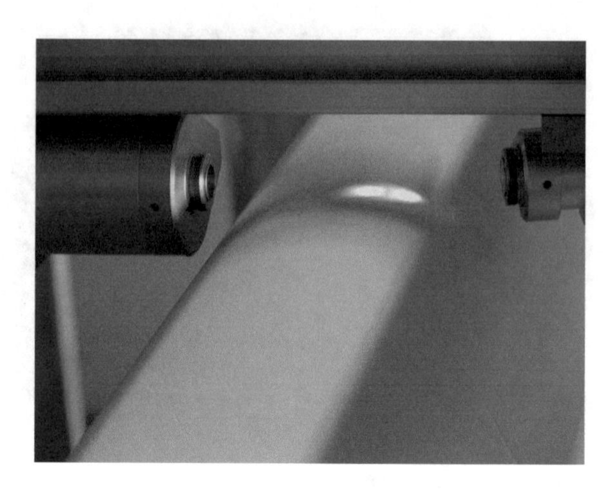

Fig. 11 Online hairiness tester

Hairiness Tester

This is an accessory which helps in controlling the singeing operation from quality as well as utility consumption point of view. It is a combination of a light source and a monochrome camera. It is fitted at a strategic position on a fabric path where it can monitor the hairiness of a moving fabric. Here, the light source illuminates the fabric surface and it is recorded by the camera. Arrangement of light source is done in such a way that the camera should not record the shadows of protruding fibres, but only reflected light is measured. The method is known as dark field measurement.

By using signals from camera, PLC can adjust intensity of flame and speed of machine as well. This will save the utility (gas) as well as improves the production (Fig. 11).

Exit Section

Spark Extinguishing Unit

The selvedge guiding ribbons are pressed against the burning selvedges and consequently extinguish them by removal of oxygen (Fig. 12).

Beating Unit

The unit has two rollers equipped with beating segments. They rotate in opposite direction of the fabric to remove brunt fibers from both fabric surface. Ash from the brunt fibers is then suck by suction duct and pass through the wet filter. The water spray nozzles are provided in wet filter to extinguish residual spark, if any.

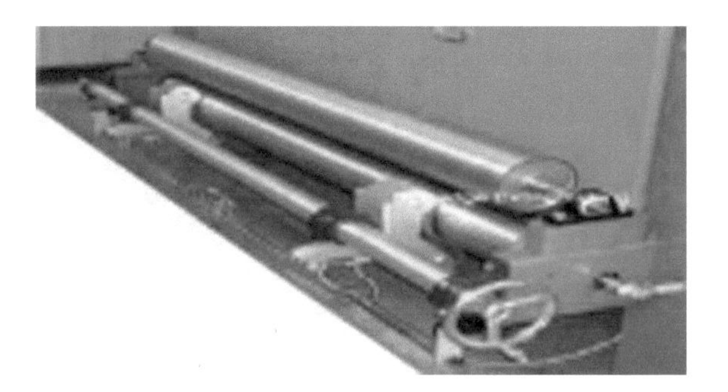

Fig. 12 Spark extinguishing unit

2.4 Desizing Range

The desizing range is usually immediately after singeing range. This way, it serves two purpose: firstly, it can use the temperature on the surface of the fabric due to exposure to live flame and secondly, the liquor in desizing tank actually act as an extinguisher apart from desizing (Fig. 13).

2.4.1 Chemical Impregnation Tank

The chemical Impregnation tanks are made up of stainless steel. It has numbers of guide rollers with special bearings at end, direct and indirect heating control, automatic level control, liquor circulation with rotary filter. Following are some of the commercially available impregnation tank (Fig. 14).

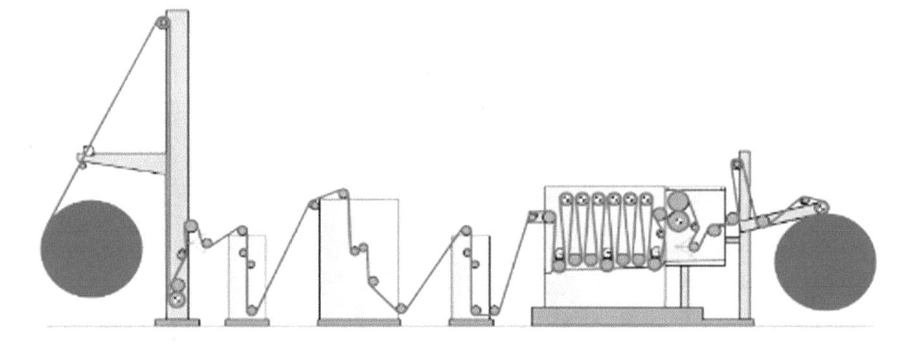

Fig. 13 Singeing and desizing range

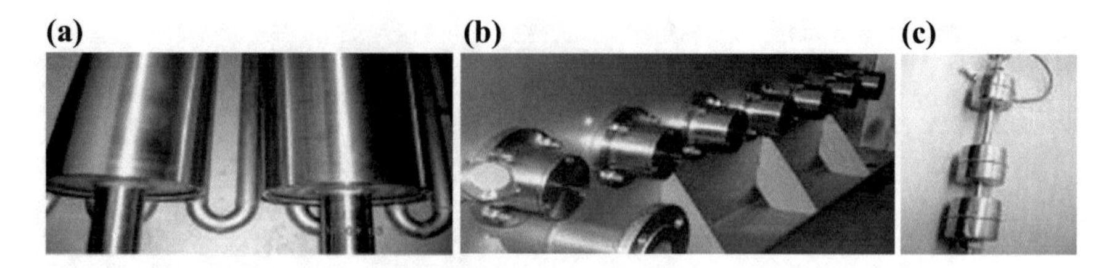

Fig. 14 Stainless steel rollers in impregnation tank (**a**); special bearings (**b**); level sensor (**c**)

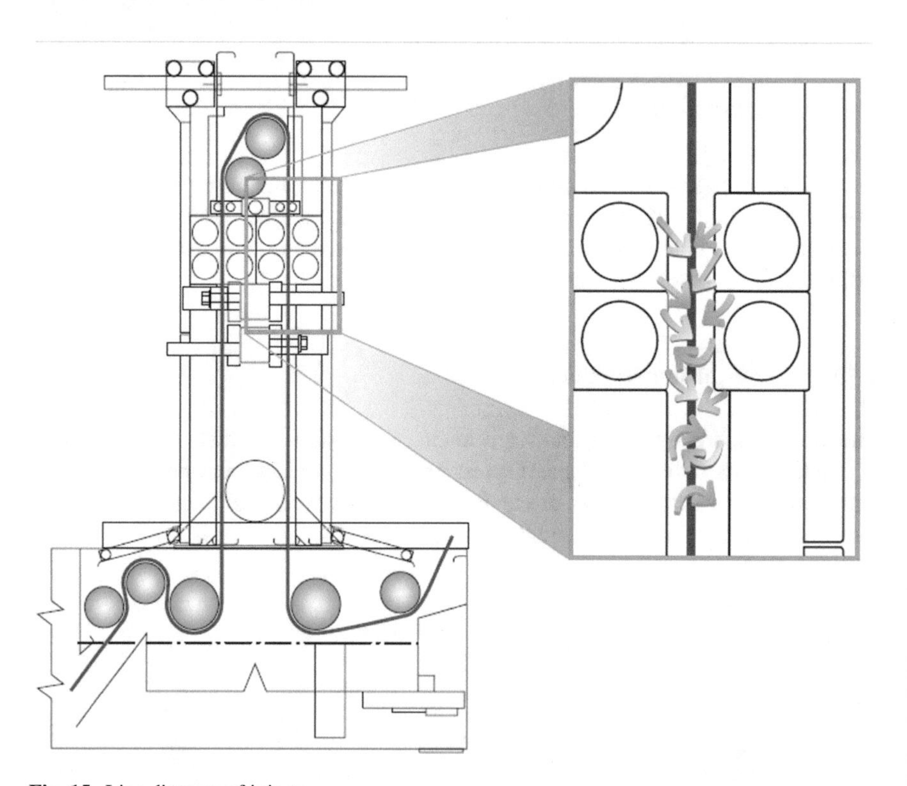

Fig. 15 Line diagram of injecta

Injecta—(Benninger)

It is the technology introduced by M/s Benninger to wash the desized fabric effectively. It uses only steam and water at higher temperature (above 100 °C) with high turbulence. The process conditions results in extremely efficient removal of size which helps in improved efficiency of subsequent processes.

The washing process is effective in removing any type of size, natural or synthetic without preswelling. Other advantages of the system includes, compact design which allows lesser space and suitability of size recovery (Fig. 15).

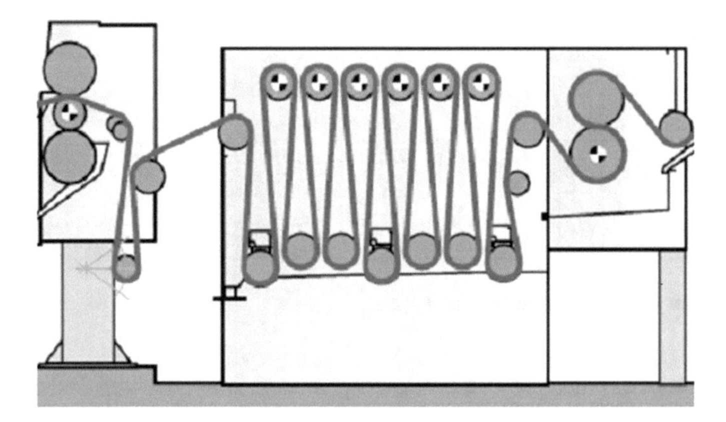

Fig. 16 Line diagram of Dip sat (Goller)

Dip Sat (Goller)

The technology developed by Goller for desizing is known as Dip sat. The set up involves a compartment in which 17-m fabric is treated. The line diagram is as shown in Fig. 16. The arrangement of rollers is in such a way that the fabric is impregnated three times. This gives good penetration of desizing liquor to the core of the fabric. The complete penetration of liquor in the fabric results in effective removal of size from the fabric resulting in even pre-treatment.

2.4.2 Squeezer

An impregnated fabric is then led between a rubber coated roller and an ebonite roller. The two rollers are pressed together to squeeze out the excess liquor from the fabric, which runs back into the impregnation tank. The squeezing effect depends on hardness of the rubber coating and squeezer pressure. Usually, pick up is in the range of 90–120% depending on the quality of fabric and size applied. The fabric treated at this stage is greige fabric and it is not absorbent enough. Thus, more the amount of desizing liquor that is applied on the fabric, more efficient desizing operation can be. Hence, usually, instead of applying pressure, only dead weight of roller is used to squeeze excess liquor.

S-Wrap Squeezer (Goller)

- Defined liquor with pick up above 100% results in excellent degree of desizing
- Slip-free and crease-free fabric transport.

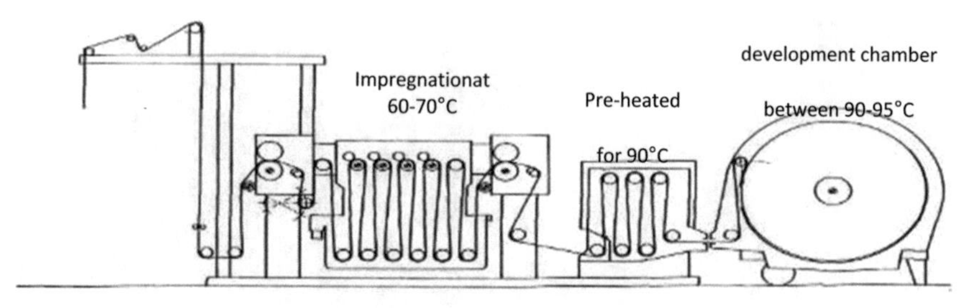

Fig. 17 Semi continuous pre-treatment range

In commercial batch process, fabric in open width is treated with desizing recipe in a suitable machine such as jigger. Usually, the recipe consists of desizing agent, preferably amylase enzyme, salt and non ionic wetting agent (Fig. 17).

In semi-continuous process, use of pad-roll installation is used. The cloth, in open width is pre-heated in a steam chamber after impregnation with desizing recipe, which contains desizing agent, salt and non-ionic wetting agent. The padded goods after steaming are batched in moveable carriages and are rotated for 8–12 h under steaming atmosphere. The material is rotated gently to prevent uneven desizing due to drainage and steam is injected into the chamber slowly to maintain required temperature.

In a continuous process, the padded fabric is steamed for 20–60 s at temperature of 95–100 °C using saturated steam. The continuous process is described in detail below.

2.5 Continuous Bleaching Range

The three Processes involved in Continuous Bleaching Range machine:

1. **Desize Wash**: (To wash the desized fabric so as to removed hydrolysed starch)
2. **Scouring**: (To remove the natural impurities present in fabric and to improve absorbency)
3. **Bleaching**: (To bring uniform whiteness to the fabric) (Fig. 18).

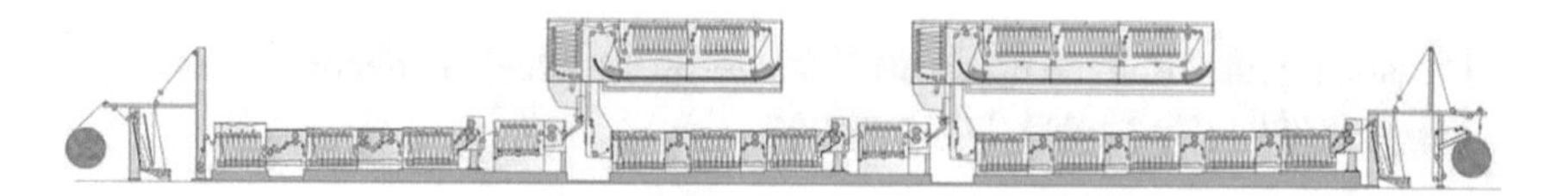

Fig. 18 Continuous pre-treatment range (Goller)

Machine consists of following subassemblies:

2.5.1 Infeed

The Infeed unit is used for the introduction of the fabric from batch carriers. The infeed drive is guaranteeing the tensionless fabric transport to the machine. The fabric tensioner is used to set the necessary fabric tension if required.

J-Scray at Inlet and Exit of Machine

J-scray stores fabric when enabled. It is use for interruption free operation. In a continuous operation, the machine continues to run with unchanged speed when a batch is changed. In this mode the fabric infeed always features a J-scray and an automatic batch change program is carried out.

2.5.2 Pre-washer (Desize Wash)

The washing module washes out the hydrolysed starch and other impurities from desized fabric. The washing process is divided into several washing steps. Different manufacturers use different technologies to improve washing efficiency. Some of the examples are,

Unipulsa

In this system, a pulsating perforated drum is used which gives high turbulent, intensive flow of liquor through fabric resulting in effective removal of hardly soluble impurities.

Vacuuset (Effecta by Goller)

Vacuum slit is used for economic water removal from fabric being processed. The technology can be used in washing off process which is used after desizing, pretreatment and dyeing. Thus, apart from extracting excess water, many other impurities such as lint, emulsified and disperse contaminants, hydrolysed dyestuffs and sizing agents can also be removed. The fresh water consumption is considerably reduced due to the increased degree of concentration that develops during the ensuing washing process. Partial disposal of this higly polluted liquor can reduce waste water contamination considerably (Fig. 19).

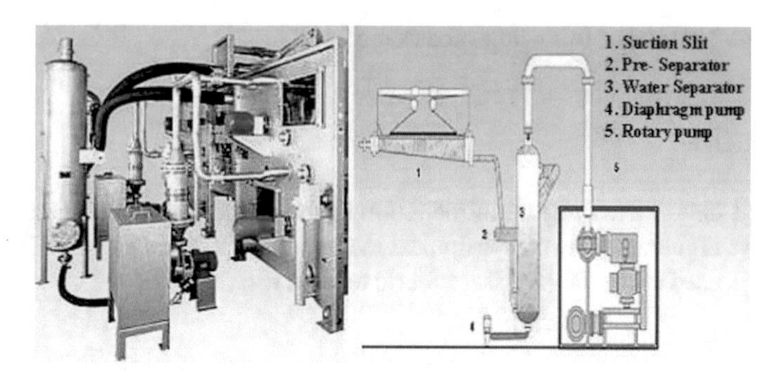

Fig. 19 Hydrovac technique of Goller

Fig. 20 Three roller squeezing device

Fig. 21 Line diagram of continuous washing range

Three Roller Squeezing Device FXT

Maximum dewatering of the fabric with evenness of dewatering over the full width. Tension on fabric is minimum with crease free running with the help of expanders (Figs. 20 and 21).

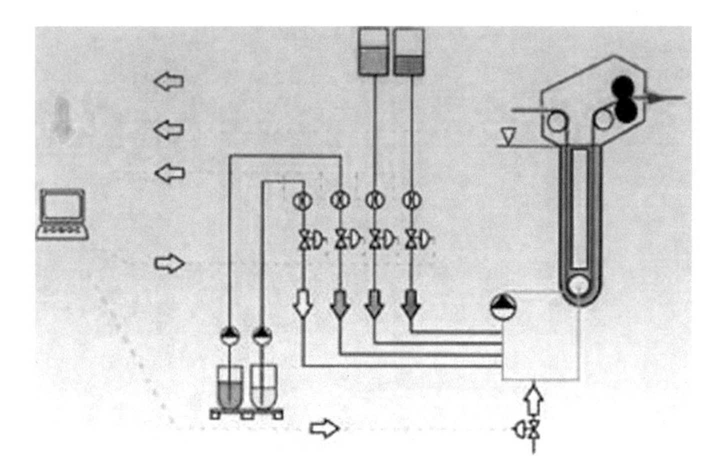

Fig. 22 Line diagram of chemical metering station

2.5.3 Chemical Metering Station

A chemical metering station is used to dose the desired chemicals and auxiliaries which are used for scouring and bleaching. In a usual pretreatment recipe, wetting agent, sequestering agent, sodium hydroxide, hydrogen peroxide and hydrogen peroxide stabiliser are used. These chemicals are dosed according to the recipe set. It is important to dose these chemicals precisely as per recipe and also every time consistently. Thus, reproducibility of dosing is extremely important for successful and consistent results. Specially designed pumps are used for this purpose which ensure uniform and reproducible results. Usually, following points are required for PLC to dispense desired chemicals uniformly: weight of fabric, capacity of tank so as to decide initial filling and dosage in ml/kg (Fig. 22).

2.5.4 Impregnation Chamber

An impregnation chamber is used to impregnate fabric with desired chemicals and auxiliaries, followed by reaction chamber, neutralisation chamber and fabric delivery section. Following are impregnation chambers manufactured by various machine manufactureres.

Impacta

The technology is developed by M/s Benninger. Important feature of the machine is a specially designed nip, which ensures uniform pick up. It is independent of speed

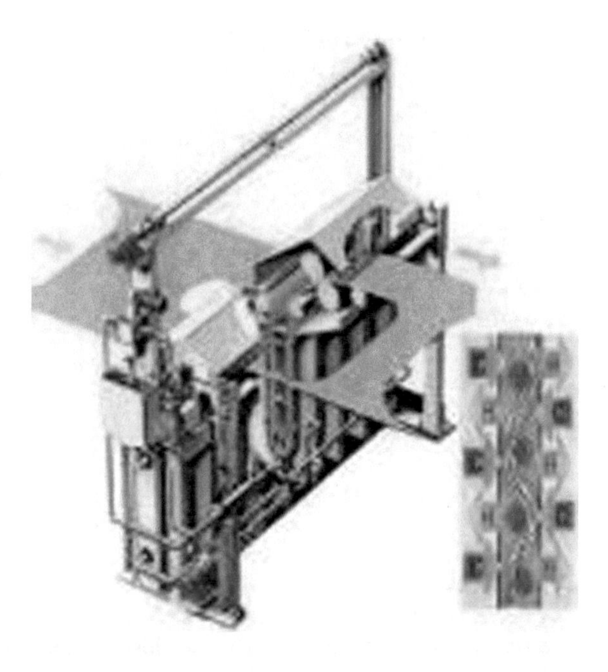

Fig. 23 Impacta—impregnation range from Benninger

and fabric quality and chemical dosing system and gives consistent application of chemicals while processing different qualities (Fig. 23).

Dip Sat (Goller)

Technology for impregnation chamber is known as dip-sat. the technology is suitable for wet on wet or dry on wet application of chemicals. The compartment has a capacity of 17 m of fabric. The threading of fabric is such that fabric is dipped in a liquor three times so that maximum penetration of liquor takes place. Usually, pick up after impregnation can go as high as 100% (Figs. 24 and 25).

2.5.5 Steaming Chamber

Steamer is an extremely crucial part of the continuous bleaching range. Each manufacturer has different designs for their reaction chambers.

Reaction chamber in a continuous bleaching range manufactured by Goller is known as Complexa. In the chamber fabric is piled so that more amount of fabric can be treated at a time. It takes 30–40 min to treat the fabric in a reaction chamber (Fig. 26).

It is fitted with a sealed housing so that steam is not leaked. The bottom of the chamber is also heated with a saturated steam so that it gives effective heating to the fabric and does not allow impurities to condense on fabric surface. The steamer is

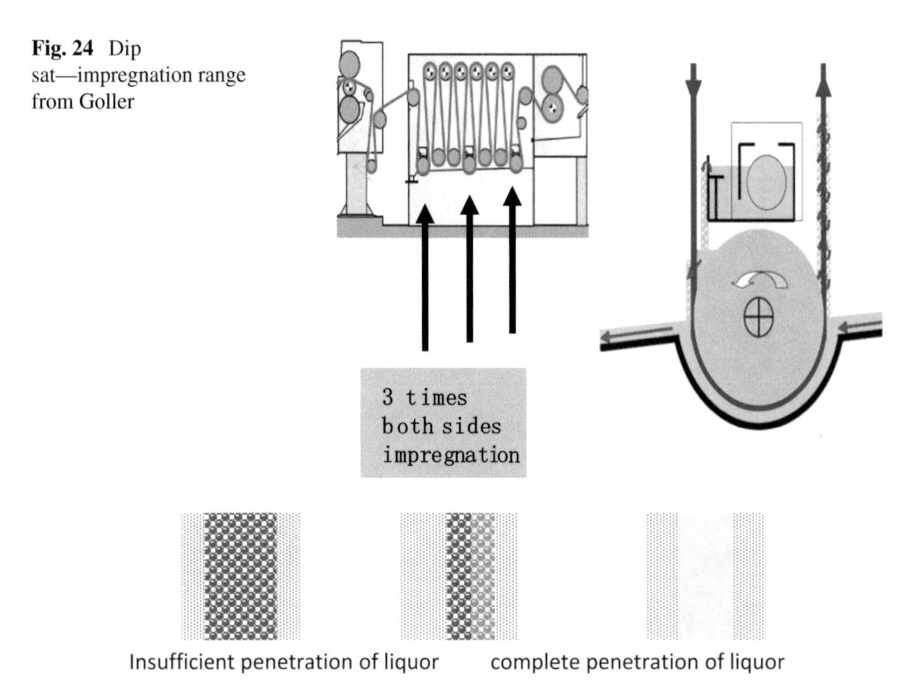

Fig. 24 Dip sat—impregnation range from Goller

Fig. 25 Pictorial representation of liquor penetration using Dip sat (Goller)

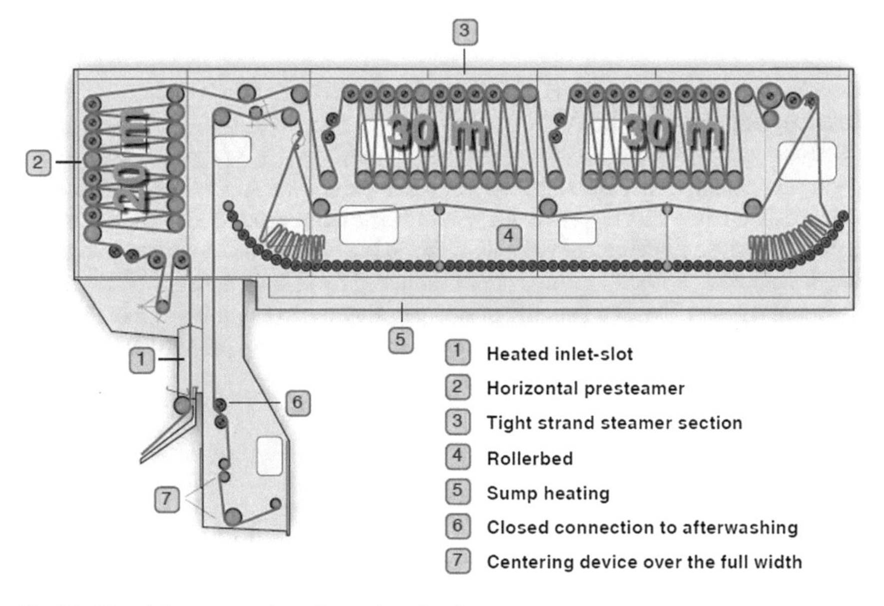

Fig. 26 Pictorial representation of steaming chamber

Fig. 27 Proportionate valve to control steam

usually fitted with frequency controlled drives and torque motors which maintains smooth flow of fabric and prevents crease formation. Load cells are used to measure fabric tension and compensator for drive control.

Roller Bed or Conveyer Bed

Steamer is equipped with roller bed or conveyer bed for transportation of fabric with required dwell time with respect to the type of fabric being processed. A plaiter is used to plait the fabric so that more amount of fabric is accommodated and desired dwell time is achieved. The strokes by plaiter is adjusted by using frequency converter.

Steam control: Desired quantity of steam is required to be introduced in the reaction chamber to achieve best results. This is achieved by using proportionate control valves (Fig. 27).

2.5.6 Post-washer

The post washing module washes out the residual chemicals from the fabric precisely and counter flow in washing tank gives high degree of efficiency. For better quality result and to minimize fresh water intake, reuse post washer liquor to pre washer by adding feed water pump having required flow rate can be used (Fig. 28).

Fig. 28 Neutralisation chamber

Neutralization Compartment

This compartment neutralizes the fabric pH by automatic acid dosing system. This compartment is equipped with pH monitoring probe to measure actual pH of liquor and control amount of acid dosing linearly by adjusting strokes of acid dosing pump. The liquor is continuously circulated by circulation pump to achieve pH consistency.

2.5.7 Vertical Drying Range

It consists of stacks of large diameter drying cylinders arranged in vertical manner to dry the fabric. Drying range are equipped with automatic moisture controller to maintain required moisture content in the fabric being processed by controlling temperature of drying cylinders precisely with the help of proportionate control valve (Fig. 29).

2.5.8 Fabric Delivery

The fabric delivery serves to wind fabric onto a batch after processing. During continuous operation, the fabric delivery with chute allows fabric changes to take place without stopping the machine, and thus delivers the optimum production result (Fig. 30).

Fig. 29 Actual photo and line diagram vertical drying range

Fig. 30 Line diagram of fabric batching unit

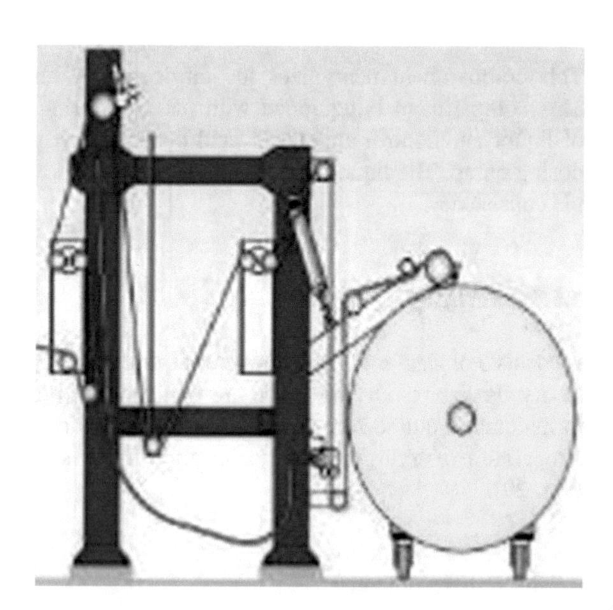

2.6 Continuous Dyeing Ranges

As the name suggests, dyeing takes place in a continuous manner. Various methods include, pad-dry-pad-steam, pad-steam, and E-control. Machine consists of three main sections which are, inlet section, chamber and exit section.

Inlet section consists of, J-Scray, cooling cylinders, padding mangle, wetting trough and pre-dryer. Chamber consists of chamber atmosphere control unit and steam injection unit. Exit section consists of cooling cylinder, J scray, batching unit.

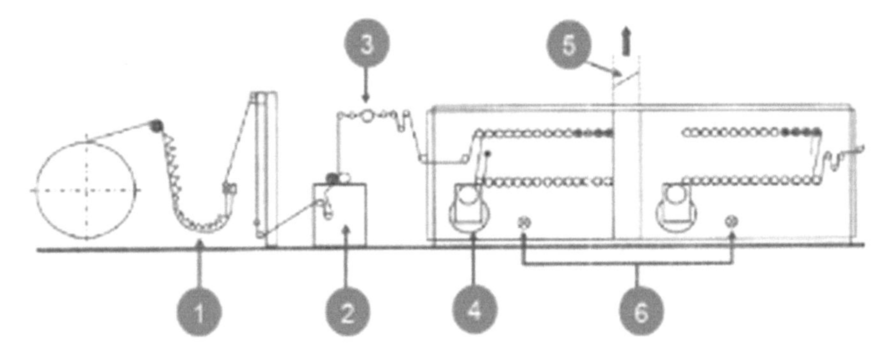

Fig. 31 Line diagram of continuous dyeing range

2.6.1 Inlet Section-J scray consists of following subassemblies:

- Rotary for tension adjustment- to neutralise uneven tension in the fabric
- Feed roll
- Traction roll with anti-lapping roll
- Optical sensor, if any lapping occurs
- Optical sensor for checking fabric presence
- Tensioner
- Brake roll
- Selvedge guider (Fig. 32).

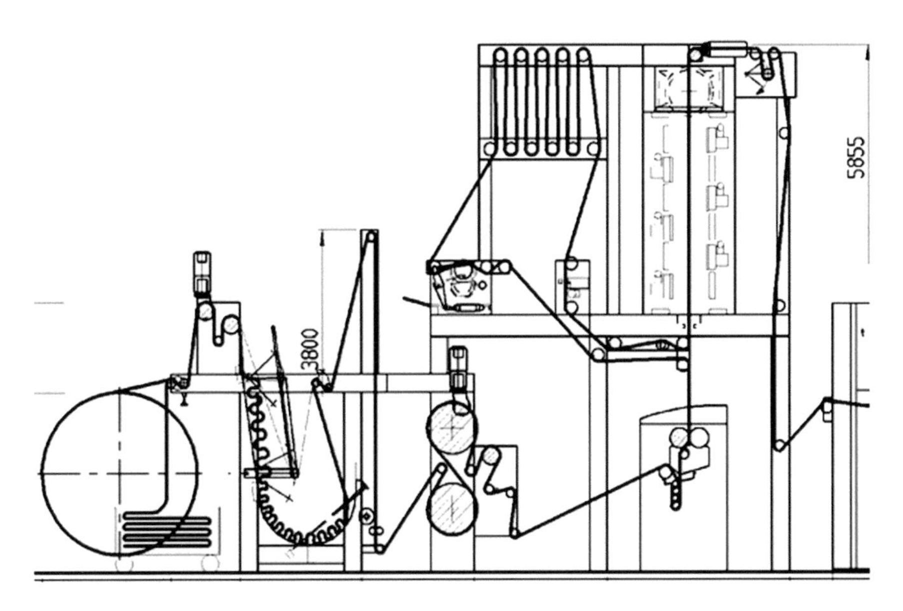

Fig. 32 Line diagram of feeding section

Cooling Cylinder

To get the even dyeing in the continuous process, the inlet fabric should be maintained at even conditions of temperature and humidity. The important function of cooling cylinders is to maintain the temperature of the infeed fabric throughout the length and width. This will ensure that there is no variation in fabric conditions as well as no change in dye bath temperature. The temperature of the trough should always be below 25° C during application of dye to the fabric.

2.6.2 Padder

The padder consists of two swimming rollers which has two pipes for hydraulic pressure. This portion is always covered with rubber sleeve. During application, hydraulic pressure portion always faces the nip. Only the rubber sleeve moves (Fig. 33). Four rollers are used to give more contact time between fabric and liquor. Padder is the most important part of semi continuous and continuous dyeing machines.

S-Roll Padder Features

- Two swimming rollers
- Hydraulic system is used at the centre of padder whereas pneumatic system is used at the edges
- Hydraulic pressure portion always faces the nip.
- Only rubber sleeve will move (Figs. 34 and 35).

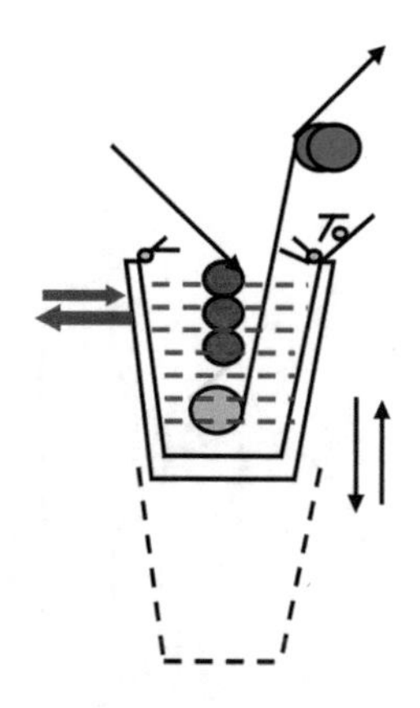

Fig. 33 Line diagram of padding mangle

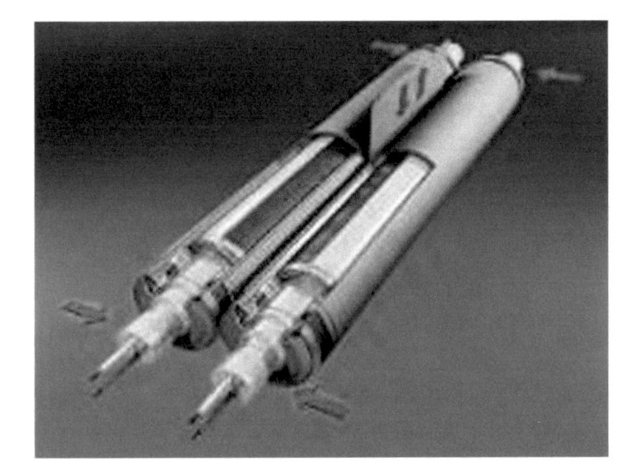

Fig. 34 S roll padding mangle

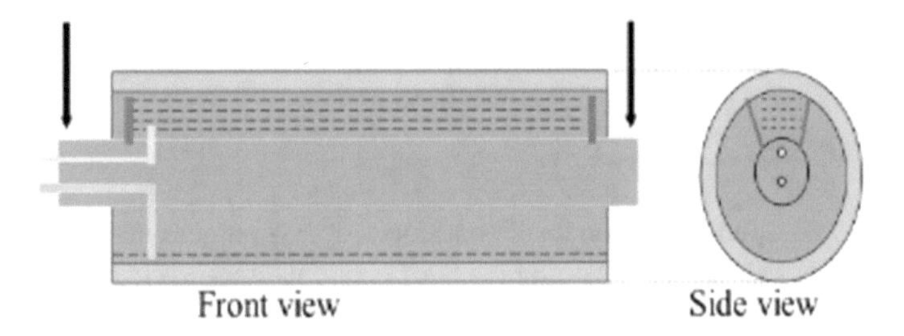

Fig. 35 Front and side view of S-roll padding mangle

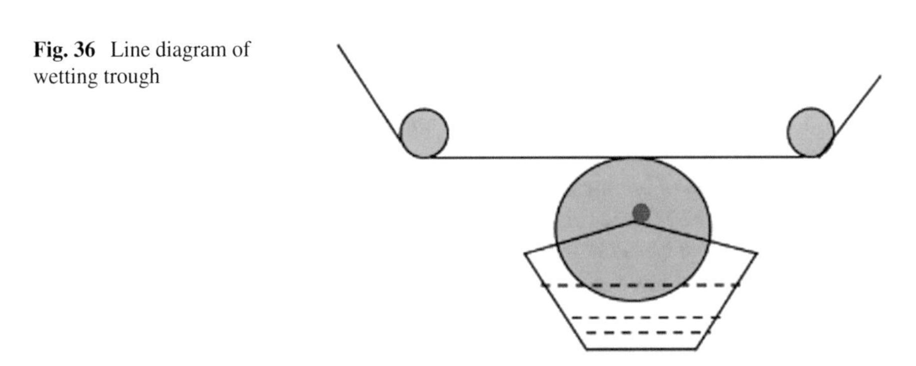

Fig. 36 Line diagram of wetting trough

2.6.3 Wetting Trough

- With rough surface takes water and makes fabric wet.
- Comes into action only when chamber requires cleaning and simulating conditions inside the chamber (Fig. 36).

2.6.4 Airing Zone

- After passing from the padding, fabric is offered time for dye
- Penetration into the fabric core for thick and tightly woven fabric substrates
- Top—5 rollers
- Bottom—6 rollers.

2.6.5 Pre-dryer

- Electrically heated or gas heated option available
- High speed evaporating efficiency which results in quick drying of fabric
- High operational safety and reliability
- Temperature range 650–950 °C
- IR modules to cater to different production requirements
- Immediate heating and cooling desired.

2.6.6 Chamber

There are few major manufacturers of continuous dyeing ranges with E-control technique. Monforts and Bruckener are the major manufacturers while Menzel has also started manufacturing the same. Some of the Indian machine manufacturers such as Dhall et al. have also have manufactured these systems, however, their ranges are only restricted to pad-dry technique.

Chamber of a continuous dyeing range has some desired features, which are as follows:

- Good insulation to minimise radiation losses
- Good even temperature distribution in the chamber and setting of different temperatures in separate chambers
- Even distribution of hot air and soft on fabric for even shade
- Dye migration free fabric travel
- Crease free run of fabric in the hot flue
- Easy cleaning and low maintenance
- Moisture management in the chamber—process control and exhaust control
- Low tensions—frequency controlled drives
- Easy fast cleaning in between shade change in hot flue.

Monforts

Major features of chamber used for E—Control technique are,

- Computer-controlled fast cleaning of the Teflon-coated fabric transport rollers.

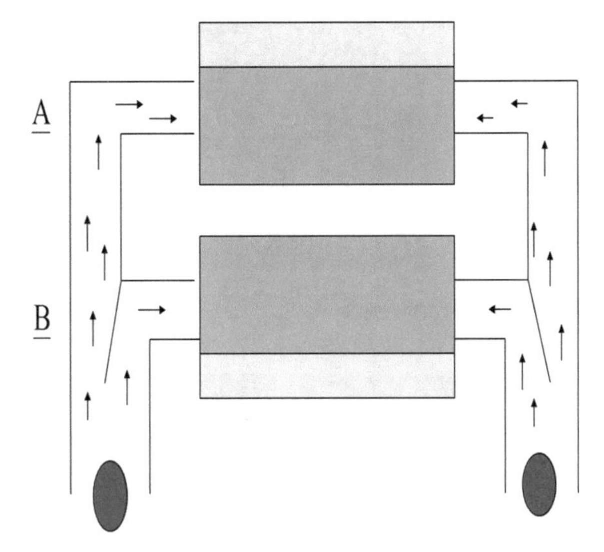

Fig. 37 Air flow in Monforts continuous dyeing chamber

- Chamber generously insulated for minimum heat radiation.
- Operating terminal with PLC system and touch screen monitor technology for set-up, operation and monitoring (Fig. 37).

The air circulation in the chamber is as shown in adjacent figure. Two blower fans are used to blow hot air from the radiator on the fabric from top as well as bottom. The patented technology used by M/s Monforts to distribute the air equally ensures the uniform development of colour on top as well as bottom surface of the dyed fabric. If the air balance is disturbed due to any malfunction, it will reflect on the dyed fabric by face—back uneven dyeing.

Another feature of Monforts chamber is known as thermo cut system. This system allows to adjust the width of the slit through which hot air is blown to desired width based on fabric being processed. This saves in heat loss as well as protects the rollers which are not covered with the fabric from direct exposure of hot air.

Up to the first chamber, brakes are provided which allows to clean the Teflon coated rollers to clean efficiently after batch is completed so that the next batch is not contaminated with colour on the rollers (Figs. 38 and 39).

Bruckener

See Fig. 40.

Main Features of Bruckner Chamber

- Symmetrical design of upper and lower pressure boxes for equal pressure in two nozzles

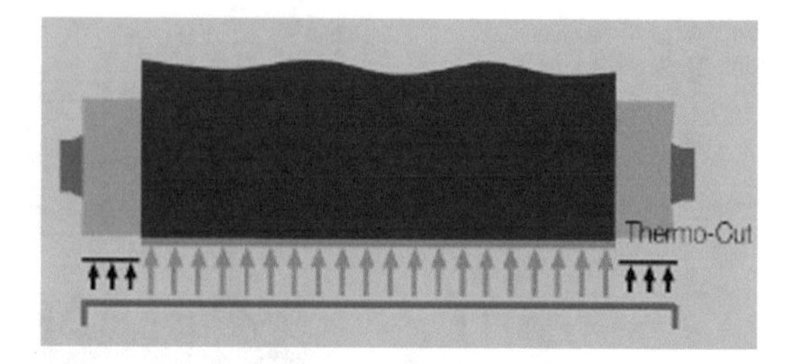

Fig. 38 Thermo cut system

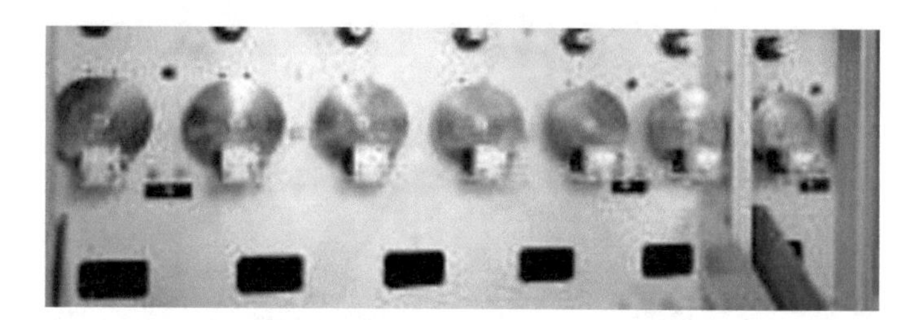

Fig. 39 Braking system in Monforts CDR

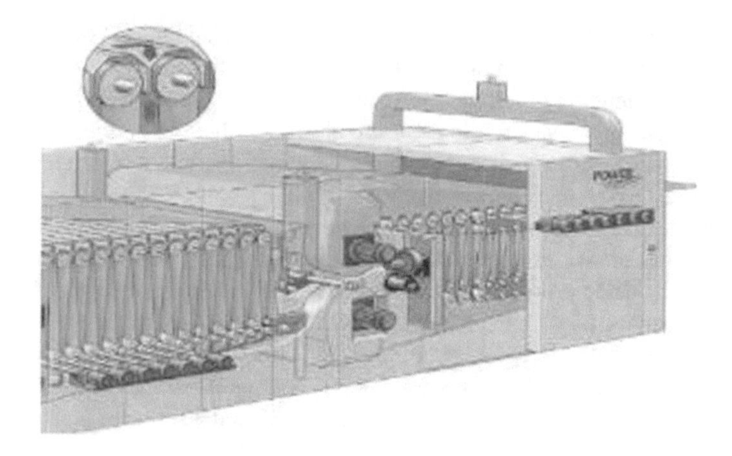

Fig. 40 Chamber design of Bruckener CDR

- Each chamber has 4 fan motors—2 for upper and 2 for lower nozzles
- Patented SPLIT FLOW air system
- Direct drive for upper rollers with individual frequency-controlled motors
- Pendulum roll at end of each chamber
- Individual drives for bottom rollers for cleaning system by differential speed (optional).

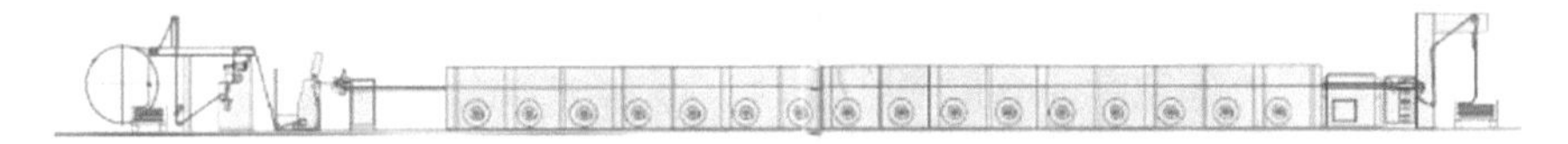

Fig. 41 Line diagram of stenter machine

2.7 *Stenter*

Stenter is a universal drying machine, which is also one of the most expensive machines. It has different assemblies for processing of woven and knitted fabrics. It is one of the most important machines in the finishing chain as it is used for various functions such as,

• Heat setting	• Width adjustment
• Application of finishing chemicals	• Dehydration
• Drying	• Curing

The line diagram of stenter for woven fabrics is as shown in Fig. 41.
Major components of stenter machine are,

• Inlet section	• Chambers
• Padder	• Weft straightener
• Radiators	• Circulation fans
• Exhaust fans	• Heat recovery
• Winder	• Clips/pins
• Cooling drums	• Electric cabinet

2.7.1 Padder

• Operator friendly design	• Pneumatic lifting and lowering trough
• Technology for even pick up	• Pneumatic valves for auto dosing of chemicals
• Rubber—ebonite roller with solid shaft	• Level control sensor
• Safety limit switches for operator safety	• Trough capacity approximately 64 lit

Fig. 42 Weft straightener

2.7.2 Weft Straightener

A fabric experiences lot of longitudinal tension during various stages of wet processing, which results in formation of bows and skews. The corrective action is to re-align weft yarns. To achieve this, defined degree of lateral tension is applied to the fabric. This is done by using different techniques. Most commonly used technique include mechanical weft straighteners and differential tenter chains. Speed of tenter chains can be varied to re-align weft or it can also be achieved by passing the material across rollers that can be pivoted around the centre point of the web.

Essential parts of a weft straightener include skew rollers, bow rollers, nose sensors, electric accuators etc. (Fig. 42).

2.7.3 Tenter Chain and Its Assemblies

The clips used by most of the manufacturers are manufactured in such a way that it can be replaced easily as many times it gets damaged during production. These are made of special aluminum alloy which are corrosion and heat resistant (Fig. 43).

Fig. 43 Pins of tenter frame

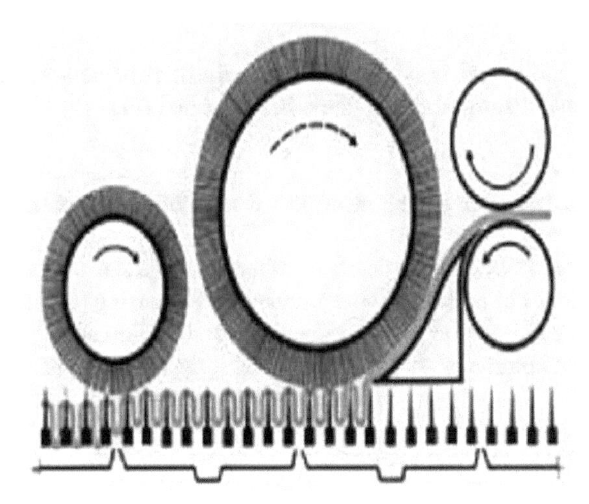

Fig. 44 Brushes to press fabric

2.7.4 Brushes After Pinning

Since the pinning shoes only bring the fabric onto the tip of the pins, the function of the subsequent brushes is to press the fabric fully onto the pins. The brushes are mounted on a pivoting lever which, after releasing a locking handle, can be moved to the non-operating position or working position (Fig. 44).

2.7.5 Selvedge Tensioning and Fabric Guiding Unit

Fabric passage along the pin rail needs to be fixed properly and efficiently on the pins. A special synchronized drive system is used for pinning the fabric efficiently with large overfeeds or having higher lycra content. Pneumatic systems are used for the efficient operation of the system. The drive of the selvedge tensioning and fabric

Fig. 45 Fabric guiding unit

guiding unit is synchronized with the drive of stenter chain so that there is no shifting and distortion of the fabric being treated (Fig. 45).

2.7.6 Selvedge Sensor and Rack/Pinion Assembly

The photoelectric detectors of the selvedge sensor operate with infrared light and are thus not affected by the ambient light. Sensing can also be performed mechanically by switch over to the mechanical sensors. Depending on the sensing signals, the rack and pinion assembly alignes the fabric selvedge parallel to the pinning devices.

2.7.7 Selvedge Unculers

This assembly is mainly used to process knitted fabric, which has a tendency to curl at the edges. The selvedge unculers may be equipped with two, three or four spindles. These can be either fixed or motor driven threaded spindles. The pivoting arrangement of the first threaded spindle means that, the wrap angle and thus the uncurling effect is infinitely variable (Fig. 46).

2.7.8 Fabric Selvedge Glueing Device

This system is typically required for processing of knitted fabric. As the selvedges of knitted fabric are prone to curling, these are required to be fixed to the tenter chain by gluing. The application of this glue is done by a device called as fabric selvedge gluing device. The device consists of a tank containing glue fitted with level sensor which automatically maintains level of glue in the tank and it also ensures even and

Fig. 46 Selvedge uncurling unit

Fig. 47 Selvedge gluing unit

steady application of the glue to the fabric. This is completely enclosed system which prevents glue from drying (Fig. 47).

2.7.9 Width Adjustment Device at Inlet

The movement of the left-hand/right-hand infeed track can be adjusted or corrected in "jog" mode at push-buttons on the small control cabinets on the chain tracks. A separate motor is used to drive each spindle. Desired width adjustment is done from the central location and can be automatically set through the entire length (Fig. 48).

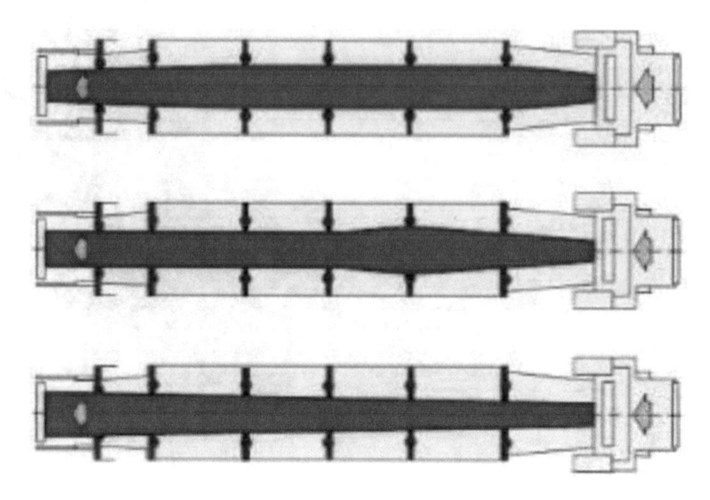

Fig. 48 Width adjustment of fabric

2.7.10 Screens

The screens prevent fluff, threads, etc. carried by the fabric from entering the air circulation system. The screens are located on the suction duct. Depending on the type and design, the machine may be equipped with fixed base screens which can be removed after opening the service doors or with auxiliary screens which can be removed during the machine operation. To ensure efficiency of the machine, lint filters should be inspected and cleaned in every shift at least once. It gives efficient cleaning of these filters if cleaned using brush or vacuum cleaners (Fig. 49).

2.7.11 Type of Radiators

Circulating Oil/Steam Heating

The circulating oil heating system supplies air heaters with heat transfer oil. The ambient air is heated to treatment temperature by the air heaters, whereby the heat transfer oil cools down.

The oil is then returned to a heating circuit in a closed-circuit loop where it is heated again. It is then pumped to the air heaters once again. The temperature regulator drives the control valve which varies the heat transfer oil supply until the preset temperature is reached (Fig. 50).

Fig. 49 Screens to filter lint

Fig. 50 Different types of radiators

Gas Fired Burner Heating System

Compressed natural gas is used as heating medium. The arrangement is as shown in (Fig. 51).

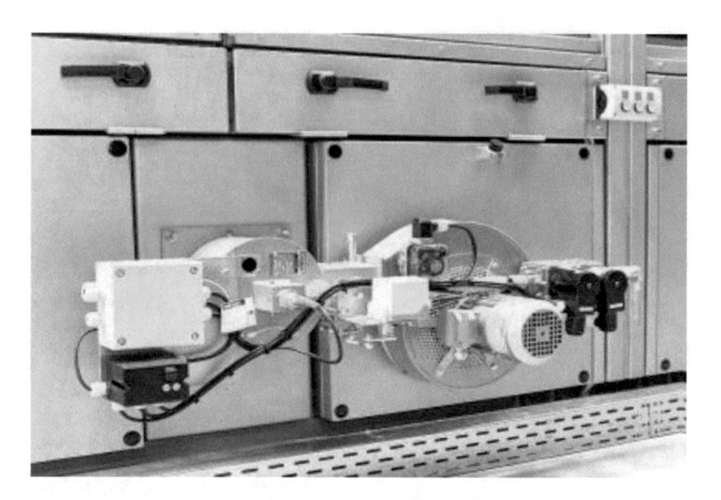

Fig. 51 Gas fired system

2.7.12 Circulation Fan

The circulation fans can be located alongside or above the roof of the treatment chambers. The air circulation fans ensure the necessary circulation of the treatment air. The air circulation fans are driven by 3-phase AC motors and variable speed inverter drives. All drive motors are of maintenance-free design.

2.7.13 Twin-Air

The technology is developed by M/s Monforts for stenter. The technology uses two independent circulation fans which gives hot air to top and bottom nozzles from radiator. The frequency of both fans can be adjusted independently. This allows completely even temperature on the fabric surface (Fig. 52).

2.7.14 Nozzle System

The nozzles are important part of the stenter machine. The air is blown on the fabric surface in such a way that maximum heat is transferred equally on both surfaces of fabric. There is a different design for woven fabric and different for knitted fabric.

2.7.15 Exhaust Air Fan

Depending on the type and configuration of the machine, the chamber is connected via a duct system to one or two exhaust fans.

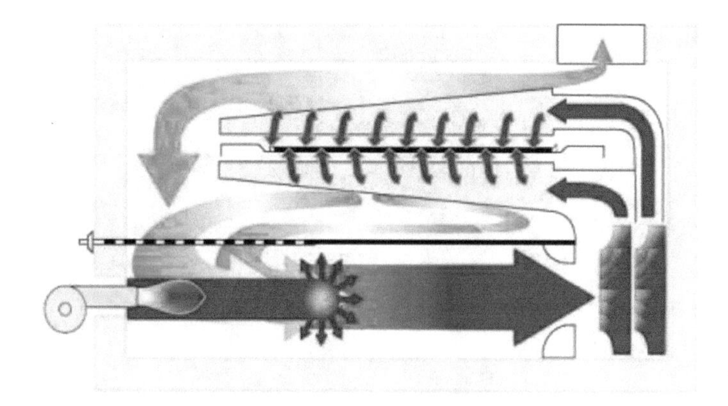

Fig. 52 Schematic representation of air flow inside chamber

The exhaust fan draws the exhaust air with the evaporated water and any finishing agent vapours out of the heat treatment chamber and discharges them into the atmosphere.

The exhausted mixture is replaced by fresh air. The exhaust air volume can be regulated via mechanically adjustable air valves.

2.7.16 Heat Recovery Unit

A heat recovery system can be installed in order to save heating energy, thereby saving on utilities.

The excess heat generated during process is directly exhausted through the exhaust air ducts. In the heat recovery system, this excess heat which is simply wasted is used to heat up the incoming process air using heat exchangers. Up to 60% of total volume of the required air can be heated in this way. The savings are in the range of 10–30%, depending on the production volume.

2.7.17 Selvedge Cutter and Suction Duct

Selvedge cutter assembly is primarily used when processing knitted fabric. The knitted fabric has a tendency to curl at the selvedges. Hence, these are glued at the selvedges using gluing device. When the process is completed, these glued portions are removed by using a combination of optical and electronic control with pneumatic support which ensure minimum selvedge cut-off. The latest technique involving laser controlled pinning and top selvedge glueing, selvedge cut-off can be reduced by up to 60% (Fig. 53).

Fig.53 Selvedge cutter and suction duct

3 Conclusion

With ever increasing pressure on textile wet processing industry to improve production without affecting quality and environment, has resulted in development of many new technologies. In the textile wet processing field, research is going on various levels and fields to improve the consistency by reducing consumption and load on effluent. Auxiliaries, colorants and process are the major areas of focus; however, machinery manufacturing is also equally important, as the wet processing of textiles involves three M's—Man, Machine and Material. Machine manufacturers continuously strive to deliver better products, which will help the user minimize problems and optimize production. Machines have become more user friendly and more sophisticated as compared to last century. In this chapter, continuous processing machines are discussed, and prime focus is given on the existing techniques used in the industry.

Printed in the United States
by Baker & Taylor Publisher Services